Chemical Bonds—Better Ways to Make Them and Break Them

stereochemistry of organometallic and inorganic compounds 3

Chemical Bonds—
Better Ways to Make Them
and Break Them

Edited by

IVAN BERNAL

Department of Chemistry, University of Houston, Houston TX 77004, U.S.A.

ELSEVIER

Amsterdam — Oxford — New York — Tokyo   1989

no anal

ELSEVIER SCIENCE PUBLISHERS B.V.
Sara Burgerhartstraat 25
P.O. Box 211, 1000 AE Amsterdam, The Netherlands .

Distributors for the United States and Canada:

ELSEVIER SCIENCE PUBLISHING COMPANY INC.
655, Avenue of the Americas
New York, NY 10010, U.S.A.

ISBN 0-444-88082-8

Printed in The Netherlands

Preface to Volume 3

As the twenty first century approaches us, a new era appears in the horizon--the era of the environmentalist-- or so it seems to me.

In this dawning era, our need to survive in an esthetically satisfying environment will place heavy demands on our future treatment of our planet, and nowhere will that demand be greatest than in our handling of our chemical industry. We can no longer live without it-- the question is "can we learn to live, safely and well, with it"?

There is a difference between an ordinary ballet performance and watching Moira Shearer dancing in Tales of Hoffmann or The Red Shoes-- it is elegance! The movements are the same, but the outcome is not, and the difference is what provides such joy to watch.

So it must be with our future chemistry-- it can no longer be only effective, it must be elegant so as to safely and efficiently provide for our most subtle needs.

The authors of the five chapters of this volume have reviewed for us how to understand the making and breaking of bonds at the dawn of what I perceive to be a more sophisticated era-- one in which we efficiently make and break bonds, but do so in a stereochemically correct manner, when such requirement is needed. These authors have summarized for us how to better understand bonds and the ways for making them and breaking more elegantly-- hopefully, some day, as elegantly and esthetically pleasing as Ms. Shearer's terpsichorean art.

Chapter 4 summarizes the fundamentals of bonding in organometallic clusters and provides the basis for the making and breaking bonds which, efficiently, do your bidding; i.e, they have, or do, the right chemistry and stereochemistry.

In Chapter 1, Braunstein and Rose show us how to create clusters which are either chemically and stereochemically correct, or correctly catalyze the formation of new bonds. The rules for understanding the making and breaking of such bonds, either for syntheses or catalyses, are discussed by Osella and Raithby in Chapter 4.

Some of the art of making and breaking bonds is discussed in Chapters 2, 3 and 5. Blackburn, Davies and Whittaker discuss the stereochemistry of a class of organometallics which the senior author has made into an effective group of smooth and efficient, stereospecific catalysts.

The use of solar energy is one of the imperatives of the new era. Thus, an understanding of photochemical processes for making and breaking bonds, in sophisticated new ways, is one of our pressing needs. Stufkens, in Chapter 3, addresses the problem of understanding the mechanism(s) of creating photoproducts which are useful or can do useful chemistry and stereochemistry for us.

Finally, one of the active participants in documenting the power and sophistication of the Sakurai Reaction kindly agreed to review it for this volume. His contribution appears in Chapter 5.

I hope readers agree that the dawn of the era of elegance are clear in these pages. Hopefully, subsequent volumes will measure to the standards of this one and educate us in the existence of sophistication in other areas of organometallic and inorganic chemistry yet unexplored by this series.

Finally, I should mention that this volume should have appeared in the Fall of 1988. This delay was caused by unexpected responsibilities and to the extent it caused anxiety to the authors, and dated by a few months their reviews, I apologize and thank them for their forbearance.

May 16th, 1989   The Editor.

TABLE OF CONTENTS

Chapter 1

Heterometallic Clusters in Catalysis

P. Braunstein and J. Rose

Chapter 2

Conformational Analysis for Ligands Bound to the Chiral Auxiliary $[(C_5H_5)Fe(CO)(PPh_3)]$

B.K. Blackburn, S.G. Davies and M. Whittaker

Chapter 3

Steric and Electronic Effects on the Photochemical Reactions of Metal-Metal Bonded Carbonyls

D.J. Stufkens

Chapter 4

Stereochemical Aspects of Organometallic Clusters. A View of the Polyhedral Skeletal Electron Pair Theory

D. Osella and P.R. Raithby

Chapter 5

The Stereochemistry of the Sakurai Reaction

Y. Yamamoto and N. Sasaki

Contributors to this Volume

B.K. Blackburn
The Dyson Perrins Laboratory, University of Oxford, South Parks Road, Oxford, OX1 3QY, United Kingdom

P. Braunstein
Département de Chimie, Institut Le Bel, Université Louis Pasteur, 4, rue Blaise Pascal,
F-67070 Strasbourg, France

S.G. Davies
The Dyson Perrins Laboratory, University of Oxford, South Parks Road, Oxford, OX1 3QY, United Kingdom

D. Osella
Dipartimento di Chimica Inorganica, Chimica Fisica e Chimica dei Materiali, Università di Torino,
Via P. Giuria 7, 10125 Torino, Italy

P.R. Raithby
University Chemical Laboratory, Lensfield Road, Cambridge CB2 1EW, United Kingdom

J. Rose
Département de Chimie, Institut Le Bel, Université Louis Pasteur, 4, rue Blaise Pascal,
F-67070 Strasbourg, France

N. Sasaki
Department of Chemistry, Faculty of Science, Tohoku University, Sendai,   Japan 980

D.J. Stufkens
Anorganisch Chemisch Laboratorium, University of Amsterdam, J.H. van 't Hoff Instituut,
Nieuwe Achtergracht 166, 1018 WV Amsterdam, The Netherlands

M. Whittaker
The Dyson Perrins Laboratory, University of Oxford, South Parks Road, Oxford, OX1 3QY, United Kingdom

Y. Yamamoto
Department of Chemistry, Faculty of Science, Tohoku University, Sendai, Japan 980

HETEROMETALLIC CLUSTERS IN CATALYSIS

P. Braunstein and J. Rose

HETEROMETALLIC CLUSTERS IN CATALYSIS

P. BRAUNSTEIN and J. ROSE

Département de Chimie, Institut Le Bel
Université Louis Pasteur
4, rue Blaise Pascal
F-67070 Strasbourg (France)

1. GENERAL INTRODUCTION

2. HETEROMETALLIC CLUSTERS IN HOMOGENEOUS CATALYSIS
   2.1. INTRODUCTION
   2.2. HYDROGENATION REACTIONS
      2.2.1. HYDROGENATION AND ISOMERIZATION OF CARBON-CARBON MULTIPLE BONDS
      2.2.2. CO HYDROGENATION
   2.3. WATER-GAS SHIFT REACTION
   2.4. HYDROFORMYLATION REACTIONS
   2.5. CARBONYLATION REACTIONS
   2.6. METHANOL HOMOLOGATION
   2.7. NORBORNADIENE DIMERIZATION
   2.8. BUTADIENE OLIGOMERIZATION
   2.9. HYDROSILATION REACTIONS
   2.10. CONCLUSION

3. HYBRID CATALYSTS PREPARED FROM MOLECULAR MIXED-METAL CLUSTERS
   3.1. INTRODUCTION
   3.2. CATALYSIS BY IMMOBILIZED MIXED-METAL CLUSTERS ON INORGANIC OXIDES
   3.3. CATALYSIS BY IMMOBILIZED MIXED-METAL CLUSTERS ON FUNCTIONALIZED SUPPORTS
   3.4. BIMETALLIC CLUSTERS IN ZEOLITES
   3.5. CONCLUSION

4. HETEROMETALLIC CLUSTERS IN HETEROGENEOUS CATALYSIS
   4.1. INTRODUCTION
   4.2. PREPARATION OF MIXED-METAL CLUSTER-DERIVED CATALYSTS
   4.3. INTERACTIONS OF MIXED-METAL CLUSTERS WITH CATALYTIC SUPPORTS
   4.4. OCCURRENCE OF HETEROMETALLIC INTERACTIONS IN MIXED-METAL CLUSTER-DERIVED CATALYSTS

4

ABBREVIATIONS

acac	acetylacetonato
Bu	butyl
COANE	cyclooctane
COD	cyclooctadiene
COENE	cyclooctene
Cp	η^5-cyclopentadienyl (η-C_5H_5)
Cy	cyclohexyl
DMF	dimethylformamide
DPAE	1,2-bis(diphenylarsino)ethane
DPPE	1,2-bis(diphenylphosphino)ethane
DPPM	bis(diphenylphosphino)methane
Et	ethyl
Me	methyl
o-MeC_6H_4	o-tolyl
MMCD	mixed-metal cluster-derived catalyst
NBD	norbornadiene
OAc	acetate
Ph	phenyl
Py	pyridine
PPN	bis(triphenylphosphine)iminium
R.T.	room temperature
THF	tetrahydrofurane
TMBA	trimethyl(benzyl)ammonium
TOF	turnover frequency
Tol	tolyl
$[M_xM'_y]$	heterogeneous catalyst prepared from a molecular cluster of metal core composition $M_xM'_y$
WGSR	water gas shift reaction

1. GENERAL INTRODUCTION

Metal cluster chemistry, which involves compounds displaying direct metal-metal interactions, provides a shining example of the fruitful interactions that can occur between various disciplines and of the subsequent cross-fertilization in the frontier areas. Cluster compounds are encountered in molecular and solid state chemistry; they belong to organometallic and inorganic chemistry and their properties arouse the interest of physicists (*e.g.*, superconductivity), of biochemists (*e.g.*, multimetallic redox centers), of academic and industrial researchers studying new materials and catalysts, and of course, of the chemists themselves. They encompass interests in main group elements or transition metals, in synthetic organic or inorganic chemistry, in theoretical, mechanistic or analytical chemistry. These areas of scientific activity do not constitute a restricted list but they already explain the exponential growth that cluster chemistry has enjoyed over the last 15 years [170].

In order for new materials to be evaluated and for new properties to emerge, the synthetic chemist must have first created the new molecules or materials of interest. Cluster chemists have displayed great skill and imagination in the synthesis and structural characterization of novel and complex molecules. Rational synthetic methods have emerged which allow, in many instances, the designed synthesis of molecules containing, *e.g.*, a specific number of different metal atoms bonded to each other or to given ligands in an unusual mode of attachment [249, 292, 303]. Beautiful achievements include tetrahedral clusters with four different metal atoms [247] (Fig. 1.1), and thus inherently chiral, and large bimetallic clusters, *e.g.*,

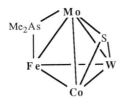

Fig. 1.1. Metal core structure of $FeCoMoW(\mu_3\text{-}S)(\mu\text{-}AsMe_2)(\eta\text{-}C_5H_5)_2(CO)_7$, the first cluster with four different metal atoms [247].

$[Ag_{12}Au_{13}Cl_6(PPh_3)_{12}]^{m+}$ [294] (Fig. 1.2), $[Ag_{19}Au_{18}Br_{11}\{P(p\text{-Tol})_3\}_{12}]^{2+}$ [295a] and $[Ni_{38}Pt_6(CO)_{48}H_{6-n}]^{n-}$ (n = 5,4) [72] (Fig.1.3), a molecular model of "cherry" crystallites studied in heterogeneous catalysis.

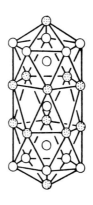

Fig. 1.2. Metal core of $[Ag_{12}Au_{13}Cl_6(PPh_3)_{12}]^{m+}$ [294].

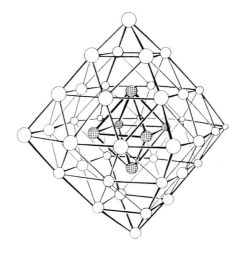

Fig. 1.3. Metal core of $[Ni_{38}Pt_6(CO)_{48}H_{6-n}]^{n-}$ (n = 5,4) [72].

Obviously, the diversity and complexity of clusters containing more than one metal type are considerable and largely justify the current efforts to better understand their synthesis, structures and physical and chemical (stoichiometric and catalytic) properties.

The goal of this chapter is to evaluate the achievements and potential in catalysis of heterometallic transition metal clusters (including heterodinuclear complexes), *i.e.*, of well-defined and characterized molecular compounds containing at least two different transition metal atoms in their core. Bimetallic catalysis has become an area of major interest and applies to homogeneous, heterogenized or heterogeneous processes [320], as shown in Scheme 1.1.

SCHEME 1.1.
Uses of Mixed-Metal Clusters in Catalysis.

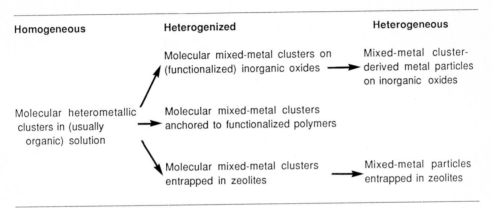

Homogeneous	Heterogenized	Heterogeneous
	Molecular mixed-metal clusters on (functionalized) inorganic oxides	Mixed-metal cluster-derived metal particles on inorganic oxides
Molecular heterometallic clusters in (usually organic) solution	Molecular mixed-metal clusters anchored to functionalized polymers	
	Molecular mixed-metal clusters entrapped in zeolites	Mixed-metal particles entrapped in zeolites

The concept of synergism in homogeneous catalysis has been discussed by Golodov [132]: "Synergism may be defined as a disproportionate increase in reaction rate observed upon mixing two catalytic systems, both of which will individually catalyze the reaction in question" [98]. However, caution must be taken when interpreting the enhanced rate effects caused by added metal complexes [291]. On the other hand, the increased diversity and availability of mixed-metal clusters, for which many high-yield syntheses have been reported and will continue to appear in series such as *Inorganic Syntheses*, have stimulated a considerable academic and industrial research interest for their catalytic properties [126]. These efforts are encouraged by the exact knowledge of the molecular structure and stereochemistry of an increasing number of mixed-metal clusters. The importance of heterometallic clusters for catalysis is based on the following reasons: (i) the adjacent metal centers offer the possibility for cooperative reactivity and the intrinsic polarity of heterometallic bonds [49] can direct the selectivity of substrate-cluster interactions [264]. This may lead to new, more active, or more selective homogeneous or heterogenized (by attachment of the clusters to solid supports) catalysts; (ii) the metal core of these clusters constitutes a well defined unit, which may be viewed as a molecular microalloy and which can be used as a precursor of novel heterogeneous catalysts. Stripping off the ancillary ligands is expected to provide catalytic materials with improved control of both particle size and composition; (iii) the knowledge of the precise structural and spectroscopic properties of atoms or ligands bound to the metal frameworks allows one to establish and evaluate conceptual and comparative models for understanding the coordination of these species, *e.g.*, H, N, NH, O, S, C, CO, CH, CH_2, CH_3, C_2H_3, to metal surfaces.

In this chapter, we shall restrict ourselves to studies in which well-characterized mixed-metal cluster compounds of the transition metals (including Cu, Ag and Au) have been used in catalysis. We will therefore not discuss situations where a mixture of homonuclear complexes of different metals has been used, except when comparisons have been made with the corresponding heterometallic precursors. We recognize that this distinction is somewhat arbitrary as mixed-metal complexes are often likely to fragment into homonuclear species under the catalytic conditions while, conversely, they may be produced *in situ* upon mixing of the individual metal complexes. Similarly, alloy formation may occur during catalysis with heterogeneous catalysts prepared from different homonuclear systems. Furthermore, it has been shown that the presence of a mixture of homonuclear, even mononuclear, complexes of different metals can be most beneficial for a given catalytic process, consecutive elementary steps occurring at each metal center. This is particularly notable, *e.g.*, in the formation of ethylene glycol from CO and H_2, catalyzed by Ru/Rh systems [97] and in the methanol homologation reaction, catalyzed by Ru/Co systems [128]. Other approaches consisting of, *e.g.*, using metal vapor chemistry for preparing metal clusters and catalysts [181, 222] will not be considered here.

The following Sections are organized according to the ways of utilizing mixed-metal clusters in catalysis, as summarized in Scheme 1.1.

2. HETEROMETALLIC CLUSTERS IN HOMOGENEOUS CATALYSIS

2. 1. INTRODUCTION

The homogeneous catalysis of chemical reactions using bimetallic molecular compounds has attracted considerable interest in the recent years [207, 215]. This is largely based on the reasonable expectations (i) that the simultaneous or consecutive activation and transformation of substrate molecules, interacting with more than one, preferably adjacent metal centers, might be easier to achieve, (ii) that the high mobility of ligands in clusters will promote reactions between different partners, and (iii) that multi-site interactions involving the ligand(s) to be activated will convey information regarding the different nature of the metal atoms involved. Metal-specific electronic and steric interactions are expected to lead to increased selectivity in the overall chemical transformation. Thus, changes in activity and/or selectivity may be attributed to the proximity of the metals, to the inherent polarity of heterometallic bonds and to the stereochemical properties of each metal center. Since we shall only consider transition metal systems, other bimetallic couples, *e.g.*, Pt-Sn, for which synergic effects have also been evidenced, will not be discussed here [17, 141, 142, 275].

The following examples illustrate specific reactivity patterns of mixed-metal clusters relevant to fundamental steps occurring in homogeneous catalysis. The selective and reversible opening of a Mn-Fe bond, concomitant with reversible CO addition, has been observed by Huttner *et al.* [159, 271] in the phosphinidene-stabilized cluster $(\eta\text{-}C_5H_5)MnFe_2(\mu_3\text{-}PR)\text{-}(CO)_8$ (eq. 1):

$$(OC)_3Fe \cdots Fe(CO)_3 \quad \xrightarrow[- CO]{CO,\ 1\ atm} \quad (OC)_3Fe \cdots Fe(CO)_4 \qquad (1)$$

The more rarely encountered reversible H_2 addition has been shown by Stone *et al.* to be possible in $Os_3Pt(\mu\text{-}H)_2(CO)_{10}(PCy_3)$ without metal-metal bond rupture, whereas this tetrahedral cluster reacts with CO to afford a butterfly, *arachno* structure [108].

    Further metal-specific reactions have been observed by Roland and Vahrenkamp [254] with the tetrahedral cluster $Ru_2Co_2(CO)_{13}$ under very mild conditions: hydrogen reacts at the ruthenium atoms to form $H_2Ru_2Co_2(CO)_{12}$ whereas alkynes insert between the cobalt atoms to form $Ru_2Co_2(\mu_4\text{-}\eta^2\text{-}C_2R_2)(CO)_{11}$ (Scheme 2.1).

SCHEME 2.1.

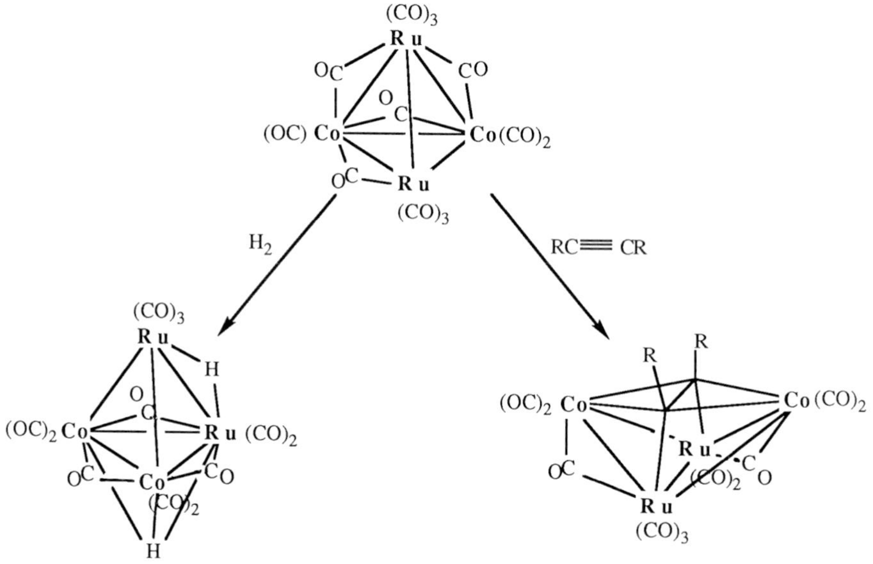

Mixed-metal cluster-mediated transformations of organic ligands have been observed, *e.g.*, with the conversion of an alkyne into a vinylidene ligand bound to a Ru/Co cluster (eq. 2) [253] or with the splitting of an alkyne into two

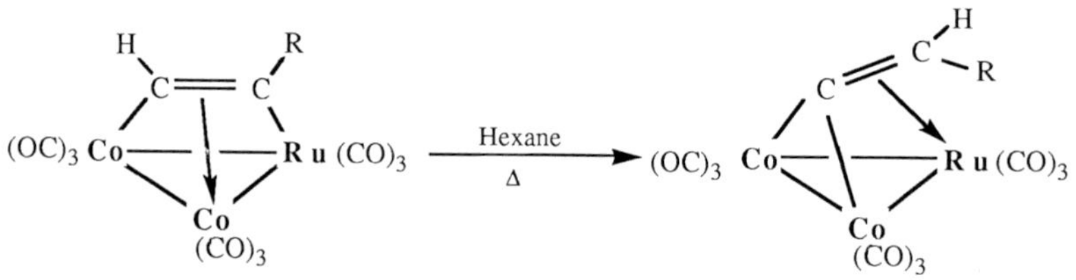

(2)

alkylidyne fragments on a tungsten-osmium cluster [223]. In studies of sulfur-capped clusters, it was found that whereas the homonuclear cluster $Ru_3(\mu_3\text{-}S)(\mu_3\text{-}CO)(CO)_9$ reacts with phenylacetylene at 98 °C to form $Ru_3(\mu_3\text{-}S)(\mu_3\text{-}HC_2Ph)(CO)_9$, head-to-tail coupling of three HC_2Ph molecules occurs with the structurally related mixed-metal cluster $Mo_2Ru(\mu_3\text{-}S)(\eta\text{-}C_5H_5)_2\text{-}(CO)_7$ (**A**) (Scheme 2.2). In the product **B**, the dimetallahexatrienyl ligand

SCHEME 2.2.

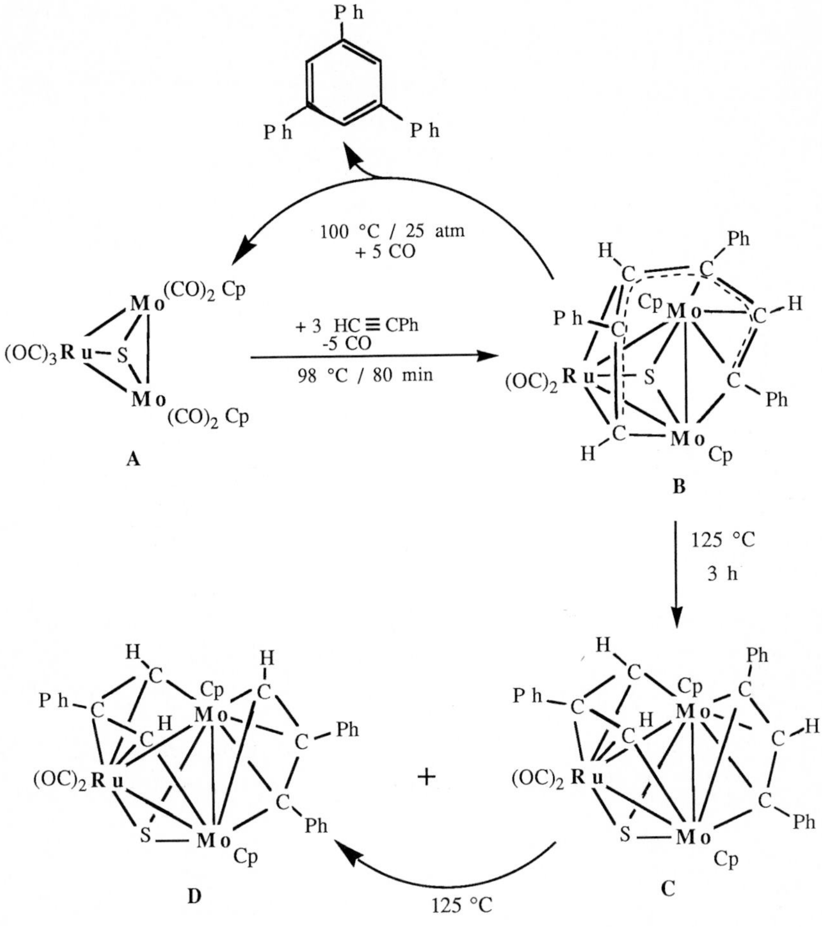

from Adams, R. D.; Babin, J. E.; Tasi, M. *Organometallics* **1987**, *6*, 2247.

is coordinated to all three metal atoms of the cluster. Thermal splitting of this ligand into two dimetallaallyl units was observed with the formation of **C** and **D** and it involves the rupture of an originally C-C triple bond in one of the alkyne molecules. The isomerization of **C** into **D** is also noteworthy. Whereas the dinuclear complex $Mo_2(\eta\text{-}C_5H_5)_2(CO)_6$ will produce the oligomerization of alkynes, no C-C bond cleavage is observed in the course of the process. Thus, the introduction of the dimolybdenum unit into **A** serves to provide an alkyne oligomerization activation site but the chemistry overall is dependent on the entire cluster functioning as a unit (Scheme 2.2).

The above mentioned transformations are not only important for understanding the molecular chemistry occurring in solution but they also provide conceptual tools for suggesting new pathways in surface chemistry and surface reaction mechanisms. These remarkable reactions are not encountered with the homonuclear clusters of the metals present in the mixed-metal compounds mentioned, emphasizing the unique reactivity that heterometallic clusters may display.

A number of tests have been suggested as criteria for catalysis by intact clusters [190, 215] and this was justified by the classical question of cluster fragmentation under the catalytic conditions, particularly if high temperatures and pressures are employed. Although it has been recently shown, for example, that Co-Rh heteronuclear metal-metal bonds might be stronger than the corresponding homonuclear ones [156], this question also applies to mixed-metal clusters and explains the efforts developed to stabilize cluster frameworks using capping ligands (*e.g.*, CR, NR, S, PR, polyphosphines) and to conduct catalytic reactions under mild conditions with the help of promoters (*e.g.*, halides and pseudohalides).

Although there is reasonable evidence for cluster-catalyzed reactions in few cases, the ultimate proof of the involvement of an intact cluster rests with asymmetric catalysis. This requires that a cluster having a rigid, chiral framework could catalyze the transformation of a substrate into isolable chiral product(s). It is important to remember here that the ligand arrangement about the cluster core results from combined electronic and steric effects. Unique effects may result from interactions between ligands bound to neighboring metal centers, a situation not available in mononuclear chemistry. Considerable efforts have been directed towards the synthesis of optically active clusters whose framework would be intrinsically chiral. Having a chiral ligand bound to a cluster would not be sufficient since it could dissociate during catalysis and be responsible for chiral induction. The first examples of such optically active clusters were provided with $FeCoM(\mu_3\text{-}S)(\eta\text{-}C_5H_5)(CO)_8$ (M = Cr, Mo, W) (Fig. 2.1) [246] and $FeCoW(\mu_3\text{-}PMe)(\eta\text{-}C_5H_5)(CO)_8$ [217].

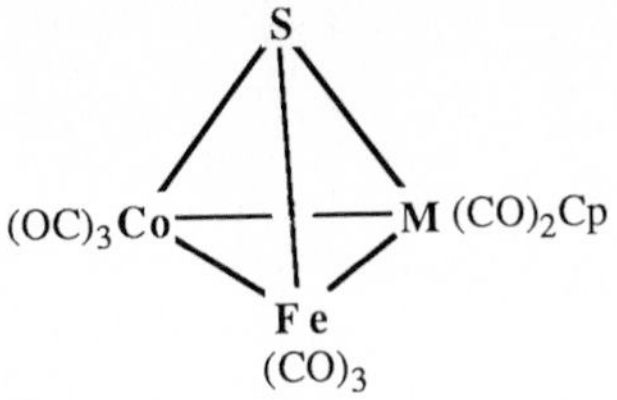

Fig. 2.1. The first intrinsically optically active clusters (M = Cr, Mo, W) [246].

Unfortunately, it was found that under catalytic conditions, these clusters racemize, as will be seen in the following Section.

2. 2. HYDROGENATION REACTIONS

2. 2. 1. Hydrogenation and Isomerization of Carbon-Carbon Multiple Bonds

The complex $CpMoRh(\mu-CO)_2(CO)(PPh_3)_2$ (**1**) in toluene solution acts as a catalyst for hydrogenation of cyclohexene to cyclohexane at ambient temperature and atmospheric pressure of dihydrogen. However, the reaction rate is significantly lower than with $RhCl(PPh_3)_3$. The nature of the catalytically active species is not determined [62].

The hydrogenation of 1- and 2-pentyne and 1,3 *cis* - and *trans* - pentadiene, catalyzed by $FeRu_2(CO)_{12}$ (**2a**) and $Fe_2Ru(CO)_{12}$ (**3a**) in toluene solution, has been studied under mild conditions [211]. The catalytic activity, lower than that of $Ru_3(CO)_{12}$, was found to decrease with increasing number of iron atoms in the dodecacarbonyls. For comparison, anchoring of **2a** or **3a** on γ-Al_2O_3 leads to a decrease in activity for pentyne hydrogenation but to increased activity for pentadiene hydrogenation (see Section 3).

The clusters **4-6**, stabilized by a capping group, were used as catalysts for the hydrogenation of styrene, α-substituted styrenes and 2-pentene. Whereas $HFe_2Co(\mu_3-PCH_3)(CO)_9$ (**5e**) was inactive and probably decomposed under the reaction conditions, the Ru-Co containing clusters **5a-c** and **6a** were found to be very active, and they could be recovered unaltered after catalysis. High turnover numbers were also observed, under mild conditions, for clusters with three different metals containing *at least* one ruthenium atom, namely **5d** and **6b**. The clusters were recovered in practically quantitative yield. The ruthenium-containing clusters were

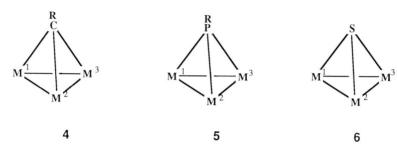

4	5	6

Cluster	R group	M^1	M^2	M^3
4 a	CH_3	$Co(CO)_3$	NiCp	$MoCp(CO)_2$
4 b	CH_3	$Co(CO)_3$	$FeH(CO)_3$	$MoCp(CO)_2$
5 a	C_6H_5	$Ru(CO)_3$	$Co(CO)_3$	$Co(CO)_3$
5 b	CH_3	$Ru(CO)_3$	$Co(CO)_3$	$Co(CO)_3$
5 c	CH_3	$Ru(CO)_3$	$Ru(CO)_3$	$CoH(CO)_3$
5 d	CH_3	$Ru(CO)_3$	$Fe(CO)_3$	$CoH(CO)_3$
5 e	CH_3	$Fe(CO)_3$	$Fe(CO)_3$	$CoH(CO)_3$
5 f	CH_3	$Fe(CO)_3$	$Co(CO)_3$	$WCp(CO)_2$
6 a	-	$Ru(CO)_3$	$Co(CO)_3$	$Co(CO)_3$
6 b	-	$Ru(CO)_3$	$Co(CO)_3$	$MoCp(CO)_2$

highly active for styrene hydrogenation, but even the most active styrene hydrogenation catalysts **5 c** and **5 d** were much less active in the hydrogenation of α-methylstyrene and totally inactive towards α-ethylstyrene under the same conditions. On the other hand, these clusters were moderatly active for hydrogenation of 1-pentene [204]. A summary of the catalytic properties of these clusters is presented in Table 2.1.

TABLE 2.1.

Catalytic Hydrogenation Activity of Mixed-Metal Clusters and Reactivity of the Organic Substrate (from ref. [204]).

Catalyst precursors				Reactivity of the substrates	
4a	-	5d	++++	2-pentene	++
4b	-	5e	-	styrene	++++
5a	+++	5f	-	α-methylstyrene	+
5b	++++	6a	++	α-ethylstyrene	-
5c	++++	6b	+		

Symbols: ++++ = very high activity; +++ = high activity; ++ = medium activity; + = low activity; - = no activity.

A remarkable activity and a very high selectivity was displayed by $H_2Ru_2Rh_2(CO)_{12}$ (**7**) in the hydrogenation of the double bond of 2-cyclohexenone, when compared with the ruthenium cluster $H_4Ru_4(CO)_{12}$ and/or the rhodium cluster $Rh_4(CO)_{12}$. Indications for the catalytic involvement of the "intact cluster" at 50 °C were provided [227].

TABLE 2.2.
Hydrogenation of 2-Cyclohexenone in the Presence of Various Catalytic Precursors Containing Ru and/or Rh (from ref. [227]).

Catalytic precursors	T(°C)	Conv.(%)	Composition of reaction products (mol %)		
			Cyclo-hexanone	2-Cyclohexenol	Cyclo-hexanol
$H_4Ru_4(CO)_{12}$	50	10.1	44.5	37.7	17.8
$Rh_4(CO)_{12}$	50	68.0	73.5	12.3	14.2
$H_4Ru_4(CO)_{12}+Rh_4(CO)_{12}$[a]	50	47.5	85.9	4.8	9.3
$H_2Ru_2Rh_2(CO)_{12}$(**7**)	40	5.8	100	0	0
	50	68.8	99.4	0.2	0.4
	60	35.8	60.1	1.2	38.7

[a] g. atom Ru/g. atom Rh =1.
Substrate 30.1 mmol; catalytic precursor 30.3×10^{-6} g.atom metal; mole substrate/g.atom metal 1003; tetrahydrofuran 9 ml; $p H_2$ 130 atm at 20 °C; reaction time 24 h.

The importance of the composition of the metal core in the low pressure hydrogenation of styrene is demonstrated by the observation that the initial hydrogenation rate with $Co_2Rh_2(CO)_{12}$ (**8**) is roughly double that of $Co_3Rh(CO)_{12}$ (**9**). This might reflect the fact that hydrogenation of styrene occurs only on the rhodium center, which is also in agreement with the inactivity of $Co_4(CO)_{12}$ alone. Additional trimethylphosphite enhanced the rate of hydrogenation except in the case of $Co_4(CO)_{12}$ which remains inactive. However, bimetallic processes were excluded [188].

The cluster $Ru_3NiCp(\mu-H)_3(CO)_9$ (**10**) is a selective hydrogenation catalyst for linear dienes such as *cis*-1,3-pentadiene, which is converted to *trans*-2-pentene, with some *cis*-2-pentene and a smaller amount of 1-pentene. However, the overall catalytic efficiency is higher for the osmium analogue $Os_3NiCp(\mu-H)_3(CO)_9$ (**11**). This is probably due to the

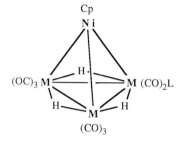

Cluster	M	L
10	Ru	CO
11	Os	CO
12 a	Os	PPh_2H
12 b	Os	$P(o\text{-}MeC_6H_4)_3$

greater stability of the nickel-osmium cluster which is recovered almost unaltered, whereas the nickel-ruthenium cluster decomposes extensively after only 40 minutes [65]. In both cases, the slowing down of the reaction rate with increased decomposition of the clusters was taken as an indication of intact cluster catalysis [66]. The conjugated cyclic diene 1,3-cyclohexadiene is selectively hydrogenated to cyclohexene in the presence of **10** [66]. Only one double bond is hydrogenated. Moreover, the hydrogenation of cyclohexene to cyclohexane occurs very slowly. With 1,4-cyclohexadiene, transformation to the 1,3-isomer occurs first, followed by hydrogenation of the latter to cyclohexene (eq. 3).

$$\text{(3)}$$

The hydrogenation of the cyclic diene is also accompanied by extensive decomposition of the Ru-Ni cluster and the reaction slows down when the cluster decomposes. Moreover, only traces of linear or of cracking products are observed. By contrast, *cluster-derived* metal particles [64] (see Section 4) readily cause C-C cleavage, and silica supported ruthenium particles derived from molecular clusters show considerable ability for C-C bond cleavage of linear and cyclic hydrocarbons [282].

The osmium-nickel cluster **11** has low activity and selectivity in the hydrogenation of *t*-butyl-alkynes or *t*-butyl-alkenes. Since this cluster can be recovered unaltered after several days of reactions at 110 °C, it might be a catalyst of low efficiency but high stability [70]. This complex is more effective in the hydrogenation of 3,3-dimethylbut-1-yne, but shows minimal activity towards the corresponding 3,3-dimethylbut-1-ene [69]. The same applies to 1-pentyne compared to 1-pentene, the former being easily hydrogenated to 1-pentene. An analogous behaviour is observed for the C_6 molecules such as cis-1,4-hexadiene, 2,4-hexadiene, 1,5-hexadiene, trans-3-hexene and 3-hexyne. In the absence of isomerization, internal triple bonds are hydrogenated more slowly than terminal ones. Conjugated or non-conjugated dienes are converted to monoenes with high efficiency and selectivity towards the terminal C=C bonds [69]. Selective hydrogenation reactions of one double bond have been observed and the efficiency toward the dienes is higher than that towards alkenes and alkynes, but the reaction stops after a few hours, probably because of cluster modifications [68]. The corresponding phosphine monosubstituted clusters $Os_3NiCp(\mu-H)_3(CO)_8L$, where L = PPh_2H (**12a**) or $P(o-MeC_6H_4)_3$ (**12b**), have comparable activity in the hydrogenation of dienes. After some hours, however, the reaction becomes considerably slower than for the unsubstituted cluster **11** and new complexes are formed [69].

On the other hand, the hexanuclear cluster $Os_3Ni_3Cp_3(CO)_9$ (**13**) was found to be more active in hydrogenation of t-butyl-alkynes and of t-butyl-alkenes, although the reduction of the former occurred more readily. During the hydrogenation reaction the cluster is decomposed to **11**, via an unidentified species probably responsible for the catalytic reaction [70].

13

The planar, triangulated clusters **14-22** were used as catalytic precursors for the hydrogenation of 1,5-cyclooctadiene (eq. 4). These clusters proved to be only moderately active hydrogenation catalyst and double bond isomerization competed effectively with hydrogenation. The yields of cyclooctane were very low in all cases. When considering only the hydrogenation products, rather selective conversions to cyclooctene are possible using the tungsten-containing clusters **16** and **17**. The molybdenum-containing clusters **14**, **15**, **18** and **19** are also effective for cyclooctadiene hydrogenation at 100 °C. It was generally found that these clusters are more active when the ligand PPh_3 is replaced by PEt_3 and that the palladium-containing clusters are somewhat more active catalyst precursors than their platinum counterparts [232].

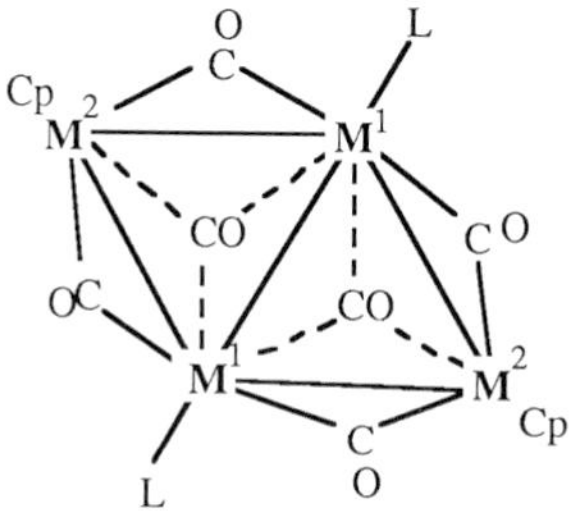

O
C
L
Cp
M2
M1
CO
CO
OC
CO
M1
M2
Cp
C
O
L

Cluster	M^1	M^2	M^3
14	Pd	Mo	PPh_3
15	Pd	Mo	PEt_3
16	Pd	W	PPh_3
17	Pd	W	PEt_3
18	Pt	Mo	PPh_3
19	Pt	Mo	PEt_3
20	Pt	W	PPh_3
21	Pt	W	PEt_3
22	Pd	Cr	PEt_3

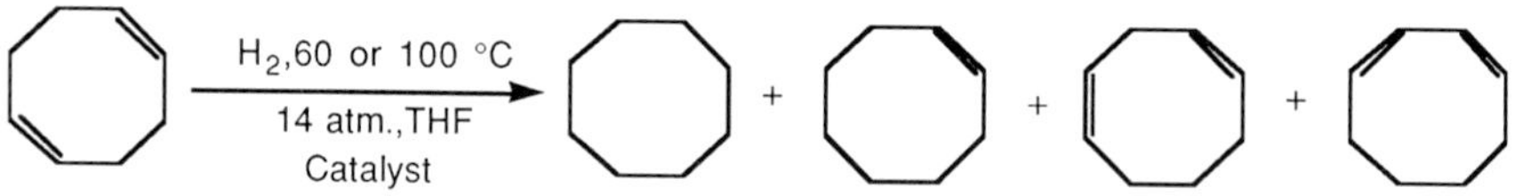

H2,60 or 100 °C
14 atm.,THF
Catalyst
+
+
+

(4)

COANE COENE 1,4-COD 1,3-COD

Phenylacetylene was hydrogenated to a mixture of styrene and ethylbenzene in the presence of clusters **14-20** (eq. 5). In each case, styrene was the major product, suggesting that these clusters might lead to efficient and selective catalysts for the alkyne reduction to alkene. Both the palladium and platinum clusters series were very active for this reaction. However, cluster **18** could not be recovered in good yields after the reaction [232].

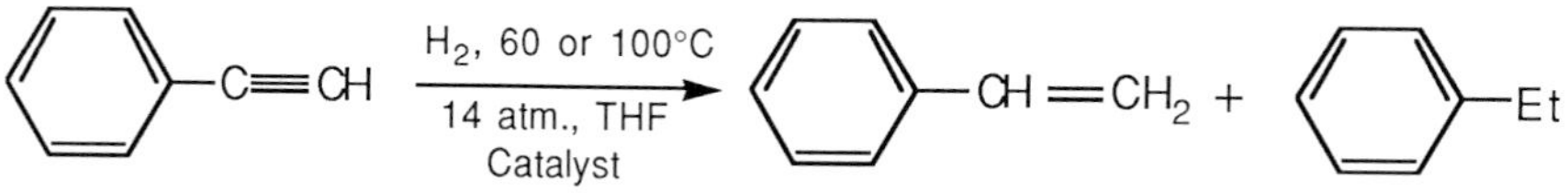

C≡CH
H2, 60 or 100°C
14 atm., THF
Catalyst
CH=CH2 +
Et

(5)

The hydrogenation of 1-hexyne (eq. 6) was also effected using clusters **14** and **15**, the linear complex *trans*-[Pt{Co(CO)$_4$}$_2$(c-C$_6$H$_{11}$NC)$_2$] (**23a**) and the triangular clusters **26** and **27** as catalyst precursors. The activity of clusters **26** and **27** is approximately the same as that of the linear complex **23a**. Cluster **15** is slightly more selective to hexene than **14**, and the tetranuclear clusters ligated with PEt$_3$ are again more active than their PPh$_3$ counterparts [232].

X(CO)$_3$M^1——Pt——M^1(CO)$_3$X

Cluster	M^1	L	X
23a	Co	c-C$_6$H$_{11}$NC	CO
24b	Fe	t-BuNC	NO
25a	Mo	c-C$_6$H$_{11}$NC	Cp

26 , L$_2$ = Ph$_2$PCH$_2$CH$_2$PPh$_2$ (DPPE)

27 , L$_2$ = Ph$_2$AsCH$_2$CH$_2$AsPh$_2$ (DPAE)

H$_2$, 60-75 °C
14-28 atm.
Catalyst

(6)

The photocatalyzed hydrogenation of 1-hexene was also studied (eq. 7). The Mo$_2$Pd$_2$ cluster **14** was a far more active catalyst than the platinum analogue Mo$_2$Pt$_2$ **18**. In fact, the latter was no more active than was cluster **14** in the absence of light [232].

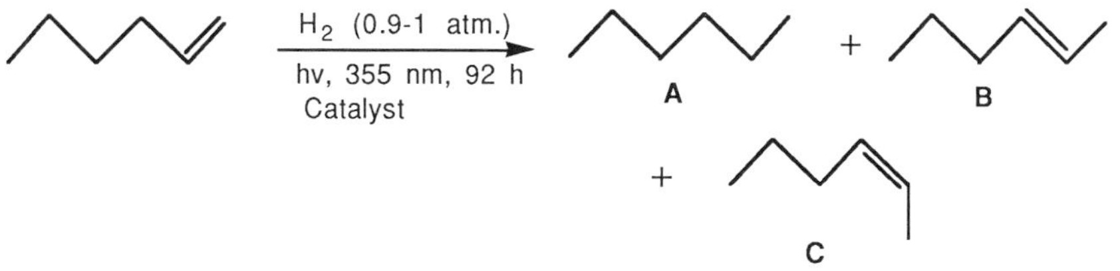

$$\text{(equation 7)}$$

Catalyst	Yields (%)		
	A	B	C
14 hv	52.1	30.2	7.9
14 dark	4.2	5.3	2.7
18 hv	3.8	1.9	1.5

A slight catalytic activity was also shown by **25a** and **28** for the hydrogenation of 1-octene, although they catalyze to a major extent the isomerization reaction [122]. Hydrogenation of terminal and internal

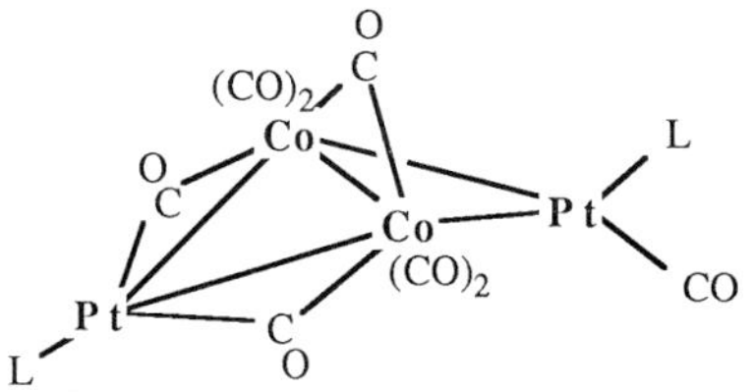

28a , L = PPh$_3$

28b , L = PEt$_3$

28c , L = AsPh$_3$

alkynes was also studied in the presence of cluster **25a, 28a** or **28c**. With terminal alkynes, a high activity was found with the linear complex **25a**, which could be recovered unaltered, while clusters **28a** and **28c** were transformed into new species [122].

With the iridium-platinum cluster Ir$_2$Pt$_2$(CO)$_7$(PPh$_3$)$_3$ (**29**), a considerable catalytic activity was found for the hydrogenation of cyclohexene, 2-methylcyclohexene, 2-butenal and 2-cyclohexenone (Table 2.3). For the last two compounds, *only* the olefinic bond was hydrogenated, the carbonyl group remaining unchanged. The homogeneous nature of the reaction mixture has been demonstrated and spectrophotometric and infrared studies did not indicate the formation of any other carbonyl complex to a detectable extent [29].

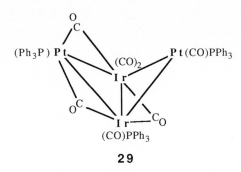

29

TABLE 2.3.

Catalytic Activity of Complex $Ir_2Pt_2(CO)_7(PPh_3)_3$ (**29**) (from ref. [29])

Reactant	Product	Turnover number[a] (h^{-1})
Cyclohexene[b]	Cyclohexane	320
2-Methylcyclohexene[b]	2-Methylcyclohexane	290
2-Cyclohexenone[c]	Cyclohexanone[d]	63
2-Butenal[c]	Butyraldehyde[e]	26
Cyclohexene[f]	Cyclohexane + Benzene	0.5

[a] (Moles of product)/(moles of catalyst x time).
[b] Reaction conditions: p H_2 (initial) = 0.68 atm; 70 °C, 0.005 mmol **29**; 25 mmol reactant; no solvent, 5 h.
[c] Conditions as in [b] but with cyclohexane as a solvent.
[d] Trace amounts of cyclohexanol.
[e] Trace amounts of n-butanol and 2-butenol.
[f] Conditions as in [b] but under vacuum in a sealed tube (10^{-3} mmHg) at 100 °C.

It was also shown [122] that high activities for the mono- and dihydrogenation of terminal alkynes can be obtained with **25a**, having the Mo_2Pt linear frame, whereas they were lower with the butterfly clusters **28**. No parallel high activity was found in the hydrogenation of internal alkynes, but selective hydrogenation of diphenylacetylene to *cis*-stilbene in relatively good yield was observed using clusters **28**. The complexes **23a**, **24b** or **25a** did not show any catalytic activity for either isomerization or hydrogenation of mono- and diolefins under all the reaction conditions investigated. While the catalyst with a Mo_2Pt core could be completely recovered after use, extensive transformation and rearrangement was found for the catalyst having the Co_2Pt_2 metallic frame [122].

The iron-palladium compound $[(OC)_4FePd(\mu-PPh_2)Cl]_2$ (**30**) is a good and selective catalyst for the hydrogenation of 1-hexyne in the presence of 1-hexene. At 175 °C, under 100 atm of H_2, 93 % of a sample of 1-hexyne in benzene was reduced to hexene and only 3 % to hexane. This is unexpected since palladium is usually an excellent catalyst for the hydrogenation of olefins [297].

30

Although isomerization reactions have been encountered above in the hydrogenation of dienes, the following systems lead more specifically to the isomerization of olefins. The $[HFeM(CO)_9]^-$ anions (**31a**, M = Cr; **31b**, M = Mo; **31c**, M = W) are found to be photocatalysts or catalyst precursors in the isomerization of external olefins into internal olefins [298]. Both 1-hexene and allylbenzene are catalytically isomerized in the presence of light. Allylbenzene has been converted into *cis*-and *trans*-propenylbenzene (*cis* : *trans* ratio *ca.* 0.1) (eq. 8).

(8)

cis   *trans*

A different dependence of catalyst activity on M is observed in the reactions with fluorescent light (Cr > Mo > W) compared to the photolysis reactions (Mo > W > Cr). The mixed-metal hydrides decompose under the photolysis conditions. The possible catalytically active fragment component $HFe(CO)_4^-$ was in fact shown to have very low activity toward olefin isomerization, in the absence of additives, whereas the $HM(CO)_5^-$ anions are inactive [298].

It was found that the tetrahedral cluster $H_2FeRu_3(CO)_{13}$ (**32**) readily catalyzes the isomerization of 1-hexene to give a mixture of *cis*- and *trans*- 2-hexenes. The *cis* : *trans* ratio was 1.0 : 3.25, and no 3-hexene was detected [117]. Approximately 60 % of the 1-hexene was isomerized, with a turnover number of *ca.* 4500 mol of hexene/mol of $FeRu_3$ cluster x hour. Most of **32** (> 95 %) was recovered unchanged.

The mixed-metal clusters $HFe_3Rh(CO)_{11}(\mu_4-\eta^2-C=CHR)$ (**33a,** R = H; **33b,** R = Ph) show specific activity under hydrogenation conditions. Thus, at 10 atm of hydrogen, isomerization of 1-heptene into *cis-* and *trans-* 2-heptene (*cis : trans* ratio *ca.* 0.3) occurs. Formation of 14 % heptane is also observed. Isomerization of 2-heptene into 3-heptene subsequently takes place at a lower rate. It thus appears that fragmentation of the cluster occurs. Under nitrogen, only a low catalytic activity was observed, and cluster decomposition occurred rapidly [10b].

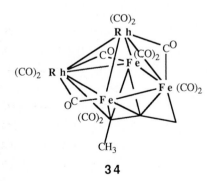

33a , R = H

33b , R = Ph

Compared to **33**, $[Ph_4P][Fe_3Rh_2(CO)_{10}(\mu-CO)_3(\mu_4-\eta^3-CH_3C=C=CH_2)]$ (**34**) was far less active toward the isomerization of alkenes. Under 10 atm of hydrogen it was necessary to heat at 60 °C to observe catalytic activity.

34

After 24 h, 25 % hydrogenation and 75 % isomerization of 1-octene into 60 % of 2-octenes and 40 % of 3-octenes were observed. At the end of the reaction, the cluster was recovered by crystallization in 60 % yield [10a]. These are two examples of mixed iron-rhodium systems in which the known reaction, the cluster was recovered by crystallization in 60 % yield   [10a]. These are two examples of mixed iron-rhodium systems in which the known hydrogenation ability of rhodium is lowered by iron, thus favoring isomerization over hydrogenation reactions.

Isomerization of mono- and dienes in the presence of Ru_3NiCp-$(\mu\text{-}H)_3(CO)_9$ (**10**) and $Os_3NiCp(\mu\text{-}H)_3(CO)_9$ (**11**) has also been investigated [66, 69], the former having higher activity for the isomerization of 1-pentene. Cluster **10** is also fairly active in the isomerization of 1,4-pentadiene to *cis* -1,3-pentadiene, in the absence of H_2, by contrast to **11**. However, cluster fragmentation was evidenced [66].

The Fe-Pd complex $[(OC)_4FePd(\mu\text{-}PPh_2)Cl]_2$ (**30**) is a selective catalyst for isomerization of 1-octene to 2-octene, no 3- or 4-isomers being formed [297].

Isomerization of 1-pentene was also studied in the presence of the complexes $H_3Ru_4\{M(PPh_3)\}(CO)_{12}$ (**35a**, M = Cu; **35b**, M = Au) and $H_2Ru_4\{M(PPh_3)\}_2(CO)_{12}$ (**36a**, M = Cu; **36b**, M = Au) [106]. Clusters **36** adopt a capped trigonal bipyramidal geometry but there is only one triply-bridging hydrido ligand in **36b**. It was clearly shown that the catalytic activity of $H_4Ru_4(CO)_{12}$ was modified by the introduction of $M(PPh_3)$ units. Thus, the two ruthenium-gold complexes clearly provide more active catalysts for 1-pentene isomerization than the parent $H_4Ru_4(CO)_{12}$, in marked contrast to the behaviour of the ruthenium-copper analogues (Table 2.4).

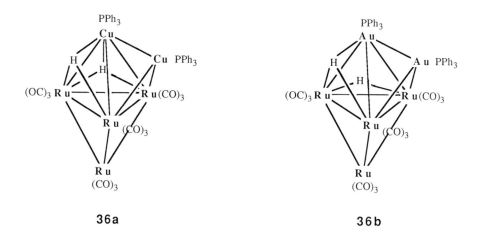

36a   **36b**

TABLE 2.4.

Isomerization and Hydrogenation of 1-Pentene[a] (from ref. [106])

Catalyst precursor	Conversion (%)	Selectivity (%) in C_5 products		
		Pentane	2-Pentene	
			trans	cis
Au(PPh$_3$)Cl	0	-	-	-
H$_4$Ru$_4$(CO)$_{12}$	32.9	1.7	19.8	11.0
[PPN][H$_3$Ru$_4$(CO)$_{12}$]	7.9	-	4.5	3.4
H$_3$Ru$_4${Au(PPh$_3$)}(CO)$_{12}$ (35b)	71.0	1.4	51.4	18.2
H$_2$Ru$_4${Au(PPh$_3$)}$_2$(CO)$_{12}$ (36b)	49.6	1.1	33.1	15.3
H$_3$Ru$_4${Cu(PPh$_3$)}(CO)$_{12}$ (35a)	29.4	0.3	18.5	10.7
H$_2$Ru$_4${Cu(PPh$_3$)}$_2$(CO)$_{12}$ (36a)	7.7	-	4.4	3.3

[a] Reaction conditions: catalyst 2.0 mol; 1-pentene, 54 mM; solvent CH$_2$Cl$_2$, 2.5 mL; 1 atm H$_2$, 35.0 ± 0.2 °C, 24 h.

The *trans : cis* ratio generally increases with overall activity. All the cluster complexes excepting **36a** were recovered unchanged after catalysis. The effects of changes in experimental conditions on the activity of **35b** are also discussed. It was observed that addition of PPh$_3$ or of a small partial pressure of CO strongly inhibited isomerization [106].

2. 2. 2. CO Hydrogenation

The homogeneous transition metal-catalyzed hydrogenation of carbon monoxide to give oxygenated products has been investigated in numerous laboratories over the last decade (eq. 9). Of special interest is the synthesis of ethylene glycol, which can be achieved with a selectivity of up to 70 % using rhodium catalysts [90, 235, 237].

$$CO + H_2 \xrightarrow[\text{> 500 atm, temp. > 200 °C}]{\text{[Rh clusters]}} HOCH_2CH_2OH + CH_3OH \qquad (9)$$

Cobalt catalysts are also active for ethylene glycol production [175, 177] whereas with ruthenium [35, 177] and iridium [176], the C_1 products methanol and methyl formate are formed with high selectivity. It is noteworthy that mixtures of rhodium and ruthenium complexes lead to significantly improved results, showing a synergy between the components of the system [97]. Furthermore, synthesis gas was also converted to ethylene glycol and its monoalkyl ether derivatives, using $Ru(acac)_3$-$Rh(acac)_3$ systems from which the mixed-metal cluster $RuRh_2(CO)_{12}$ (37), was generated and isolated [182].

An homogeneous process for converting CO and H_2 to methanol (CO/H_2 ratio was 1 : 1, THF, 275 °C, 1200 atm) in the presence of a cluster compound of formula $Ru_6(ML)_2C(CO)_{16}$ (L = organonitrile; 38a, M = Cu; 38b, M = Ag; 38c, M = Au) has been investigated [236]. It is interesting to note that no hydrocarbon was formed, which would have indicated the presence of ruthenium metal since the latter is known to catalyze the formation of hydrocarbons from synthesis gas.

38a

CO hydrogenation using a rhodium-platinum cluster as the catalyst has also been studied [257]. In the case of $[PPN][Rh_5Pt(CO)_{15}]$ (39b), ethylene glycol and methanol are the favoured products. Similar results were obtained with the rhodium cluster $[PPN][Rh_5(CO)_{15}]$ or with the mixture $Rh(CO)_2(acac)/Pt(acac)_2/[PPN][OAc]$. Therefore, it does not seem critical whether the different catalyst components are added separetely or as the mixed cluster compound. In contrast, the catalytic properties of the Rh/Pt cluster $[PPN]_2[Rh_4Pt(CO)_{12}]$ (40), as well as the yield in ethylene glycol, are greatly reduced, although the amount of rhodium involved in the reaction was the *same* as with the Rh_5Pt or Rh_5 clusters. This low activity was correlated with the enhanced amount of $[PPN]^+$ present rather than with the increased Pt/Rh ratio. Examination of product yields with reference to

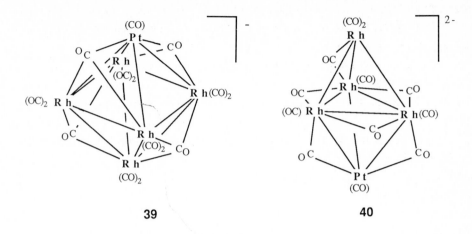

39   40

catalysts composed of mononuclear rhodium and platinum species appears to indicate cluster catalysis. Since platinum alone does not catalyze ethylene glycol but rather methanol formation [177], this result led the authors to suggest that mixed molecular clusters are probably involved in the equilibria which lead to the still unknown catalytic species [257].

Monovalent metal salts having general formula $M_2[Rh_{12}(CO)_{30}]$, where M is Cu (**41a**), Ag (**41b**), Au (**41c**) or Ir (**41d**) have also been evaluated as catalyst precursors in the reaction of CO and H_2 to produce oxygenated compounds such as methanol, ethylene glycol, glycerol and 1,2-propylene glycol. The clusters were recovered from the filtrate by recrystallization [310a]. Similarly, divalent metal salts of general formula $M'[Rh_{12}(CO)_{30}]$ where M' is Mn (**42a**), Co (**42b**), Zn (**42c**) or Pt (**42d**) as well as trivalent metal salts of the type $M''_2[Rh_{12}(CO)_{30}]_3$ in which M" is Ir (**43a**) or Re (**43b**) catalyze the synthesis of organic compounds from CO and H_2 [310b, 310c]. Salts of Sr, Mg or Ca containing the anions $[m_y m'_{12-y}(CO)_{30}]^{2-}$, m and m' being different metals, either Co, Ir or Rh, are also useful catalysts in the reaction of CO and H_2 to produce oxygenated compounds, such as ethylene glycol [51]. Similarly, Al, Ga, Ir, Se, Y or Re salts of the anions $[m_y m'_{12-y}(CO)_{30}]^{2-}$, such as $[Rh_6 Ir_6 (CO)_{30}]^{2-}$ (**44**), $[Rh_9 Ir_3 (CO)_{30}]^{2-}$ (**45a**) or $[Rh_9 Co_3 (CO)_{30}]^{2-}$ (**45b**), catalyze the reaction between CO and H_2. Oxygenated compounds, such as methanol, ethylene glycol, glycerol and 1, 2-propylene glycol, are produced [51].

2. 3. WATER-GAS SHIFT REACTION

The water-gas shift reaction (WGSR) (eq. 10) is of considerable industrial importance, since it provides means of adjusting the H_2/CO ratio in gas for, *e.g.*, Fischer-Tropsch synthesis.

$$H_2O(g) + CO(g) \longrightarrow H_2(g) + CO_2(g) \qquad (10)$$

The reaction is extremely slow in the absence of catalyst. It appears that catalysis of the WGSR can be effected by a surprisingly wide range of metal complexes and under markedly different reaction conditions. A result of particular interest is the higher catalytic activity of mixed-metal carbonyl clusters compared to homonuclear metal carbonyls alone.

In the mixed-metal Fe-Ru catalysis, one of the remarkable features is the observation that $H_2FeRu_3(CO)_{13}$ (**32**) is more active than either ruthenium carbonyl or iron carbonyl individually (Table 2.5). However, comparable activity was observed with a mixture of $Fe(CO)_5$ and $Ru_3(CO)_{12}$ in alkaline solution. Spectral characterization as well as isolation of various reaction components indicate the presence of several mixed-metal clusters, including **32**, in these solutions. The synergic catalysis by the mixed-metal carbonyl system is consistent with catalytic cycles in which the rate-limiting path is reductive elimination of H_2 from a cluster hydride species, and with mixed-metal clusters being more reactive toward H_2 elimination [114, 301]. Independently, the mixed tetranuclear cluster $H_4FeRu_3(CO)_{12}$ has been reported as being considerably less stable than the $H_4Ru_4(CO)_{12}$ homologue toward loss of H_2 [185].

TABLE 2.5.

WGSR Activities of Iron-Ruthenium Carbonyl Catalysts in Alkaline Aqueous Ethoxyethanol Solution[a] (from ref. [301])

Initial complex[b]	H_2 activity[c]	CO_2 activity[c]
$H_2FeRu_3(CO)_{13}$ (**32**)	10.3	10.9
$Fe(CO)_5$	1.0	1.1
$H_2Ru_4(CO)_{13}$	4.4	4.0
$H_4Ru_4(CO)_{12}$	3.7	3.3
$Ru_3(CO)_{12}$	2.8	2.7

[a] T = 100 °C, p CO = 0.9 atm; reaction carried out in all-glass vessels, stirred magnetically.
[b] Solutions prepared from 0.04 mmol of complex (0.012 M), 2 mmol of KOH (0.6 M), 0.02 mol of H_2O (6 M) and 3 mL of 2-ethoxyethanol.
[c] Activity is mol gas (mol complex)$^{-1}$ (day)$^{-1}$.

The clusters $FeRu_2(CO)_{12}$ (**2a**), $Fe_2Ru(CO)_{12}$ (**3a**), their phosphine and phosphite derivatives (Table 2.21), as well as $H_2FeRu_3(CO)_{13}$ (**32**) were also examined under basic conditions for the water-gas shift reaction [307]. Both parent clusters **2a** and **3a** were found to be more active catalysts precursors than the homonuclear clusters $Fe_3(CO)_{12}$ and $Ru_3(CO)_{12}$ alone, in accordance with the results of Ungermann et al. [301]. On the other hand, a pyridine solution of a mixture of $Fe(CO)_5$ and $Ru_3(CO)_{12}$ catalyzes the WGSR at a turnover frequency comparable to that of $FeRu_2(CO)_{12}$. The results for mixed clusters show a clear decrease in turnover frequency as the iron content of the cluster decreases. Monosubstitution by phosphine or phosphite enhances the activity of the clusters with a high iron content, e.g., **3b**, by stabilizing the parent cluster. Disubstitution of $Fe_2Ru(CO)_{12}$ (**3a**) seems to diminish the stability of the cluster under the basic WGSR conditions specified in the footnotes of Table 2.6 [307].

The precursors $Fe_2Os(CO)_{12}$ (**47**) and $H_2FeOs_3(CO)_{13}$ (**48**) produced a catalytically inactive system at low CO pressures [307]. These results are not surprising since neither $Fe(CO)_5$ nor $Os_3(CO)_{12}$ is active at low CO pressures.

Under relatively mild conditions, $HFeCo_3(CO)_{12}$ (**49**) is an inactive catalyst precursor, as $Fe_3(CO)_{12}$ or $Co_4(CO)_{12}$ alone. With $Na_2[FeIr_4(CO)_{15}]$ (**50**) or a mixture of $Fe(CO)_5$ and $Ir_4(CO)_{12}$ as catalyst precursors, moderate activity was observed, showing a weak synergic effect of the two metals. $Ir_4(CO)_{12}$ alone produced only a weakly active system [307].

The catalytic activity of Ru-Co and Ru-Rh clusters as catalyst precursors has also been reported [307]. Only $H_3Ru_3Co(CO)_{12}$ (**51**) was observed to initiate a catalytic system, the activity of which is comparable to that of $Ru_3(CO)_{12}$ (Table 2.6). Similar activities were also noted for mixtures of $Ru_3(CO)_{12}$ and $Co_4(CO)_{12}$ with various Ru/Co ratios, indicating that cobalt carbonyls do not significantly affect the activity of $Ru_3(CO)_{12}$. Under basic conditions, $RuCo_2(CO)_{11}$(**52**) and $HRuCo_3(CO)_{12}$(**53**) were practically inactive. Cluster **51** decomposes easily under CO atmosphere to yield $Ru_3(CO)_{12}$, which seems to be responsible for the catalytic activity of this system. Formation of active ruthenium species is probably hindered in the systems based on **52** or **53**.

A pyridine solution of the Ru-Rh mixed-metal clusters catalyzes the WGSR at a significant activity which, however, is lower than that observed for the most active rhodium catalyst precursor $Rh_2(CO)_4Cl_2$ under similar conditions (Table 2.6). Both in the cases of mixed-metal clusters and of metal carbonyl mixtures, ruthenium does not increase the catalytic activity of the rhodium species. The mixture of $[Ru(CO)_3Cl_2]_2$ and $Rh_2(CO)_4Cl_2$ produces a catalytic system showing the same activity as that initiated by $Rh_2(CO)_4Cl_2$ alone [307].

TABLE 2.6.

Catalysis of the WGSR by Ruthenium-Cobalt and Ruthenium-Rhodium Carbonyl Catalyst Precursors[a] (adapted from ref. [307]).

Precursor compounds	Amount of catalyst precursor (μ mol)	Heating time (h)	Turnover frequency[b]	
			$A\,H_2$	$A\,CO_2$
$RuCo_2(CO)_{11}$ (52)	40	2	0	5
$HRuCo_3(CO)_{12}$ (53)	20	17	0	2
$H_3Ru_3Co(CO)_{12}$ (51)	20	17	12	11
$Ru_3(CO)_{12}$	20	17	15	15
$Ru_3(CO)_{12}/Co_4(CO)_{12}$	40/20	17	18	15[c]
$Ru_3(CO)_{12}/Co_4(CO)_{12}$	40/40	17	19	15[c]
$Ru_3(CO)_{12}/Co_4(CO)_{12}$	20/40	17	20	16[c]
$Co_4(CO)_{12}$	38	17	0	0
$Rh_2(CO)_4Cl_2$	20	2	280	270
$HRuRh_3(CO)_{12}$ (54)	24	17	47	47
$HRuRh_3(CO)_{12}$ (54)	24	2	110	130
$[PPN][RuRh_5(CO)_{16}]$ (55)	20	17	49	42
$H_2Ru_2Rh_2(CO)_{12}$ (7)	20	17	16	15
$Ru_3(CO)_{12}/Rh_4(CO)_{12}$	40/40	2	28	26
$Ru_3(CO)_{12}/Rh_2(CO)_4Cl_2$	20/20	2	130	110
$[Ru(CO)_3Cl_2]_2/Rh_2(CO)_4Cl_2$ [d]		2	280	250

[a] Conditions: base / solvent = pyridine; T = 100 °C; p CO = 0.45-0.50 atm; pyridine 3.0 mL; H_2O 0.36 ml.
[b] Turnover frequency = mol gas (mol catalyst precursor x 24)$^{-1}$.
[c] Activities were determined per mol of $Ru_3(CO)_{12}$.
[d] Activity has been determined as an average for seven mixtures with different Ru/Rh contents. $A\,H_2$ and $A\,CO_2$ were determined per mol of $Rh_2(CO)_4Cl_2$.

Under basic WGSR conditions, the precursors $Co_2Rh_2(CO)_{12}$ (8) and $Co_3Rh(CO)_{12}$ (9) produce also a highly active catalytic system. The cluster $Co_2Ir_2(CO)_{12}$ (56) is only a weakly active catalysts precursor and no synergic effect was observed between cobalt and iridium (Table 2.7). Among the cobalt group metals, only the rhodium compounds showed high catalytic activity whereas the cobalt carbonyls are totally inactive. The possibility of synergic behaviour in the Co-Rh mixed-metal carbonyls has been excluded under the present conditions.

TABLE 2.7.

Catalysis of the WGSR by Cobalt-Rhodium, Cobalt-Iridium and Rhodium-Iridium Carbonyl Catalyst Precursors[a] (adapted from ref. [307]).

Precursor cluster compounds	Amount of catalyst precursor (μ mol)	Heating time (h)	Turnover frequency[b]	
			A H_2	A CO_2
$Co_3Rh(CO)_{12}$ (9)	20	2.5	180	150
$Co_3Rh(CO)_{12}$ (9)	10	2.5	350	310
$Co_2Rh_2(CO)_{12}$ (8)	0.5	22	700	570
$Co_4(CO)_2$	38	17	0	0
$Co_4(CO)_{12}/Rh_4(CO)_{12}$	37/10	2	140	120[c]
$Rh_4(CO)_{12}$	20	17	110	100
$Co_2Ir_2(CO)_{12}$ (56)	12	17	0	3
$Rh_2(CO)_4Cl_2$	20	2	280	270
$Rh_2(CO)_4Cl_2/[Ir(CO)_3Cl]_2$	0.7/0	19	590	460
$Rh_2(CO)_4Cl_2/[Ir(CO)_3Cl]_2$	7.2/20	2	120	110
$[Ir(CO)_3Cl]_2$	42	17	3	4

[a] Conditions: base/solvent = pyridine; T = 100 °C; p CO = 0.45 -0.60 atm; pyridine 3.0 ml; H_2O 0.36 ml.
[b] Turnover frequency = mol gas (mol catalyst precursor x 24 h)$^{-1}$.
[c] Activities were determined per mol of $Co_4(CO)_{12}$ + $Rh_4(CO)_{12}$. Turnover frequencies per mol of $Rh_4(CO)_{12}$ are A H_2 = 640 and A CO_2 = 550.

General results obtained with carbonyl precursors of the Groups 8 and 9 are summarized in Table 2.8. The synergic effect in these mixed-metal carbonyl systems was evident in the cases of the Fe/Ru and Fe/Ir systems but was much weaker in the latter case.

TABLE 2.8.

Catalytic Activity of Mixed-Metal Carbonyl Precursors of the Groups 8 and 9 in the WGSR (from ref. [307]).

	Fe	Ru	Os	Co	Rh	Ir
Fe	5	250	0	0	30	32
Ru		15	c	36	140	24[a]
Os			0	c	c	3[b]
Co				0	700	1
Rh					590	130
Ir						3

[a] Activity of a mixture of $RuCl_3$ and $[Ir(CO)_3Cl_2]_2$ as a catalyst precursor.
[b] Activity of a mixture of $Os_3(CO)_{12}$ + $[Ir(CO)_3Cl_2]_2$.
[c] Not studied.

2. 4. HYDROFORMYLATION REACTIONS

The mixed-metal complex $Cp_2Zr(CH_2PPh_2)_2RhH(PPh_3)$ **(57)** was found to be an active hydroformylation catalyst at 80 °C and 20 atm ($H_2/CO = 1:1$). A high rate of conversion of 1-hexene was observed, but the normal/branched selectivity decreased during the course of the reaction to reach a ratio of *ca.* 2. However, a significant improvement of the catalytic performance is achieved when using a mixture of Rh and Zr complexes, higher conversion rates and an increased ratio of *n/iso* aldehyde being observed [82].

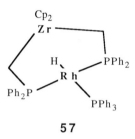

57

Similarly, the heterobimetallic complex **58a** catalyzes the hydroformylation of 1-hexene at 80 °C and 5 atm CO + H_2. Its enhanced activity compared to the analogous complex **58b** is attributed to the Zr center inducing more electron density on rhodium. The complexes **58a** and **58b** could be recovered after catalysis [276, 277].

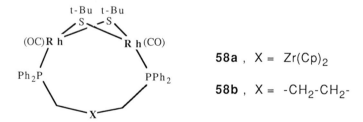

58a , X = $Zr(Cp)_2$

58b , X = $-CH_2-CH_2-$

A synergic effect was observed when the mixed-metal cluster $[Et_4N][FeCo_3(CO)_{12}]$ **(59b)** was used as catalyst in the hydroformylation of cyclohexene to yield cyclohexanecarbaldehyde (CHC) (eq. 11) (Table 2.9). In comparison, $[Et_4N][Fe_3Co(CO)_{13}]$ **(60b)**, which has also a tetrahedral structure, exhibited little activity. Addition of $Fe_3(CO)_{12}$, with or without

[Bu$_4$N]Cl, to Co$_2$(CO)$_8$ did not improve the activity [150].

(11)

TABLE 2.9.

Hydroformylation of Cyclohexene with Fe-Co and Ru-Co Mixed-Metal Catalysts[a] (adapted from ref. [150, 151]).

Catalyst	Ru/Co or Fe/Co ratio	Yield of CHC (%)[b]	Initial rate [c]
Co$_2$(CO)$_8$	-	14	1.0
Ru$_3$(CO)$_{12}$[d]	-	3	0.3
[Et$_4$N][FeCo$_3$(CO)$_{12}$] (59b)	0.33	27	2.3
Co$_2$(CO)$_8$ + Fe$_3$(CO)$_{12}$ + [Bu$_4$N]Cl	0.35	10	1.1
Co$_2$(CO)$_8$ + Fe$_3$(CO)$_{12}$	0.91	9.2	1.0
[Et$_4$N][Fe$_3$Co(CO)$_{13}$] (60b)	3.0	0.2	0.0
Co$_2$(CO)$_8$ + Ru$_3$(CO)$_{12}$	0.34	32	3.6
Co$_2$(CO)$_8$ + Ru$_3$(CO)$_{12}$	0.95	52	5.9
Co$_2$(CO)$_8$ + Ru(CO)$_3$(PPh$_3$)$_2$[e]	0.89	10	1.1
Co$_2$(CO)$_8$ + Ru(CO)$_4$(PPh$_3$)[f]	0.99	35	3.3
HRuCo$_3$(CO)$_{12}$ (53) + HC(PPh$_2$)$_3$	0.33	37	3.0

[a] Reaction conditions: Co, 2 x10^{-4} g. atom; cyclohexene, 80 mmol; THF, 10 mL; T = 110 °C; reaction time, 4 h ; CO/H$_2$, 40 : 40 atm (initial pressure at room temp.)
[b] The yield of cyclohexanecarbaldehyde is based on the starting cyclohexene. It includes the 2,4,6-tricyclohexyl-1,3,5-trioxane formed, because the latter quantitative'·' decomposes to CHC under the conditions of direct GC analysis.
[c] Relative to Co$_2$(CO)$_8$ alone.
[d] Ru, 6 x 10^{-4} g. atom.
[e] T = 130 °C.
[f] Benzene as solvent, 10 mL.

In the ruthenium-cobalt bimetallic systems presented in Table 2.9, a similar synergic effect was also observed. The activity of HRuCo$_3$(CO)$_{12}$ (53) was investigated in the presence of triphenylphosphine or of the assembling ligand tris(diphenylphosphino)methane, expected to maintain the metals in close proximity. The results were similar to those obtained for the FeCo$_3$ cluster 59b. However, the activity was greatly improved when increasing the Ru/Co ratio [150, 151]. The hydroformylation by the bimetallic catalysts was remarkably influenced by the nature of the solvent used, and alcohols such as methanol and ethanol were the best among the solvents employed [151].

In iron-rhodium bimetallic systems, no specific effect due to the presence of iron was observed. For example, in the hydroformylation of 1-pentene, the cluster $HFe_3Rh(CO)_{11}\{\mu_4\text{-}\eta^2\text{-}C=CH(Ph)\}$ (**33b**) showed the same activity as $Rh_4(CO)_{12}$ whereas in olefin isomerization reactions, the mixed-metal system showed specific activity (see Section 2.2.1). It thus appears that under catalytic conditions, fragmentation of the Fe_3Rh cluster occurs, resulting in the formation of $Rh_4(CO)_{12}$ [10b]. The cluster $[Ph_4P][Fe_5RhC(CO)_{16}]$ (**107**) was also used as catalyst precursor for the hydroformylation of 1-pentene. Whereas 70 % of 1-pentene were converted into a 2.7 : 1 mixture of hexanal and 2-methylpentanal, the remaining 30 % was converted into 2-pentene. Under these catalytic conditions (60 atm CO + H_2 at 100 °C), the cluster **107** was transformed and the new neutral cluster $Fe_4Rh_2C(CO)_{16}$ (**108**) and the known $[Ph_4P][Fe_4RhC(CO)_{14}]$ (**109**) were isolated in 18 % and 74 % yield, respectively. Cluster **108** showed a better activity than **107**, and at the end of the reaction, the infrared spectrum of the solution only gave evidence of the starting cluster. Cluster **109** showed an activity similar to **107**, but was transformed during catalysis into $[Ph_4P][Fe_3Rh_3C(CO)_{15}]$ (**110**). This cluster was also shown to be a catalyst for the hydroformylation of 1-pentene, with an activity similar to that of **108**. Cluster **110** could be recovered in 70-80 % yield after catalysis [10c].

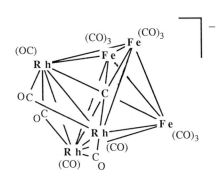

110

A cobalt-rhodium system, promoted by triphenylphosphine, enables a selective hydroformylation of dicyclopentadiene under relatively mild conditions (eq. 12). Bimetallic species, such as $CoRh(CO)_6(PPh_3)$ (**61**), were thought to play a key role in this reaction [123]. A comparison of the catalytic behaviour of the mixed *vs.* the corresponding homometallic systems clearly points to a synergic effect in the former case (Table 2.10).

(12)

TABLE 2.10.
Synergic Effect of Cobalt and Rhodium in the Catalyzed Hydroformylation of Dicyclopentadiene (after ref. [123])^a

	Selectivity %		
Catalyst precursor	Monoaldehydes	Dialdehydes	Ketones
$Co_2(CO)_8$	60.95	36.37	1.25
$Co_2(CO)_8 + Rh_4(CO)_{12}$	5.50	94.50	-
$Rh_4(CO)_{12}$	15.45	67.97	8.94

^{a} [M]/PPh$_3$ = 2; P = 40 atm; T = 110 °C.

Similarly, a synergic effect was found by Ojima [220a] in the hydroformylation of pentafluorostyrene. Recent evidence points to the involvement of the coordinatively unsaturated bimetallic compound $CoRh(CO)_7$ (**62**) as the active species [220b]. Excellent regioselectivities were also achieved in the hydroformylation step of the hydroformylation-amidocarbonylation of trifluoropropene (eq. 13) using a $Co_2(CO)_8/Rh_6(CO)_{16}$ system in which **62** could be generated as well.

62

$$CF_3CH = CH_2 + CO + H_2 + CH_3CONH_2 \longrightarrow$$

$$(13)$$

A mixed-metal $Co_xRh_y(CO)_z$ species has also been postulated previously as an active catalyst in the hydrocarbonylation of diketene [289].

In the hydroformylation of cyclohexene, under mild conditions (50 °C, 1 atm $CO+H_2$), the bimetallic cluster $Co_2Rh_2(CO)_{12}$ (**8**) shows an activity greater than that of $Rh_4(CO)_{12}$ and addition of $P(OPh)_3$ to **8** enhances the activity. When considering rhodium as the only active metal center in these systems, the observed results suggest the occurrence of a synergic effect, which apparently decreases on increasing the $P(OPh)_3/Co_2Rh_2(CO)_{12}$ molar ratio [73a]. This contrasts with the hydroformylation of 1-pentene and of styrene for which cluster **8** or $Rh_4(CO)_{12}$ alone is poorly active. A similar behaviour is observed for mixtures of $Rh_4(CO)_{12}$ and $Co_2(CO)_8$. However, addition of phosphines lead to active systems and high hydroformylation rates of, *e.g.*, styrene are observed. The ligand inducing the highest activity is triphenylphosphine and the branched 2-phenylpropanal is the isomer prevailing in the reaction mixture [73a] (Table 2.11). The combined addition of PPh_2H and of an additional tertiary phosphine greatly affects the nature of the species present in solution, and generates very active systems [73b]. The catalytic activities for 1-hexene hydroformylation of the mixed cluster $Co_3Rh(CO)_{12}$ (**9**) were compared with those of $Rh_4(CO)_{12}$ and $Rh_6(CO)_{16}$ [311b].

TABLE 2.11.
Hydroformylation of Styrene[a] (adapted from ref. [73a]).

Compound[a]	L	L[c] M$_4$(CO)$_{12}$	T (°C)	Rate[d]	Iso /n [e]
Co$_2$Rh$_2$(CO)$_{12}$ (8)	-	-	25-50	ca. 0	-
Co$_2$Rh$_2$(CO)$_{12}$ (8)	PPh$_3$	3	25	4.3	8.6
Co$_2$Rh$_2$(CO)$_{12}$ (8)	PPh$_3$	3	50	25.6	1.6
Rh$_4$(CO)$_{12}$	-	-	25-50	ca. 0	-
Rh$_4$(CO)$_{12}$	PPh$_3$	3	25	10.2	14.6
Rh$_4$(CO)$_{12}$	PPh$_3$	5	25	43.0	14.5
Rh$_4$(CO)$_{12}$	PPh$_3$	5	50	110.6	1.6

[a] Reactions conditions: toluene at 25 °C and 1 atm of CO / H$_2$ (1 : 1)mixture
[b] Concentration 7 x 10^{-3} M.
[c] Molar ratio.
[d] Initial hydroformylation rate expressed as mol aldehyde per mol cluster per h.
[e] 2-Phenylpropanal / 3-phenylpropanal molar ratio.

In the hydroformylation of formaldehyde, the bimetallic cluster Co$_2$Rh$_2$(CO)$_{12}$ (8) provided an unexpected result: although it shows an activity only slightly greater than expected for a system formed from equimolar amounts of Co$_4$(CO)$_{12}$ and Rh$_4$(CO)$_{12}$ behaving independently, it greatly alters the product distribution and produces both glycolaldehyde and ethylene glycol with an overall molar selectivity of ca. 50% (Table 2.12). Variable mixtures made by combining either Co$_2$(CO)$_8$ or Co$_4$(CO)$_{12}$ with Rh$_4$(CO)$_{12}$ were also found to produce C$_2$-oxygenated species, however with a lower overall selectivity and with preferential formation of ethylene glycol over glycolaldehyde [206]. Although these results seem to imply involvement of bimetallic Co-Rh species in catalysis, it was found that Co$_2$Rh$_2$(CO)$_{12}$ in DMF and under carbon monoxyde is not stable and, in a matter of minutes, quantitatively disproportionates acoording to eq. 14.

$$5 \text{ Co}_2\text{Rh}_2\text{(CO)}_{12} \xrightarrow{\text{DMF, CO}} 2 \text{ [Rh}_5\text{(CO)}_{15}]^- + 4 \text{ Co}^{2+} + 6 \text{ [Co(CO)}_4]^- + 6 \text{ CO} \qquad (14)$$

It has also been established recently by Horvath et al. [157b] that the Co-Rh mixed system Co$_2$Rh$_2$(CO)$_{12}$ (8) reacts with carbon monoxide in a facile and reversible reaction to give CoRh(CO)$_7$ (62) (eq. 15) whose unusual stability (see eq. 16 and eq. 17) is strongly influenced by the nature of the ligands

TABLE 2.12

Effect of Combining Cobalt and Rhodium Precursors on Activity and Selectivity[a] in the Hydroformylation of Formaldehyde (after ref. [206])

Precursor Complexes	Co/Rh ratio	P_i (atm)	Time (h)	Activity[b]	Selectivity[c]	
					$HOCH_2CHO$	$HOCH_2CH_2OH$
$Co_2Rh_2(CO)_{12}$ (8)	1	120	9	83.0	41.4	6.6
$Co_2(CO)_8$ + $Rh_4(CO)_{12}$	1	120	6	82.1	1.5	8.2
$Co_4(CO)_{12}$ + $Rh_4(CO)_{12}$	0.5	100	8	81.9	0.3	17.6
CoI_2 + $Rh_4(CO)_{12}$	1	100	8	73.2	15.7	10.4
$CoCl_2·6H_2O$ + $Rh_4(CO)_{12}$	1	100	8	65.1	1.7	25.2
$[PPN][Co(CO)_4]$ + $Rh_4(CO)_{12}$	1	98	3	84.0	-	4.7
CoI_2 + $[TMBA]_2$-$[Rh_{12}(CO)_{30}]$	0.13	120	8	82.4	1.3	12.5
$CoCl_2·6H_2O$ + $RhCl_3·3H_2O$	1	87	6	35.0	18.3	0.6

[a] $[M' + M''] = 4 \times 10^{-2}$ M; solvent = DMF; volume of the solution = 10 ml; [HCHO] = 4 M; P_i = initial pressure at 25 °C ; T = 110 °C.
[b] Activity = moles of converted HCHO x 100/moles of starting HCHO.
[c] Selectivity = moles of product x 100/moles of converted HCHO.

present [157a]. This is in sharp contrast to other mixed clusters, such as $HRu_3Co(CO)_{13}$ and $H_2FeRu_3(CO)_{13}$ (32) which decompose easily to the corresponding homonuclear species under 1 atm of CO at 25-70 °C [116]. Furthermore, 62 has been show to activate molecular hydrogen at low temperature in the presence of carbon monoxide [157c].

$$Co_2Rh_2(CO)_{12} + 2\,CO \xrightarrow[\text{hexane 0 °C, 2 atm CO}]{} 2\,CoRh(CO)_7 \quad (15)$$
$$(> 98\ \%\ \text{transformation})$$

$$CoRh(CO)_7 + CO \xrightarrow[\text{hexane, 0 °C, 160 atm CO}]{} CoRh(CO)_8 \quad (16)$$
$$(50\ \%)$$

$$4\,CoRh(CO)_7 \xrightarrow{\text{R.T., 10 atm CO, 5 days}} Co_2(CO)_8 + Rh_4(CO)_{12} \quad (17)$$
$$(50\ \%\ \text{transformed})$$

The mixed-metal clusters **4a, 63a-c, 65** and **66** have been tested as catalysts for the hydroformylation of 1-pentene and of styrene [233b, 244].

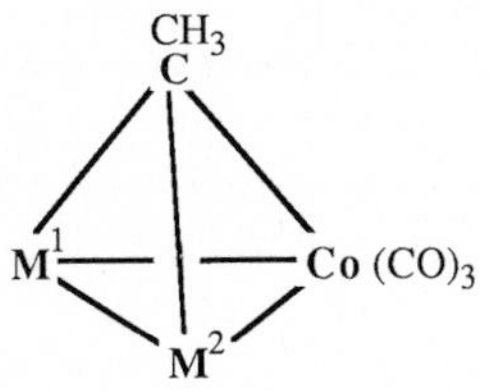

Cluster	M^1	M^2
4a	Mo(CO)$_2$Cp	NiCp
63a	Co(CO)$_3$	NiCp
63b	Co(CO)$_3$	Mo(CO)$_2$Cp
63c	Co(CO)$_2$P(OMe)$_3$	Mo(CO)$_2$Cp

Hydroformylation of 1-pentene to hexanal and 2-methylpentanal was achieved under mild batch conditions with all the clusters examined with the exception of **65**. Isomerization of 1-pentene competes with hydroformylation and is especially pronounced in the case of clusters **4a**, **63a** and **66** (Table 2.13).

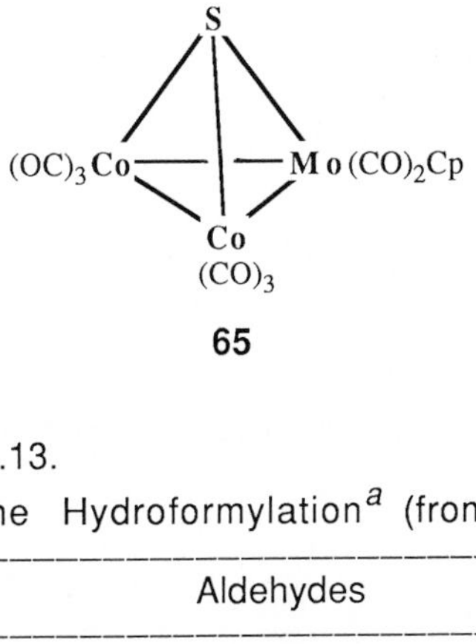

65

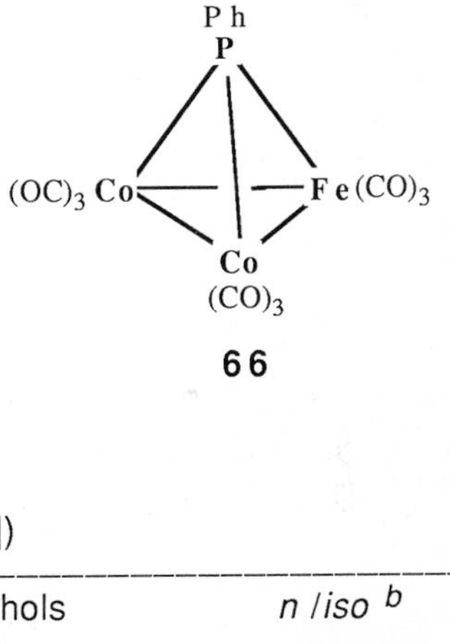

66

TABLE 2.13.
1-Pentene Hydroformylation[a] (from ref. [244])

Catalyst	Aldehydes	Alcohols	n /iso [b]
4a	8.0	-	3.2
63a	87.5	11.4	0.6
63b	13.5	-	·2.4
63c	25.5	-	2.9
65	2.4	-	c
66	89.4	1.2	1.4
Co$_2$(CO)$_8$	91.6	1.0	1.5

[a] Reaction conditions: 0.01mmol of cluster, 130 °C, THF, 40 atm H$_2$ / CO (1 : 1). Yields based on pentene consumed.
[b] Ratio of hexanal to 2-methylpentanal.
[c] Hexanal only.

Hydroformylation of styrene was achieved under mild conditions with moderate to high branched to normal selectivity when cluster **4a**, **63a** or **63c** was employed. The conditions used are shown in eq. 18.

$$\text{(styrene)} \xrightarrow[\substack{60\ ^\circ\text{C ,THF} \\ \text{Catalyst}}]{56\ \text{atm}\ \ \text{CO/H}_2(1:1)} \text{(branched CHO)} + \text{(linear CHO)} \qquad (18)$$

With the exception of **66** used in 1-pentene hydroformylation, all the clusters employed could be recovered in high yield (> 90 %) after catalysis [244]. Other optically active clusters were studied in order to examine the process of asymmetric induction in cluster-catalyzed reactions. Thus, in order to check the catalytic properties of clusters **67a** and **67b**, the hydroformylation of styrene was studied [201]. Cluster **67a** is an effective catalyst, giving a conversion of 25 %, the ratio of branched to linear aldehydes being 60 : 40 whereas **67b** did not yield any detectable hydroformylation products. Thus, the replacement of cobalt in **67a** by molybdenum in **67b** results in complete loss of activity. In both cases, only a small amount of the organometallic precursor could be recovered. These experiments do not yet prove the reality of asymmetric synthesis catalyzed by an intrinsically chiral cluster [201].

$$
\begin{array}{l}
\text{L-X} = \mu\text{-1,2-}\eta^2\text{-C(Me):N(Ph)} \\
\textbf{67a} \ , \quad M = Co(CO)_3 \\
\textbf{67b} \ , \quad M = Mo(CO)_2Cp
\end{array}
$$

Hydroformylation of 1-pentene has also been examined in the presence of palladium- or platinum-containing mixed-metal compounds [232] (Table 2.14). Only those systems which contained cobalt were active as catalysts at the low temperature employed (60-100 °C). For example, $Mo_2Pd_2Cp_2(CO)_6$-$(PPh_3)_2$ (**14**) and $Mo_2Pt_2Cp_2(CO)_6(PPh_3)_2$ (**18**) were totally inactive as

TABLE 2.14.

Hydroformylation of 1-Pentene Catalyzed by Mixed-Metal Clusters[a] (from ref. [232]).

Catalyst		Temp. (°C)	Time (h)	Conversion[b]	Products[b]	
No.	Formula				Hexanal	2-methyl pentanal
14	$Mo_2Pd_2Cp_2(CO)_6(PPh_3)_2$	95	17	0	-	-
18	$Mo_2Pt_2Cp_2(CO)_6(PPh_3)_2$	95	17	0	-	-
26	$Co_2Pt(DPPE)(CO)_8$	60	15	0	-	-
26	$Co_2Pt(DPPE)(CO)_8$	80	21	39	30.7	8.2
27	$Co_2Pt(DPAE)(CO)_6$	80	22	0	-	-
23 a	trans -[Pt{Co(CO)_4}_2(C_6H_{11}NC)_2]	62	20	0	-	1
23 a	trans -[Pt{Co(CO)_4}_2(C_6H_{11}NC)_2]	100	20	39.3	30	9.3
28 a	$Co_2Pt_2(CO)_8(PPh_3)_2$	62	17	trace	trace	trace
28 a	$Co_2Pt_2(CO)_8(PPh_3)_2$	75	18	16.5	14	2.5
28 a	$Co_2Pt_2(CO)_8(PPh_3)_2$	100	17	85.4[c]	63.5	14.6

[a] Reaction conditions : ≈ 56 atm, H_2/CO (1 : 1) 9 mmol 1-pentene, 0.01 mmol catalyst, benzene (4ml), with the exception of the reactions using 28a where toluene (4ml) was used.

[b] Mol % based on 1-pentene charged.

[c] 7.3 % of 1-hexanol was also present.

hydroformylation catalysts at 95 °C. In contrast, cobalt-containing clusters, such as the triangular $Co_2Pt(DPPE)(CO)_7$ (**26**), the linear $Co_2Pt(CO)_8L_2$ (L = c-$C_6H_{11}NC$) (**23a**) and the tetranuclear $Co_2Pt_2(CO)_8(PPh_3)_2$ (**28a**) were quite active catalysts within the 80-100 °C range but not at 60-65 °C. A remarkable ligand effect was noticed when comparing the catalytic activity of **26** *vs.* that of its isostructural analogue $Co_2Pt(DPAE)(CO)_8$ (**27**). The former was an active hydroformylation catalyst at 80 °C whereas the DPAE-ligated **27** was inert. This is a particularly interesting phenomenon since the chelating ligand is not bonded to the active cobalt atoms.

The activity of **28a** for 1,3-butadiene hydroformylation is higher than that of the corresponding mononuclear Co-P and Pt-P complexes. This was related to the electron distribution within the mixed-metal cluster [94].

2. 5. CARBONYLATION REACTIONS

The carbonylation of 1-octene was studied under mild conditions (75°C, 50 atm) using the iron-palladium complex $[(OC)_4FePd(\mu\text{-}PPh_2)Cl]_2$ (**30**), (eq. 19).

$$CH_3(CH_2)_5CH = CH_2 \quad \xrightarrow[\substack{4 \text{ \% HCl, EtOH} \\ \text{Catalyst}}]{CO} \quad CH_3(CH_2)_5CH \diagup\diagdown \substack{CH_3 \\ CO_2Et} \quad + \quad CH_3(CH_2)_7CO_2Et \quad (19)$$

The total yield of esters was ten times greater for the mixed-metal catalyst than for $PdCl_2(PPh_3)_2$. The distribution of isomer products is *ca.* 1 : 1, with a tendency for the *iso* to predominate. Complex **30** was also an effective catalyst for the carbonylation of 1,5-cyclooctadiene (eq. 20), but not as good as $PdCl_2(PPh_3)_2/FeCl_3$, which was thought to possibly lead to $[PdCl(PPh_3)_2]^+$ $[FeCl_4]^-$ [297].

$$\text{(cyclooctadiene)} \quad \xrightarrow[\text{Catalyst}]{CO \text{ , EtOH}} \quad \text{(cyclooctene-}CO_2Et\text{)} \quad \longrightarrow \quad \text{diesters} \quad (20)$$

Preliminary data indicate that the manganese-palladium complex $MnPdBr(CO)_3(\mu\text{-}DPPM)_2$ (**105a**) produces ethylformate from a CO_2/H_2 mixture (1 : 1, 12 atm) in the presence of ethanol and triethylamine at 130 °C. The corresponding iodo complex **105b** has been tested in methanol homologation reaction (see next section) [158].

2. 6. METHANOL HOMOLOGATION

In the presence of the manganese-palladium complex $MnPdI(CO)_3(\mu-DPPM)_2$ (**105b**) and aqueous HI, methanol reacted with CO/H_2 (1 : 1, 14 atm) at 130 °C to afford Me_2O, AcOH, AcOMe and $MeCH(OMe)_2$ (55 % molar selectivity) [158].

Salts of $[FeCo_3(CO)_{12}]^-$ (**59**) [100a, 152, 300] or $[Et_4N][Fe_3Co(CO)_{13}]$ (**60b**) [152, 300] promoted with methyl iodide have proven to be catalysts for the methanol homologation reaction (eq. 21), depending on the reaction conditions.

$$CH_3OH \xrightarrow[-H_2O]{CO + H_2} CH_3CHO \xrightarrow{H_2} CH_3CH_2OH \qquad (21)$$

The reaction was carried out at relatively high pressures and moderate temperatures. It is possible to obtain acetaldelyde in 80 % selectivity at methanol conversions of 75 %. Under such conditions, only small amounts of ethanol were observed. For an optimum yield of ethanol, long reaction times at high temperature are required, but this results, for some catalysts, in decomposition along with the production of relatively large amounts of methane and of high boiling condensation products derived from acetaldehyde [100a] (Table 2.15). Infrared evidence has shown that the iron-cobalt clusters **49** and **59** do not remain intact during the reaction, but the nature

TABLE 2.15.
Methanol Homologation with Fe-Co Complexes[a] (from ref. [100a]).

Catalyst	Methanol Conversion (%)	Selectivity (%)[b]	
		Ethanol	Acetaldehyde[c]
$Co_2(CO)_8$	9	30	40
$Fe(CO)_5$	2	30	60
$Co_2(CO)_8 + PPh_3$	29	34	52
$Co_2(CO)_8 + Fe(CO)_5$	11	50	40
$HFeCo_3(CO)_{12}$ (**49**)	16	47	43
$[Bu_4N][FeCo_3(CO)_{12}]$ (**59c**)	75	73	10

[a] Reaction conditions: ≈ 270 atm, 220 °C, H_2 / CO (60 : 40), 6 h, methanol / metal (2200 : 1), metal/iodide (1 : 2).
[b] Liquid product only.
[c] Acetaldehyde plus dimethylacetal.

of the active catalyst species is not known. A higher selectivity for ethanol at lower conversion of methanol was obtained with $[Et_4N][Fe_3Co(CO)_{13}]$ (60b) at 40 atm CO, 80 atm H_2, 180 °C [300].

Catalysts employing ruthenium in addition to cobalt are preferred for ethanol production, presumably owing to their ability to readily hydrogenate acetaldehyde, thereby eliminating the undesired by-products [128]. Thus, high ethanol selectivity at high methanol conversion was observed with the mixed Ru-Co compound $CpRu(PPh_3)_2Co(CO)_4$ (68) as catalyst precursor. However, a mixture of $CpRu(PPh_3)_2Cl$ and $Co_2(CO)_8$ provided the same activity. This mixture is unexpectedly superior to the anticipated additive effects of the individual metal complexes. This reaction is also temperature dependent [99, 100b].

Improved catalytic activities have been observed with the mixed clusters $HRuCo_3(CO)_{12}$ (53) and $M[RuCo_3(CO)_{12}]$ (69a-e) promoted by methyl iodide, compared with those for $Co_2(CO)_8$ or $Ru_3(CO)_{12}$ alone [99, 151-153, 300]. The yield of ethanol depends upon the nature of the cation employed [152]. Another Ru-Co mixed cluster, 70a, gave lower yield of ethanol under the same conditions [152, 153]. Although consistent with consecutive transformations (eq. 21), the synergy resulting from a mixed-metal system has been identified in the case of Ru-Co carbonyl clusters. However, these could not be recovered intact after catalysis.

69a, M = Na   69d, M = Ph_4P

69b, M = Cs   69e, M = PPN   70a

69c, M = Et_4N 69f, M = $[Ph_3P]_2Au$

Similarly, high activity and good selectivity were observed for other Ru-Co clusters such as 69f, $RuCo_3(AuPPh_3)(CO)_{12}$ (71) or $[Et_4N][RuCo_3(CO)_{10}^-$

(C$_2$Ph$_2$)] **(72)**. However, no improvement due to the presence of additional gold in **69f** and **71** , or of the alkyne in **72** was noticed [256].

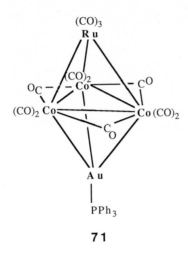

71

High activity has also been achieved with [Et$_4$N][OsCo$_3$(CO)$_{12}$] **(73)**, but selectivity in ethanol was much lower than that observed with Ru-Co systems [300].

The synergic effect of ruthenium and rhodium was observed at 100 atm synthesis gas pressure, where Ru or Rh chloride alone is inactive for ethanol synthesis. When the metal chlorides were replaced by the cluster compounds (Ru$_3$(CO)$_{12}$ + Rh$_6$(CO)$_{16}$, 200 °C, 100 atm), a slight decrease in the formation of ethanol could be observed. No enhancement of ethanol production was observed with the mixed-metal compounds H$_2$Ru$_2$Rh$_2$(CO)$_{12}$ **(7)**, HRuRh$_3$(CO)$_{12}$ **(54)**, HRuRh$_3$(CO)$_{10}$(PPh$_3$)$_2$ **(74)** and [PPN][RuRh$_5$(CO)$_{16}$] **(55)** as catalyst precursors. This is consistent with the cluster decomposition found to occur in all the experiments. Similar catalytically active mononuclear species are probably formed from both the metal chlorides and the cluster compounds [239].

High activities were also observed for the Rh-Co clusters Co$_2$Rh$_2$(CO)$_{12}$ **(8)** and Co$_3$Rh(CO)$_{12}$ **(9)** but selectivities in ethanol are always lower than those obtained for the RuCo$_3$ systems precedently described [153, 300]. Moreover, these systems do not exceed the activity of the well known Ru-Co catalysts for methanol homologation (Table 2.16).

TABLE 2.16.

Homologation of Methanol Catalyzed by Mixed-Metal Clusters[a] (adapted from ref. [153])

Catalyst Precursor	Conversion (%)	Selectivity (%)[b]	
		Total C_2	EtOH
$Co_2(CO)_8$[c]	46	31	1.4
$Co_4(CO)_{12}$[d]	48	32	1.2
$Co_2Pd(DPPE)(CO)_7$ (75)[c]	61	39	1.3
$Co_2Pt(DPPE)(CO)_7$ (26)[c]	49	36	3.3
$Co_3Rh(CO)_{12}$ (9)	52	39	1.6
$Na[RuCo_3(CO)_{12}]$ (69a)	46	41	30
$Cs[RuCo_3(CO)_{12}]$ (69b)	43	51	41
$[Et_4N][RuCo_3(CO)_{12}]$ (69c)	41	54	51
$[Et_4N][Ru_3Co(CO)_{13}]$ (70a)[d]	40	26	23
$Ru_3(CO)_{12}$	18	11	10

[a] Unless otherwise noted: catalyst (0.13 mol), methanol (500 mmol), MeI(5 mmol), CO (40 atm) and H_2 (80 atm), 3 hours at 180 °C.
[b] The selectivity is defined as [product (mmol)/MeOH reacted (mmol)] x 100.
[c] The amount of catalyst was 0.2 mmol.
[d] The amount of catalyst was 0.1 mmol.

In the case of the cobalt-palladium cluster $Co_2Pd(DPPE)(CO)_7$ (75), the rate of methanol carbonylation to acetaldehyde increased compared with $Co_2(CO)_8$ or $Co_4(CO)_{12}$ but the reduction of the latter to ethanol was not enhanced. On the other hand, use of the corresponding cobalt-platinum cluster 26 slightly increased the yield of ethanol [153, 300].

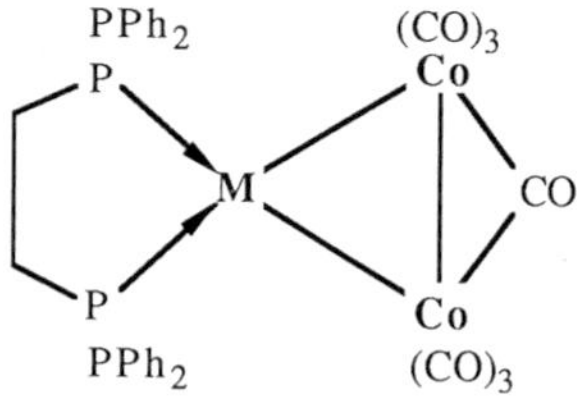

26 , M = Pt

75 , M = Pd

2. 7. NORBORNADIENE DIMERIZATION

Several cobalt- or iron-containing complexes are capable of catalytic, stereospecific dimerization of norbornadiene (NBD) (eq. 22).

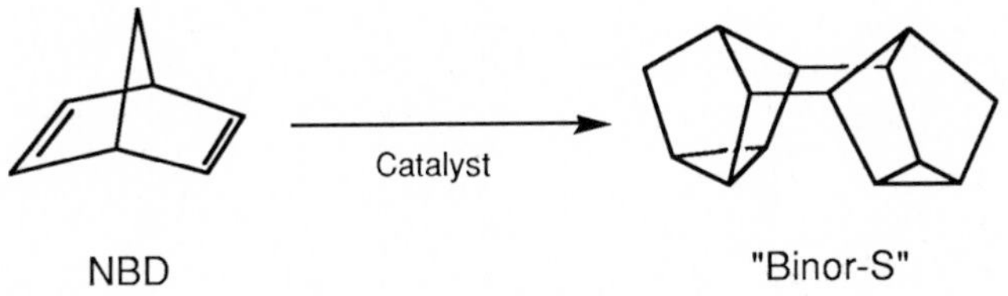

(22)

NBD   "Binor-S"

Thus, the "head-to-head" dimer 1,2,4:5,6,8-dimetheno-s-indacene, "Binor-S", is formed in quantitive yield by dimerization of NBD with the bifunctional transition metal catalyst $Zn[Co(CO)_4]_2$ (**76**), with or without a Lewis acid cocatalyst [273]. A transition state in which two substrate molecules could come sufficiently close for bond formation, giving rise to the "Binor-S", is shown below:

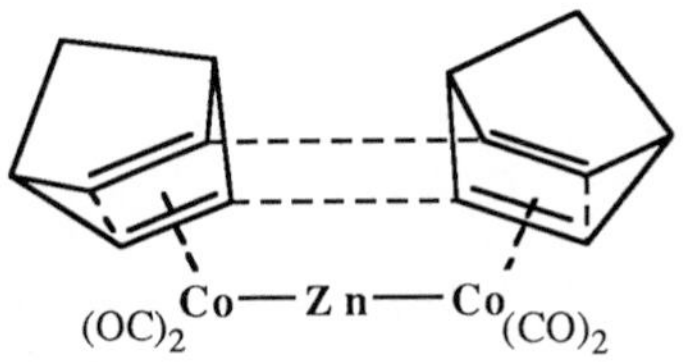

Although $Co_2(CO)_8$ is a dimer and hence potentially a binuclear catalyst, it was found to dimerize norbornadiene only to a mixture characteristic of mononuclear catalysts. On the other hand, **76** and $Cd[Co(CO)_4]_2$ (**77**), gave mixtures of dimers, including "Binor-S", under the same conditions. The mercury analogue $Hg[Co(CO)_4]_2$ (**78**) dimerizes norbornadiene to a mixture of four dimers but no "Binor-S" is formed. In the presence of Lewis acids, however, this catalyst gives "Binor-S" exclusively [272].

It was also found that $Co_2[ZnCo(CO)_4]_2(CO)_7$ (**79**) is an efficient catalyst for the stereospecific dimerization of norbornadiene. The main difference observed when comparing **79** with $Zn[Co(CO)_4]_2$ (**76**) was that the induction period typically observed for the latter was greatly reduced for **79** [61].

79

The anionic cluster $[Me_4N][FeCo_3(CO)_{12}]$ (**59a**) showed a greater catalytic activity than the corresponding tetranuclear neutral clusters given in Table 2.17. Thus, availability of electron density in the cluster framework appears to be an important factor [71].

TABLE 2.17.

Relative Rates for the Dimerization of Norbornadiene to "Binor-S " (T = 46 °C) (from ref. [71]).

Catalyst	Rate x 10^3 (mmol/min)
$[Me_4N][FeCo_3(CO)_{12}]$ (**59a**)	73.1 ± 5.7
$HFeCo_3(CO)_{12}$ (**49**)	27.2 ± 2.4
$Co_4(CO)_{12}$	4.96 ± 0.85
$Co_2(CO)_8$	16.44 ± 0.18

A similarly high catalytic activity with stereospecific dimerization of NBD to "Binor-S" was also obtained with catalysts having a Co_2Pt (**23**) or a Fe_2Pt (**24**) linear framework, a butterfly core, as in $Co_2Pt_2(CO)_8L_2$ (L =

Cluster	M	L	X
23a	Co	c-$C_6H_{11}NC$	CO
23b	Co	t-BuNC	CO
24a	Fe	c-$C_6H_{11}NC$	NO
24b	Fe	t-BuNC	NO
25a	Mo	c-$C_6H_{11}NC$	Cp
25b	Mo	t-BuNC	Cp
80a	W	c-$C_6H_{11}NC$	Cp
80b	W	t-BuNC	Cp
81a	Mn	c-$C_6H_{11}NC$	$(CO)_2$
81b	Mn	t-BuNC	$(CO)_2$

PPh$_3$, **28a**; L = PEt$_3$, **28b**) or a trigonal bipyramidal structure, as in Pt$_3$Co$_2$(CO)$_9$ L$_3$ (**82**), in the presence of a Lewis acid [14].

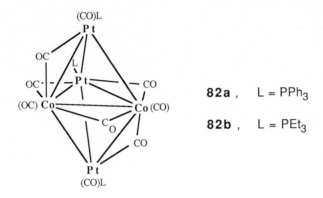

82a ,   L = PPh$_3$

82b ,   L = PEt$_3$

It was also shown that the Fe$_2$Pt complex (**24**), *without* a Lewis acid, selectively leads to the "exo-trans-exo" isomer.

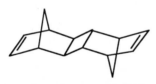

"exo-trans-exo" dimer

The other complexes **25** and **80** led to polymerization of norbornadiene, whereas no catalytic reaction was observed with the linear Mn$_2$Pt complexes **81** [14].

2. 8. BUTADIENE OLIGOMERIZATION

The tetranuclear M$_2$Pd$_2$(CO)$_6$Cp$_2$(PPh$_3$)$_2$ clusters with M = Mo, **14** or M = W, **16**, catalyze butadiene oligomerization to form low molecular weight polymers and a mixture of 4-vinylcyclohexene, 1,5-cyclooctadiene and cyclododecatriene [232]. The catalytic properties of the butterfly cluster Co$_2$Pt$_2$(CO)$_8$(PPh$_3$)$_2$ (**28a**) were also studied for 1,3-butadiene cyclo-oligomerization [94].

2. 9. HYDROSILATION REACTIONS

The photocatalyzed 1-pentene hydrosilation was performed in the presence of the clusters **14-19** as catalytic precursors at room temperature (Table 2.18) (eq. 23).

$$\text{[1-pentene]} \xrightarrow[\substack{\text{R.T., hv 355nm} \\ \text{Catalyst}}]{\text{Et}_3\text{SiH}} \text{[product]}\text{SiEt}_3 + \text{[product]}$$

$$+ \quad \text{[product]} \quad + \quad \text{[product]} \tag{23}$$

Only clusters **18** and **19** exhibited a little hydrosilation activity. The predominant reaction was olefin isomerization. Similarly, the platinum-containing linear catalyst **23a** and the triangular cluster **26** gave no detectable hydrosilation nor isomerization upon irradiation at 355 nm. Cluster conversion or fragmentation probably occurs during the reaction [232].

Photoinitiated hydrosilation of acetophenone (eq. 24) with triethyl silane was performed in the presence of the clusters **4a, 63a-c, 66, 83** and **84**.

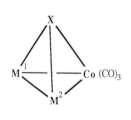

Cluster	X	M^1	M^2
4a	CMe	Mo(CO)$_2$Cp	NiCp
63a	CMe	Co(CO)$_3$	NiCp
63b	CMe	Co(CO)$_3$	Mo(CO)$_2$Cp
63c	CMe	Co(CO)$_2$P(OMe)$_3$	Mo(CO)$_2$Cp
66	PPh	Fe(CO)$_3$	Co(CO)$_3$
83	CH	Co(CO)$_3$	W(CO)$_2$Cp
84	S	Fe(CO)$_3$	Mo(CO)$_2$Cp

$$\underset{\text{O}}{\overset{\text{O}}{\text{Ph}-\overset{\|}{\text{C}}-\text{CH}_3}} + \text{Et}_3\text{SiH} \longrightarrow \underset{\text{A}}{\text{Ph}-\overset{\text{OSiEt}_3}{\underset{|}{\text{CH}}}-\text{CH}_3} + \underset{\text{B}}{\text{Ph}-\overset{\text{OSiEt}_3}{\underset{|}{\text{C}}}=\text{CH}_2} + \text{H}_2 \tag{24}$$

TABLE 2.18.

Product Distributions in Photocatalyzed (λ = 355 nm) Hydrosilation of 1-Pentene with Triethylsilane, in the Presence of the Planar, Triangulated Clusters **14-19**[a] and **26** (from ref. [232]).

N°	Catalyst precursor Formula	L	Irradiation time (h)	Dark time (h)	Product distribution[b] (%)			2-Pentene	
					$C_5H_{11}SiEt_3$	Pentane		trans -	cis -
14	$Mo_2Pd_2(CO)_6Cp_2L_2$	PPh$_3$	48	-	-	2.9		14.3	5.5
15	$Mo_2Pd_2(CO)_6Cp_2L_2$	PEt$_3$	48	-	-	0.4		35.9	4.8
16	$W_2Pd_2(CO)_6Cp_2L_2$	PPh$_3$	48	-	-	4.6		12.9	7.9
17	$W_2Pd_2(CO)_6Cp_2L_2$	PEt$_3$	-	48	-	0.7		2.3	1.0
17	$W_2Pd_2(CO)_6Cp_2L_2$	PEt$_3$	12	36	-	1.3		5.4	3.3
17	$W_2Pd_2(CO)_6Cp_2L_2$	PEt$_3$	24	24	-	2.2		9.9	3.7
17	$W_2Pd_2(CO)_6Cp_2L_2$	PEt$_3$	48	-	-	4.8		15.0	4.8
17	$W_2Pd_2(CO)_6Cp_2L_2$	PEt$_3$	80	-	-	7.6		18.3	14.0
18	$Mo_2Pt_2(CO)_6Cp_2L_2$	PPh$_3$	48	-	3.0	0.7		8.2	1.2
19	$Mo_2Pt_2(CO)_6Cp_2L_2$	PEt$_3$	48	-	4.0	?		?	?
26	$Co_2Pt(CO)_7L$	DPPE	16	-	-	-		-	-

a In each run 3.66 mmol of 1-pentene and triethylsilane and 0.01 mmol of catalyst were used.
b Mol % based on 1-pentene charged.

All these clusters were found to be suitable catalysts for this reaction. In most cases, only **B** was produced (Table 2.19). Furthermore, after photolysis, the clusters **4a**, **63a-c**, **66**, **83** and **84** could be recovered (90-98 %) by chromatography on neutral silica gel [233a].

TABLE 2.19.
Photoinitiated Hydrosilation Catalyzed by Mixed-Metal Clusters[a]
(from ref. [233a]).

Cluster	λ=254 nm		λ=355 nm	
	ϕ^b	A : B	ϕ^b	A : B
4a	-	-	0.03	c
63a	0.1	c	0.01	c
63b	0.048	c	0.016	c
63c	0.037	c	0.014	c
66	0.21	c	0.03	1.0:1
83	0.028	c	0.034	c
84	0.45	0.68:1	0.17	0.5:1

[a] Reaction after degassing in hermetically sealed Pyrex ampoules (λ = 355 nm) or in vacuum-tight quartz cells (λ = 254 nm) at 25 °C. Cluster concentration: 0.01 mol/L; acetophenone / triethylsilane / cluster = 366 : 366 : 1. No additional solvent.
[b] ϕ is the number of moles of product(s) (based on acetophenone consumed) per incident photon.
[c] Only **B** was produced.

Asymmetric hydrosilation with the optically active cluster (+)-**84** would provide irrefutable evidence for involvement of the intact chiral SMoFeCo tetrahedron during catalysis. When this reaction was carried out under the conditions given in Table 2.19 (λ = 355 nm), **B** (48 %), racemic **A** (16 %) and racemic cluster **84** were isolated after 12 days. It has been subsequently established that the photoracemization of (+)-**84** at λ = 355 nm proceeds substantially faster than the hydrosilation reaction [233a].

2. 10. CONCLUSION

In this Section, we have examined the various reactions in which mixed-metal cluster compounds have been used as precursors of homogeneous catalysts. In many instances, a synergic effect has been noted when comparing the performances of the heterometallic cluster with those of the corresponding homometallic components. Valuable comparisons can only be made for systems studied under strictly analogous conditions, which turns out to be only rarely possible, partly because of the diversity of systems studied and of the increasing number of research groups engaged in this field. Much remains to be learned about the stoichiometric and catalytic reactivity of mixed-metal cluster compounds, particularly about the site selectivity induced by the presence of different metals [264]. Furthermore, attempts to correlate the catalytic activity with the structure of the cluster core should be encouraged. Although one could anticipate that more open structures, *e.g.*, chains or butterflies, should be more reactive and lead to more selective pathways than the corresponding *closo* structures, *e.g.*, triangular or tetrahedral, this has not yet been established. The hetero-metallic associations examined for a given catalytic reaction are presented in Table 2.20 and a listing of the heterometallic clusters which have been used in homogeneous catalysis is given in Table 2.21.

In order to address the question of cluster fragmentation during catalysis, chiral clusters have been employed as catalysts but found to racemize under the catalytic conditions. This approach remains of fundamental importance for a improved knowledge of the reaction mechanisms involving cluster compounds. With the hope of *i.a.*, stabilizing cluster cores and of decreasing their tendency to fragment into mononuclear species, important efforts have been directed towards their immobilization onto various supports. This will now be examined in Section 3.

TABLE 2.20

Heterometallic Associations in Homogeneous Catalysis with Transition Metal Clusters.

Reaction	Heterometallic Associations
Hydrogenation/Isomerization of Carbon-Carbon Multiple Bonds	Cr-Fe, Cr-Pd Mo-Fe, Mo-Rh, Mo-Pd, Mo-Pt W-Fe, W-Pd, W-Pt Fe-Ru, Fe-Co, Fe-Rh, Fe-Pd, Fe-Pt Ru-Co, Ru-Rh, Ru-Ni, Ru-Cu, Ru-Au Os-Ni Co-Rh, Co-Pt Ir-Pt Mo-Fe-Co, Mo-Ru-Co, Mo-Co-Ni W-Fe-Co Fe-Ru-Co
CO Hydrogenation	Mn-Rh Re-Rh Ru-Rh, Ru-Cu, Ru-Ag, Ru-Au Co-Rh Rh-Ir, Rh-Pt, Rh-Cu, Rh-Ag, Rh-Au, Rh-Zn
Water-Gas Shift Reaction	Fe-Ru, Fe-Os, Fe-Co, Fe-Ir Ru-Co, Ru-Rh Co-Rh, Co-Ir
Hydroformylation Reactions	Zr-Rh Mo-Co, Mo-Pd, Mo-Pt Fe-Co, Fe-Rh Ru-Co Co-Rh, Co-Ni, Co-Pt Mo-Co-Ni
Carbonylation Reactions	Mn-Pd Fe-Pd
Methanol Homologation	Mn-Pd Fe-Co Ru-Co, Ru-Rh Os-Co Co-Rh, Co-Pd, Co-Pt Ru-Co-Au

Norbornadiene Dimerization	Mo-Pt
	W-Pt
	Mn-Pt
	Fe-Co, Fe-Pt
	Co-Pt, Co-Zn, Co-Cd, Co-Hg
Butadiene Oligomerization	Mo-Pd
	W-Pd
	Co-Pt
Hydrosilation Reactions	Mo-Co, Mo-Pd, Mo-Pt
	W-Co, W-Pd
	Fe-Co
	Co-Ni, Co-Pt
	Mo-Fe-Co, Mo-Co-Ni

TABLE 2.21.

Utilization of Molecular Mixed-Metal Clusters in Homogeneous Catalysis

Bimetallic Couple [a]	Precursor Cluster	Ref. Synth.	Catalyzed Reactions	Ref. Catal.
Zr-Rh	$Zr(\eta\text{-}C_5H_5)_2(CH_2PPh_2)_2RhH(PPh_3)$ (57)	82	Hydroformylation of 1-hexene	82
	$Rh_2(S\text{-}t\text{-}Bu)_2[(Ph_2PCH_2)_2Zr(\eta\text{-}C_5H_5)_2](CO)_2$ (58a)	81	Hydroformylation of 1-hexene	276
Cr-Fe	$PPN[HCrFe(CO)_9]$ (31a)	8	Isomerization of alkenes	298
Cr-Pd	$Cr_2Pd_2(\eta\text{-}C_5H_5)_2(CO)_6(PEt_3)_2$ (22)	20	Hydrogenation of 1,5-COD	232
Mo-Fe	$PPN[HMoFe(CO)_9]$ (31b)	8	Isomerization of alkenes	298
Mo-Co	$MoCo_2(\eta\text{-}C_5H_5)(CO)_8(\mu_3\text{-}CMe)$ (63b)	27	Hydroformylation of 1-pentene and styrene	233b, 244
			Hydrosilation of acetophenone	233a
	$MoCo_2(\eta\text{-}C_5H_5)(CO)_7[P(OMe)_3](\mu_3\text{-}CMe)$ (63c)	27	Hydroformylation of 1-pentene and styrene	244
			Hydrosilation of acetophenone	233a
	$MoCo_2(\eta\text{-}C_5H_5)(CO)_8(\mu_3\text{-}S)$ (65)	303	Hydrosilation of acetophenone	244
	$MoCo_2(\eta\text{-}C_5H_5)(CO)_6[C(Me):N(Ph)](\mu_3\text{-}S)$ (67b)	201	Asymmetric hydroformylation of styrene	201
Mo-Rh	$MoRh(\eta\text{-}C_5H_5)(\mu\text{-}CO)_2(CO)(PPh_3)_2$ (1)	62	Hydrogenation of cyclohexene	62
Mo-Pd	$Mo_2Pd_2(\eta\text{-}C_5H_5)_2(CO)_6L_2$ L = PPh₃ (14)	22	Hydrosilation of 1-pentene	232
			Oligomerization of butadiene	232
			Hydrogenation of 1,5-COD and alkynes	232
			Hydroformylation of 1-pentene	232
	L = PEt₃ (15)	22	Hydrogenation of 1,5-COD and alkynes	232
			Hydrosilation of 1-pentene	232
Mo-Pt	$Mo_2Pt(CO)_6(\eta\text{-}C_5H_5)_2(c\text{-}C_6H_{11}NC)_2$ (25a)	15	Hydrogenation of alkenes and alkynes	12
			Dimerization of norbornadiene	14
	$Mo_2Pt(CO)_6(\eta\text{-}C_5H_5)_2(t\text{-}BuNC)_2$ (25b)	15	Dimerization of norbornadiene	14

Mo-Pt	$Mo_2Pt_2(\eta\text{-}C_5H_5)_2(CO)_6L_2$ $\quad$ L = PPh$_3$ (18)	21, 23	Hydrosilation of 1-pentene	232
			Hydrogenation of 1,5-COD and phenylacetylene	232
			Hydroformylation of 1-pentene	232
	$\quad$ L = PEt$_3$ (19)	21, 23	Hydrosilation of 1-pentene	232
W-Fe	PPN[HFeW(CO)$_9$] (31c)	8	Isomerization of alkenes	298
W-Co	$WCo_2(\eta\text{-}C_5H_5)(CO)_8(\mu_3\text{-}CH)$ (83)	25	Hydrosilation of acetophenone	233a
W-Pd	$W_2Pd_2(\eta\text{-}C_5H_5)_2(CO)_6L_2$ $\quad$ L = PPh$_3$ (16)	22	Hydrogenation of 1,5-COD and phenylacetylene	232
			Hydrosilation of 1-pentene	232
			Oligomerization of butadiene	232
	$\quad$ L = PEt$_3$ (17)	22	Hydrogenation of 1,5-COD and phenylacetylene	232
			Hydrosilation of 1-pentene	232
W-Pt	$W_2Pt(CO)_6(\eta\text{-}C_5H_5)_2(c\text{-}C_6H_{11}NC)_2$ (80a)	15	Dimerization of norbornadiene	1
	$W_2Pt(CO)_6(\eta\text{-}C_5H_5)_2(t\text{-}BuNC)_2$ (80b)	15	Dimerization of norbornadiene	14
	$W_2Pt_2(\eta\text{-}C_5H_5)_2(CO)_6L_2$ $\quad$ L = PPh$_3$ (20)	23	Hydrogenation of 1,5-COD and phenylacetylene	232
	$\quad$ L = PEt$_3$ (21)	23	Hydrogenation of 1,5-COD	232
Mn-Rh	$Mn[Rh_{12}(CO)_{30}]$ (42a)	310b	CO Hydrogenation	310b
Mn-Pt	$Mn_2Pt(CO)_{10}(c\text{-}C_6H_{11}NC)_2$ (81a)	15	Dimerization of norbornadiene	14
	$Mn_2Pt(CO)_{10}(t\text{-}BuNC)_2$ (81b)	15	Dimerization of norbornadiene	14
Mn-Pd	$MnPdX(CO)_3(\mu\text{-}DPPM)_2$ $\quad$ X = Br (105a)	158	Carbonylation reaction	158
	$\quad$ X = I (105b)	158	Homologation of methanol	158
Re-Rh	$Re_2[Rh_{12}(CO)_{30}]_3$ (43b)	310c	CO Hydrogenation	310c
Fe-Ru	$FeRu_2(CO)_{12}$ (2a)	316	Hydrogenation 2-pentyne and 1,3 pentadienes	211
	$FeRu_2(CO)_{11}L$ $\quad$ L = Phosphines (2b)	308	WGSR	307
	$FeRu_2(CO)_{10}L_2$ $\quad$ L = Phosphines (2c)	308	WGSR	307
	$Fe_2Ru(CO)_{12}$ (3a)	316	Hydrogenation 2-pentyne and 1,3-pentadienes	211
	$Fe_2Ru(CO)_{11}L$ $\quad$ L = Phosphines (3b)	308	WGSR	307
	$Fe_2Ru(CO)_{10}L_2$ $\quad$ L = Phosphines (3c)	308	WGSR	307

Fe-Ru	$H_2FeRu_3(CO)_{13}$ **(32)**	129	WGSR	114, 301, 307
Fe-Os	$Fe_2Os(CO)_{12}$ **(47)**	214	Isomerization of 1-hexene	117
	$H_2FeOs_3(CO)_{13}$ **(48)**	84, 129a, 234	WGSR	307
			WGSR	307
Fe-Co	$FeCo_2(CO)_9(\mu_3\text{-}PPh)$ **(66)**	217, 245	Hydroformylation of 1-pentene and styrene	244
			Hydrosilation of acetophenone	233a
	$HFe_2Co(CO)_9(\mu_3\text{-}PMe)$ **(5e)**	217	Hydrogenation of alkenes	204
	$HFeCo_3(CO)_{12}$ **(49)**	76	WGSR	307
	$M[FeCo_3(CO)_{12}]$ M = Et$_4$N **(59b)**	76	Dimerization of norbornadiene	71
			Hydroformylation of cyclohexene	150
			Homologation of methanol	100a, 300 / 152
	M = Me$_4$N, Et$_4$N, Bu$_4$N **(59a-c)**	76	Dimerization of norbornadiene	71
			Hydroformylation of cyclohexene	150
			Homologation of methanol	152, 300
	$(Et_4N)[Fe_3Co(CO)_{13}]$ **(60b)**	290		
Fe-Rh	$HFe_3Rh(CO)_{11}(\mu_4\text{-}\eta^2\text{-}C=CH(R))$ R = H **(33a)**; Ph **(33b)**	10b	Hydroformylation of alkenes	10b
			Isomerization of 1-heptene	10b
	$[Ph_4P][Fe_3Rh_2(CO)_{13}(\mu_4\text{-}\eta^3\text{-}CH_3C=C=CH_2)]$ **(34)**	10a	Isomerization of 1-octene	10a
	$[Ph_4P][Fe_4RhC(CO)_{14}]$ **(109)**	10c	Hydroformylation of 1-pentene	10c
	$[Ph_4P][Fe_3Rh_3C(CO)_{15}]$ **(110)**	10c	Hydroformylation of 1-pentene	10c
	$Fe_4Rh_2C(CO)_{16}$ **(108)**	10c	Hydroformylation of 1-pentene	10c
	$[Ph_4P][Fe_5RhC(CO)_{16}]$ **(107)**	74c	Hydroformylation of 1-pentene	10c
Fe-Ir	$Na_2[FeIr_4(CO)_{15}]$ **(50)**	164	WGSR	307
Fe-Pd	$[(OC)_4Fe(\mu\text{-}PPh_2)Pd(\mu\text{-}Cl)]_2$ **(30)**	24	Hydrogenation of alkynes	297
			Isomerization of alkenes	297
			Carbonylation of 1,5-COD	297

System	Cluster	Reaction	Ref.
Fe-Pt	$Fe_2Pt(CO)_6(NO)_2(c\text{-}C_6H_{11}NC)_2$ (24a)	Dimerization of norbornadiene	1, 15
	$Fe_2Pt(CO)_6(NO)_2(t\text{-}BuNC)_2$ (24b)	Hydrogenation of alkynes	232, 15
		Hydrogenation of alkenes and dienes	122
		Dimerization of norbornadiene	14
Ru-Co	$RuCo(\eta\text{-}C_5H_5)(CO)_4(PPh_3)_2$ (68)	Homologation of methanol	99, 99
	$RuCo_2(CO)_{11}$ (52)	WGSR	307, 252
	$RuCo_2(CO)_9(\mu_3\text{-}PPh)$ (5a)	Hydrogenation of alkenes	204, 251
	$RuCo_2(CO)_9(\mu_3\text{-}PMe)$ (5b)	Hydrogenation of alkenes	204, 251
	$RuCo_2(CO)_9(\mu_3\text{-}S)$ (6a)	Hydrogenation of alkenes	204, 255
	$HRu_2Co(CO)_9(\mu_3\text{-}PMe)$ (5c)	Hydrogenation of alkenes	204, 217
	$HRuCo_3(CO)_{12}$ (53)	Hydroformylation of alkenes	99, 151, 48, 152
		WGSR	307
	$M[RuCo_3(CO)_{12}]$ M = Na, Cs, NEt$_4$, Ph$_4$P, PPN (69a-e)	Homologation of methanol	152, 153, 48, 46, 152
	$[Et_4N][RuCo_3(CO)_{10}(C_2Ph_2)]$ (72)	Homologation of methanol	300, 46
	$H_3Ru_3Co(CO)_{12}$ (51)	WGSR	256, 131
	$[Et_4N][Ru_3Co(CO)_{13}]$ (70a)	Homologation of methanol	307, 290
Ru-Rh	$RuRh_2(CO)_{12}$ (37)	CO Hydrogenation	182, 182
	$HRuRh_3(CO)_{12}$ (54)	Homologation of methanol	239, 242
		WGSR	307
	$HRuRh_3(CO)_{10}(PPh_3)_2$ (74)	Homologation of methanol	239, 242
	$H_2Ru_2Rh_2(CO)_{12}$ (7)	Hydrogenation of 2-cyclohexenone	227, 240, 241
		Homologation of methanol	239
		WGSR	307
	$PPN[RuRh_5(CO)_{16}]$ (55)	Homologation of methanol	239, 243
		WGSR	307
Ru-Ni	$H_3Ru_3Ni(\eta\text{-}C_5H_5)(CO)_9$ (10)	Hydrogenation of alkynes and dienes	66, 70, 192
		Isomerization of dienes	66, 263

Metals	Complex	Reaction	Ref.
Ru-Cu	H3Ru4{Cu(PPh3)}(CO)12 (35a)	Isomerization of 1-pentene	260, 106
	H2Ru4{Cu(PPh3)}2(CO)12 (36a)	Isomerization of 1-pentene	118, 119, 106
	Ru6(CuL)2C(CO)16 L = RCN (38a)	CO Hydrogenation	36, 236
Ru-Ag	Ru6(AgL)2C(CO)16 L = RCN (38b)	CO Hydrogenation	236, 236
Ru-Au	H3Ru4{Au(PPh3)}(CO)12 (35b)	Isomerization of 1-pentene	107, 261, 106
	H3Ru4{Au(PPh3)}2(CO)12 (36b)	Isomerization of 1-pentene	118, 119, 106
	Ru6(AuL)2C(CO)16 L = RCN (38c)	CO Hydrogenation	236, 236
Os-Co	[NEt4][OsCo3(CO)12] (73)	Homologation of methanol	183, 300
Os-Ni	H3Os3Ni(η-C5H5)(CO)9 (11)	Hydrogenation of alkenes	70, 192, 66
		Hydrogenation of alkynes	280, 70
		Hydrogenation of alkenes and dienes	68, 69
		Isomerization of dienes	66, 69
	H3Os3Ni(η-C5H5)(CO)8(PPh2H) (12a)	Hydrogenation of dienes	265, 69
	H3Os3Ni(η-C5H5)(CO)8{P(o-MeC6H4)3} (12b)	Hydrogenation of dienes	265, 69
	Os3Ni3(η-C5H5)3(CO)9 (13)	Hydrogenation of alkenes and alkynes	262, 70
Co-Rh	CoRh(CO)7 (62)	Hydroformylation of diketene	157b, 289
	CoRh(CO)6(PPh3) (61)	Hydroformylation of pentafluorostyrene	220
		Hydroformylation-amidocarbonylation of trifluorostyrene	220
	Co2Rh2(CO)12 (8)	Hydroformylation of dicyclopentadiene	157a, 289, 123
		Hydrogenation of alkenes	209, 188
		Hydroformylation of alkenes	73
		Hydroformylation of formaldehyde	206
		Homologation of methanol	153, 300
		WGSR	307

Metal	Compound	Ref.	Reaction	References
Co-Rh	Co$_3$Rh(CO)$_{12}$ (9)	209	Hydrogenation of alkenes	188
			Hydroformylation of alkenes	311b
			Homologation of methanol	153, 300
			WGSR	307
	Co[Rh$_{12}$(CO)$_{30}$] (42b)	310b	CO Hydrogenation	310b
	M$_2$[Co$_3$Rh$_9$(CO)$_{30}$]$_3$ (45b)	51	CO Hydrogenation	51
Co-Ir	Co$_2$Ir$_2$(CO)$_{12}$ (56)	189	WGSR	307
Co-Ni	Co$_2$Ni(η-C$_5$H$_5$)(CO)$_6$(μ_3-CMe) (63a)	27	Hydroformylation of 1-pentene and styrene	244
			Hydrosilation of acetophenone	233a
Co-Pd	Co$_2$Pd(DPPE)(CO)$_7$ (75)	153	Homologation of methanol	153, 300
Co-Pt	Co$_2$Pt(CO)$_8$(c-C$_6$H$_{11}$NC)$_2$ (23a)	15	Dimerization of norbornadiene	1
	Co$_2$Pt(CO)$_8$(t-BuNC)$_2$ (23b)	15	Hydrogenation of alkenes and dienes	122, 232
			Hydroformylation of 1-pentene	122, 232
			Dimerization of norbornadiene	14
	Co$_2$Pt(CO)$_7$(DPPE) (26)	95	Hydrosilation of 1-pentene	122, 232
			Hydrogenation of alkenes, dienes and alkynes	122, 232
			Hydroformylation of 1-pentene	122, 232
			Homologation of methanol	153, 300
	Co$_2$Pt(CO)$_7$(DPAE) (27)	95	Hydrogenation of 1-hexyne	232
			Hydroformylation of 1-pentene	232
	Co$_2$Pt$_2$(CO)$_8$L$_2$ L = PPh$_3$ (28a)	39	Hydrogenation of olefins and alkynes	94, 122, 232
			Hydroformylation of alkenes	94, 232
			Dimerization of norbornadiene	14
			Butadiene oligomerization	94
	L = PEt$_3$ (28b)	39	Dimerization of norbornadiene	14
	L = AsPh$_3$ (28c)	39	Hydrogenation of alkynes	122
	Co$_2$Pt$_3$(CO)$_9$L$_3$ L = PPh$_3$ (82a)	16	Dimerization of norbornadiene	14
	L = PEt$_3$ (82b)	16	Dimerization of norbornadiene	14

		Ref	Reaction	Ref
Co-Zn	$Zn[Co(CO)_4]_2$ (76)	272	Dimerization of norbornadiene	272, 273
	$Co_2[ZnCo(CO)_4]_2(CO)_7$ (79)	61	Dimerization of norbornadiene	61
Co-Cd	$Cd[Co(CO)_4]_2$ (77)	272	Dimerization of norbornadiene	272
Co-Hg	$Hg[Co(CO)_4]_2$ (78)	154	Dimerization of norbornadiene	272
Rh-Ir	$Ir_2[Rh_{12}(CO)_{30}]$ (41d)	310c	CO Hydrogneation	310c
	$Ir_2[Rh_{12}(CO)_{30}]_3$ (43a)	310a	CO Hydrogenation	310a
	$M_2[Rh_6Ir_6(CO)_{30}]_3$ M = Al (44)	51	CO Hydrogenation	51
	$M_2[Rh_9Ir_3(CO)_{30}]_3$ M = Al (45a)	51	CO Hydrogenation	51
Rh-Pt	$[PPN]_2[Rh_4Pt(CO)_{12}]$ (40)	121	CO Hydrogenation	257
	$PPN[Rh_5Pt(CO)_{15}]$ (39b)	120, 121	CO Hydrogenation	257
	$Pt[Rh_{12}(CO)_{30}]$ (42d)	310b	CO Hydrogenation	310b
Rh-Cu	$Cu_2[Rh_{12}(CO)_{30}]$ (41a)	310a	CO Hydrogenation	310a
Rh-Ag	$Ag_2[Rh_{12}(CO)_{30}]$ (41b)	310a	CO Hydrogenation	310a
Rh-Au	$Au_2[Rh_{12}(CO)_{30}]$ (41c)	310a	CO Hydrogenation	310a
Rh-Zn	$Zn[Rh_{12}(CO)_{30}]$ (42c)	310b	CO Hydrogenation	310b
Ir-Pt	$Ir_2Pt_2(CO)_7(PPh_3)_3$ (29)	29	Hydrogenation of alkenes	29

Trimetallic System	Precursor Cluster	Ref. Synth.	Catalyzed Reaction	Ref. Catal.
Mo-Fe-Co	$HMoFeCo(\eta\text{-}C_5H_5)(CO)_8(\mu_3\text{-}CMe)$ (4b)	246, 248	Hydrogenation of 2-pentene and styrenes	204
	$MoFeCo(\eta\text{-}C_5H_5)(CO)_8(\mu_3\text{-}S)$ (84)	255	Hydrosilation of acetophenone	233a
Mo-Ru-Co	$MoRuCo(\eta\text{-}C_5H_5)(CO)_8(\mu_3\text{-}S)$ (6b)	255	Hydrogenation 2-pentene	204
Mo-Co-Ni	$MoCoNi(\eta\text{-}C_5H_5)_2(CO)_5(\mu_3\text{-}CMe)$ (4a)	26	Hydrogenation of 2-pentene and styrenes	204
			Hydroformylation of 1-pentene and styrene	244
			Hydrosilation of acetophenone	233a
W-Fe-Co	$WFeCo(\eta\text{-}C_5H_5)(CO)_8(\mu_3\text{-}PMe)$ (5f)	217, 218	Hydrogenation of 2-pentene and styrenes	204
Fe-Ru-Co	$HFeRuCo(CO)_9(\mu_3\text{-}PMe)$ (5d)	205	Hydrogenation of 2-pentene and styrenes	204
Ru-Co-Au	$RuCo_3Au(PPh_3)(CO)_{12}$ (71)	47	Homologation of methanol	256
	$[(Ph_3P)_2Au][RuCo_3(CO)_{12}]$ (69f)	47	Homologation of methanol	256

a The metals are listed with increasing number of their group.

3. HYBRID CATALYSTS PREPARED FROM MOLECULAR MIXED-METAL CLUSTERS

3. 1. INTRODUCTION

The study of new catalysts prepared by immobilizing mixed-metal clusters on solid supports such as organic polymers or inorganic oxides, functionalized or not, has been strongly developed in recent years. The original incentive to produce such systems was to overcome the limitations and difficulties encountered in both homogeneous catalysis, *i.e.*, catalyst recovery and cluster fragmentation, and in heterogeneous catalysis, *i.e.*, limited tuning due to frequently ill-defined active sites, reduced effectiveness of components of a multimetallic catalyst and severe reaction conditions. At the same time, it was hoped that the advantages inherent to homogeneous catalysis, *i.e.*, better known mechanisms and catalytic cycles, mild reaction conditions, high efficiency of the metal atoms, and easier tuning of the electronic and steric properties, could be combined with those characterizing heterogeneous catalysis, *i.e.*, higher stability of the catalyst, applicability to a wide range of reactions, technological versatility, and ready separation from the reaction products. Further advantages of these new hybrid catalysts could be to stabilize unsaturated, reactive clusters that are normally unstable and to prevent cluster fragmentation as well as formation of larger aggregates under catalytic conditions. Many review articles have appeared on the immobilization of transition-metal carbonyls, and although most of the information available concerns mononuclear complexes and homonuclear clusters, general considerations and requirements about the preparation, characterization and properties of these catalytic materials are germane to the mixed-metal systems reviewed here [12, 18, 52, 105, 125, 136, 145, 146, 162a, 167, 169, 202, 238, 318, 320]. We shall therefore not go into the details of, *e.g.*, how clusters interact with various supports: relatively little is known about the specific features associated with heterometallic clusters.

In the anchoring process, the cluster should ideally maintain its integrity. This should be checked by control experiments [19], in order to obtain a catalytic material of known nuclearity. Two main routes may be discerned for the preparation of hybrid catalysts :

(i) bonding to inorganic oxides and

(ii) linkage to functionalized organic or inorganic supports. These will be examined in the following. We advise the reader to consult excellent recent review articles on the characterization by physical methods of metal clusters in and on supports [135, 184, 222, 225].

3. 2. CATALYSIS BY IMMOBILIZED MIXED-METAL CLUSTERS ON INORGANIC OXIDES

Various impregnating methods are available for immobilizing a mixed-metal cluster onto inorganic oxides such as SiO_2, Al_2O_3, MgO.... They can be derived from those established for simpler systems [12, 87, 155, 165, 184], provided they respect the conditions of the chemical stability peculiar to the cluster considered. Two main bonding modes between the molecular species and the support have been identified: ionic bonding between a negatively charged cluster and a surface metal cation, generally occurring via a carbonyl ligand (Scheme 3.1, a) and covalent bonding between the cluster and the surface (Scheme 3.1, b).

SCHEME 3.1.

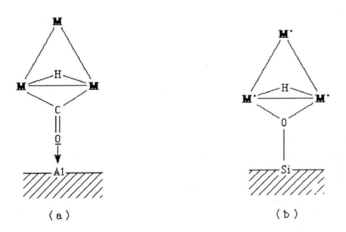

(a)　　　　　　　　　　(b)

A variety of chemical reactions may be involved in the chemisorption process [18, 105, 162a, 238]. They are, in most cases, related to basic elementary steps known in organometallic chemistry.

Recently, the interaction of the mixed hydrido carbonyl clusters $HFeCo_3(CO)_{12}$ (49) and $H_2FeM_3(CO)_{13}$ (M = Ru, 32; M = Os, 48) with Al_2O_3 or MgO has been studied [77, 80, 140, 186]. These clusters transfer a proton to a basic site on the support, as anticipated from the known acidity of these clusters. The resulting anionic clusters $[FeCo_3(CO)_{12}]^-$ and $[HFeM_3(CO)_{13}]^-$ (M = Ru, Os), respectively, are immobilized on the surface through ionic bonding. The pretreatment of the oxide support affects the base strength and therefore plays an important role in the formation of the metal carbonyl surface species. The magnesia-supported anionic cluster $[FeCo_3(CO)_{12}]^-$ starts to decompose to CO and H_2 at 323 K, showing lower stability than $[HFeM_3(CO)_{13}]^-$ (M = Ru, Os) [77]. The alumina-supported cluster $Al^+[HRuOs_3(CO)_{13}]^-$, formed by the surface reaction shown in eq. 25 [59, 274], was inferred to give $[Al]^+[H_3RuOs_3(CO)_{12}]^-$ when heated.

$$Al\text{-}OH + H_2RuOs_3(CO)_{13} \xrightarrow{-H_2O} [Al]^+[HRuOs_3(CO)_{13}]^- \qquad (25)$$

This new surface species was shown to be relatively stable at temperatures up to 373 K under vacuum, but it decomposes at higher temperatures. In contrast, greater thermal stability was observed in the presence of CO + H_2 [59]. This cluster was catalytically active for 1-butene isomerization but it decomposed during ethylene hydrogenation at 340 K [274]. After catalysis, $[Al]^+[H_3RuOs_3(CO)_{12}]^-$ was the only detectable metal carbonyl species.

Anchoring of $Fe_3(CO)_{12}$, $Fe_2Ru(CO)_{12}$ (3a), $FeRu_2(CO)_{12}$ (2a) and $Ru_3(CO)_{12}$ on γ-Al_2O_3 produced catalysts which were less active towards hydrogenation of 1- and 2-pentyne and more active towards hydrogenation of 1,3-*cis*- and 1,3-*trans*- pentadiene compared to their homogeneous toluene solutions, examined in Section 2.2.1 [211]. Catalytic activity was found to decrease with increasing number of Fe atoms in the dodecacarbonyls. A complete disruption of the clusters to give metal particles was ruled out. In most cases, however, the immobilized mixed-metal cluster is thermally decomposed under the catalytic conditions. This leads to metal particles which are no longer molecular species. The catalytic properties of these cluster-derived particles will be examined in Section 4.

Surface metal ions may also be involved in bonding interactions with an organometallic complex, possibly leading to heterometallic species on inorganic oxides. Indeed, it has been found recently that a very active bimetallic, heterogeneous olefin metathesis catalyst is formed by reaction of a reduced Philipps catalyst with Fischer-type molybdenum or tungsten carbene complexes [311a]. The carbene complex is thought to interact with the Si-O-Cr-O-Si spacer as shown in Scheme 3.2.

SCHEME 3.2.

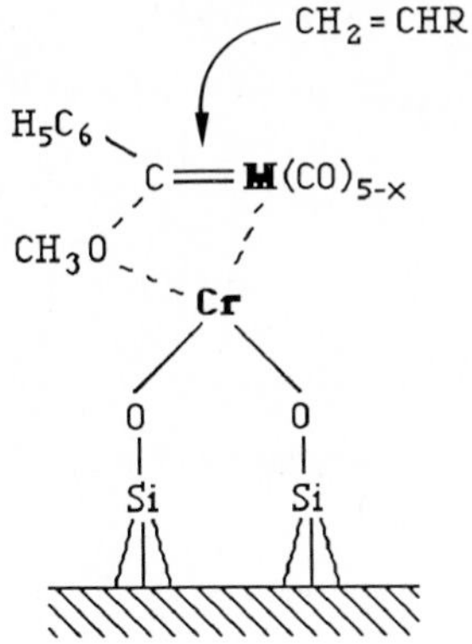

It is noteworthy that the olefin polymerization activity of the reduced Philipps catalyst is lost in this new bimetallic catalyst [311a].

3. 3. CATALYSIS BY IMMOBILIZED MIXED-METAL CLUSTERS ON FUNCTIONALIZED SUPPORTS.

Recently, the use of chemically modified supports has been developed with two main objectives: (i) the replacement of oxygen-donating ligands, the only ones available on the surface of oxides, with groups having a greater affinity for low oxidation-state metal carbonyls, and (ii) the availability of a larger set of model compounds whose spectroscopic properties are well studied and may be used for comparison with those of the immobilized material. As far as inorganic oxides are concerned, pendant ligands may be introduced via the reaction of surface hydroxyl groups with molecules of the silane type (eq. 26).

$$-\overset{|}{\underset{|}{Si}}-OH \xrightarrow{R(CH_2)_nSiX_3} -\overset{|}{\underset{|}{Si}}-O-\overset{X}{\underset{X}{Si}}-(CH_2)_n-R \qquad (26)$$

$n = 0, 1, 2,...$ R = organofunctional group
e.g., PPh_2, NH_2, Cp
X = a hydrolyzable group
e.g., Cl, NH_2, OR, OCOR

The cluster may be anchored to the functionalized surface by deprotonation (eq. 27) or by ligand exchange (eq. 28). For example, amino functions on a silica gel matrix were used as anchoring groups via an acid-base reaction with the cluster $HFeCo_3(CO)_{12}$ (**49**) [149] :

$$SIL\text{-}[CH_2]_3\text{-}NH_2 \ + \ HFeCo_3(CO)_{12} \longrightarrow [SIL\text{-}(CH_2)_3\text{-}NH_3]^+[FeCo_3(CO)_{12}]^- \qquad (27)$$

Infrared spectroscopy indicated deprotonation but preservation of the cluster framework. It was also possible to regenerate **49** from the supported cluster catalyst. The catalyst prepared by decarbonylation-activation of the supported cluster in a stream of hydrogen at atmospheric pressure and 473 K was higly active in Fischer-Tropsch synthesis. At atmospheric pressure, 20 % conversion of synthesis gas was observed at 513 K and the unusually narrow product distribution showed a maximum at C6 (Figure 3.1).

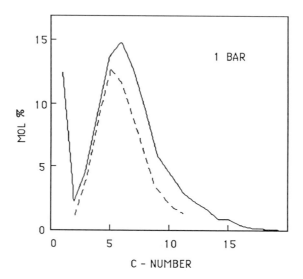

Fig. 3.1. Product distribution in Fischer-Tropsch synthesis using $[SIL(CH_2)_3NH_3]^+$ $[FeCo_3(CO)_{12}]^-$ as the catalyst precursor at 1 atm of CO/H_2 (1 : 1); —, total hydrocarbons; --, alkenes (adapted from ref. [149]).

Ligand exchange between the cluster and the functional support is another common method for preparing immobilized clusters (eq. 28).

$$SIL\text{---}PPh_2 + HOs_3Au(CO)_{10}PPh_3 \longrightarrow HOs_3Au(CO)_{10}(Ph_2P\text{---}SIL) + PPh_3 \quad (28)$$

The catalyst $HOs_3Au(CO)_{10}(Ph_2P\text{---}SIL)$ was found to be active for 1-butene isomerization at 383 K, the corresponding homonuclear system $[H_2Os_3(CO)_9-(Ph_2P\text{---}SIL)]$ being however ca. 10 times more active [229]. On the other hand, when studying the catalytic properties of $HOs_3Au(CO)_{10}(Ph_2P\text{---}SIL)$ and $ClOs_3Au(CO)_{10}(Ph_2P\text{---}SIL)$ (Fig. 3.2.), prepared by anchoring the PPh_3 clusters **86** and **87**, respectively, onto phosphine-functionalized silica, Knözinger et al. [313] found no catalytic activity for 1-butene isomerization and ethylene hydrogenation below 383 K. The stability of the "$ClOs_3Au$" system was much lower than that of the "HOs_3Au" one.

The synthesis of an anchored heteronuclear cobalt-palladium complex was also performed by the reaction of an anchored cobalt complex SIL-$(CH_2\text{-}CH_2\text{-}CH_2PCy_2)_3Co_3(CO)_7$ with a benzene solution of PdL_4. The catalytic activity of the resulting system for the hydroformylation of propylene, under mild conditions (40-100 °C, 1 atm of CO), greatly exceeds the performances, under similar conditions, of individual homonuclear cobalt and palladium complexes [213].

Functional organic polymers have also been widely used to tether mixed-metal clusters [127]. Thus, with phosphine-functionalized poly-(styrene-divinylbenzene) as a support, it was possible to generate the species $Fe_2Pt(CO)_8(Ph_2P\text{---}\textbf{P})_2$ and $RuPt_2(CO)_5(Ph_2P\text{---}\textbf{P})_3$ [**P** = poly(styrene-divinylbenzene)] [229]. Both polymer-bound clusters catalyzed ethylene hydrogenation under mild conditions. The similarity in the kinetics of the catalytic reaction for the two systems allowed the authors to suggest that the mechanisms were similar and that the clusters, and not aggregated metal, provided the active sites for hydrogenation [229]. The functionalized poly(styrene-divinylbenzene) support was also used to tether the cluster "$ClOs_3Au$"(**87**) and produce an active catalyst for ethylene hydrogenation at 1 atm and temperatures less than 373 K [228]. This catalyst was found to be significantly more stable than when **87** is anchored onto a phosphine-functionalized silica [313]. In contrast, cluster "HOs_3Au" (**86**) on this polymer led to an inactive system [228]. This different behaviour was incorrectly assigned to tetrahedral vs. butterfly structures for these clusters, the more open butterfly being associated with a more reactive cluster. In fact, it has now been established that both clusters $XOs_3Au(CO)_{10}(PPh_3)$ are of the butterfly type [45], however, with 58 and 60 electrons for X = H, **86** [171] and X = Cl, **87** [34], respectively (Fig. 3.2.). A possible explanation for the different reactivity of the supported clusters could be found in the ability of chloride to function as a one (terminally bound) or three (bridging) electron donor ligand.

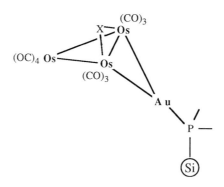

Fig. 3.2. Structure of Os$_3$Au(μ-X)(CO)(Ph$_2$P $\sim$ SIL); X = H, butterfly angle: 109.8 ° and X = Cl, butterfly angle: 121.4 °.

The cluster H$_2$Os$_3$Rh(CO)$_{10}$(acac) (**100**) was supported on poly(styrene-divinylbenzene) support by ligand association reaction, giving [H$_2$Os$_3$Rh(CO)$_{10}$(acac)Ph$_2$P — **P**]. It was tested in the hydrogenation of ethylene and the isomerization of 1-butene. Although the cluster was initially anchored intact to the support, it broke up to give catalytically active species, rhodium aggregating into crystallites responsible for hydrogenation and osmium remaining in the form of trinuclear clusters accounting for the isomerization activity [194].

An active catalyst for olefin hydrogenation was produced by immobilizing the butterfly cluster Co$_2$Pt$_2$(CO)$_8$(PPh$_3$)$_2$ (**28a**) on a phosphine functionalized polymer [124] (eq. 29):

$$\text{(P)} - \text{PPh}_2 + \quad \text{Co}_2\text{Pt}_2(\text{CO})_8(\text{PPh}_3)_2 \longrightarrow \text{Co}_2\text{Pt}_2(\text{CO})_8(\text{PPh}_2 - \text{(P)})_2 \quad + \; 2 \text{ PPh}_3 \tag{29}$$

Its slightly improved stability compared to that of **28a** under homogeneous conditions is of interest since in the latter case the mixed-metal cluster rapidly transformed into the homonuclear cluster $Pt_5(CO)_6(PPh_3)_4$ [122].

The use of ion exchange resins constitutes a further method for anchoring anionic mixed-metal carbonyl clusters, although little has been described yet in the literature. This method is especially attractive since there exists a large number of fully characterized anionic clusters. Polymeric supports presenting R_3N^+ functional groups were used to attach the anionic cluster $[H_3RuOs_3(CO)_{12}]^-$ [208]. The resulting catalyst displayed activity for the hydroformylation and isomerization of 1-hexene. It was observed that its activity was greater than that of corresponding homometallic systems, suggesting a synergic enhancement. Although the supported cluster remained partially intact after catalysis, some metal leaching occurred because of the presence of traces of water in the support, converting the anionic cluster into the neutral and more soluble cluster $H_4RuOs_3(CO)_{12}$.

Amine functionalized resins have been used to support Co-Rh clusters, such as $Rh_xCo_{4-x}(CO)_{12}$ (x = 0-2) [101, 147]. These systems were tested in olefin hydroformylation reactions [147].

Further studies have shown that an amberlite anion exchange resin may be used as the support material to immobilize $[Rh_5Pt(CO)_{15}]^-$ **(39)** [28]. This supported mixed-metal cluster showed catalytic activity with respect to aromatic ring hydrogenation reactions, observed with phenol, anisole and toluene but neither with aniline nor nitrobenzene. Under comparable conditions, catalytic activity was absent in supported catalysts obtained from homonuclear clusters of either metal ($[Pt_5(CO)_{30}]^{2-}$ or $[Rh_{12}(CO)_{30}]^{2-}$). It was shown that in this case, the anchored species is very different from the parent cluster and exhibits unique catalytic properties [28].

The heterometallic couples examined for a given catalytic reaction are presented in Table 3.1 and a listing of the heterometallic clusters which have been used to prepare hybrid catalysts is given in Table 3.2.

TABLE 3.1.

Heterometallic Couples used in Hybrid Catalysis by Immobilized Mixed-Metal Clusters.

Reaction	Heterometallic Couples
Hydrogenation of Carbon-Carbon Multiple Bonds	Fe-Ru, Fe-Pt Ru-Os, Ru-Pt Os-Rh, Os-Au Co-Pt Rh-Pt
Isomerization of Carbon-Carbon Multiple Bonds	Ru-Os Os-Rh, Os-Au
CO Hydrogenation	Fe-Co
Hydroformylation of Olefins	Ru-Os Co-Rh

3. 4. BIMETALLIC CLUSTERS IN ZEOLITES

Metal containing zeolites are attracting considerable academic and industrial interest. The preparation of bimetallic zeolites has been achieved recently by successive incorporation of the different metals. EXAFS methods may be used to ascertain short range heterometallic interactions [225, 295c]. This area has been reviewed recently [167]. Although Ni-Pd and Ni-Pt complexes have been used as precursors [102], no information appears available on the incorporation in zeolites of bimetallic clusters, as defined in this chapter. The study of the catalytic properties of the corresponding mixed-metal clusters entrapped in zeolites should be of great interest.

TABLE 3.2.

Immobilized Mixed-Metal Cluster Catalysts.

Bimetallic couple [a]	Precursor cluster	Ref. Synth.	Support [b]	Catalyzed Reaction	Ref.Catal.
Fe-Ru	$FeRu_2(CO)_{12}$ (**2a**)	316	γ-Al_2O_3	Hydrogenation of pentynes and pentadienes	211
	$Fe_2Ru(CO)_{12}$ (**3a**)	316	γ-Al_2O_3	Hydrogenation of pentynes and pentadienes	211
Fe-Co	$HFeCo_3(CO)_{12}$ (**49**)	76	SIL-NH_2	CO Hydrogenation	149
Fe-Pt	$Fe_2Pt(CO)_8PPh_3$ (**88**)	55	P-PPh_2	Hydrogenation of ethylene	229
Ru-Os	$H_2RuOs_3(CO)_{13}$ (**85**)	60, 129a, 274	γ-Al_2O_3	Hydrogenation of ethylene	59
				Isomerization of 1-butene	59, 274
	$H_3RuOs_3(CO)_{12}^-$ (**106**)	60, 274	γ-Al_2O_3	Isomerization of 1-butene	274
			P-NR_3^+	Hydroformylation of 1-, and 2-hexene	208
Ru-Pt	$RuPt_2(CO)_5(PPh_3)_3$ (**89**)	56	P-PPh_2	Hydrogenation of ethylene	229
Os-Rh	$H_2Os_3Rh(CO)_{10}(acac)$ (**100**)	109	P-PPh_2	Hydrogenation of ethylene	194
				Isomerization of 1-butene	194
Os-Au	$HOs_3Au(CO)_{10}PPh_3$ (**86**)	171	SIL-PPh_2	Isomerization of 1-butene	229, 313
				Hydrogenation of ethylene	313
			P-PPh_2	Hydrogenation of ethylene	228
	$ClOs_3Au(CO)_{10}PPh_3$ (**87**)	34	SIL-PPh_2	Isomerization of 1-butene	229, 313
				Hydrogenation of ethylene	313
			P-PPh_2	Hydrogenation of ethylene	228

Co-Rh	Co$_2$Rh$_2$(CO)$_{12}$ (**8**)	209	Resin	Hydroformylation of alkenes	147
	Co$_3$Rh(CO)$_{12}$ (**9**)	209	Resin	Hydroformylation of alkenes	147
Co-Pt	Co$_2$Pt$_2$(CO)$_8$(PPh$_3$)$_2$ (**28a**)	39	**P**-PPh$_2$	Hydrogenation of alkenes	124
Rh-Pt	Na[Rh$_5$Pt(CO)$_{15}$] (**39a**)	121	Resin	Hydrogenation of aromatic rings	28

a The metals are listed with increasing number of their group.

b Abbreviations used:

P = Poly(styrene-divinylbenzene)

SIL = silica gel matrix

Resin = amberlite anion exchange resin IRA 401

3. 5. CONCLUSION

When inorganic supports are used to anchor mixed-metal cluster compounds, a careful study of the resulting surface organometallic chemistry is necessary. Detailed spectroscopic studies on the chemisorption products of molecular clusters onto various supports can prove very valuable [2, 184]. There is clearly a need for more such studies on mixed-metal clusters. It is often tempting but dangerous to place too much confidence on infrared data as many very different carbonyl clusters have spectra looking similar, even for the pure molecular species, in solution, in their finger-print region (ca. 2100- 1700 cm^{-1}). This tends to be even more so with high nuclearity clusters, precisely those which are used as models for the surface species which may be generated on the support. Spectral resolution is of course not enhanced upon adsorption of a complex molecule, so that severe mistakes may be made. However, by combining different spectroscopic methods ("One method is no method"!), one could obtain important information regarding the reactivity of the metal core and of the coordinated ligands and follow the structural changes eventually occurring during attachment to the support and, subsequently, during catalysis. The supported systems often have a high lability, particularly under the thermal conditions required for the catalysis, which will limit their life-time. The lack of stability of supported clusters seems to be quite general. Formation of metallic particles often results and such cluster-derived heterogeneous catalysts may actually display unique properties, as discussed in Section 4.

A wider range of functional inorganic and organic supports may be used. By varying their nature and that of the molecular cluster to be immobilized, a collection of new catalysts may be prepared. They often display the anticipated advantages such as improved stability, mentioned in the Introduction to this Section. However, synthetic limitations have been encountered, *e.g.*, the interaction of $H_2FeRu_3(CO)_{13}$ with phosphine-functionalized silica led to cluster fragmentation and aggregated metal resulted from attempts to anchor $Fe_2Pt(CO)_8(PPh_3)_2$ [229].

4. HETEROMETALLIC CLUSTERS IN HETEROGENEOUS CATALYSIS

4. 1. INTRODUCTION

Bimetallic systems represent an important class of heterogeneous catalysts [283, 286]. Just to give one example, alumina-supported platinum-rhenium is one of the most important commercial bimetallic catalyst, which is used in the catalytic reforming of petroleum naphthas [266]. As a rule, the present industrial heterogeneous catalysts are multicomponent. Sinfelt and coworkers [284-287] have pioneered the development and study of catalytic materials composed of very small bimetallic particles highly dispersed over the surface of an oxide support. The properties of these catalysts are largely determined by the homogeneous distribution of the components in the bulk and on the surface of the particles and also by their mutual interaction, necessary for the occurrence of active centers having unique characteristics. Therefore, an optimal method of preparation of the catalyst is required, including the selection of the precursors and of the support, the conditions of their interaction and of the catalyst activation. These catalytic materials customarily have been prepared by simultaneous or sequential impregnation of a separate precursor for each metal, a method which, in general, does not allow a fine tuning of intermetallic stoichiometry in the resulting particles.

The use of transition metal carbonyls for the preparation of highly dispersed metallic catalysts was suggested back in 1964 by Parkyns, who studied the decomposition of $Ni(CO)_4$ on an alumina surface, giving metallic nickel [224]. Ten years later, Anderson and Mainwaring drew attention in a pioneering work to the utilization of *molecular bimetallic clusters* as specific precursors for the generation of dispersed bimetallic catalysts [6]. Obviously the idea of studying cluster-derived catalysts must have been in the air as the use of carbonyl cluster compounds was independently suggested at about the same time by Robertson and Webb [250], and Smith *et al.* [288]. Using appropriate bimetallic cluster compounds could, in principle, allow a better control of individual particle composition and structure. The description of cluster core structures as fragments of metal surfaces or of the bulk has become very popular and of great heuristic value [75, 103, 216]. There is indeed a striking analogy between the core structure of large molecular clusters such as $[Ni_{38}Pt_6(CO)_{48}H_{6-n}]^{n-}$ [72] (Fig. 1.2) and that of small metal particles. High nuclearity metal clusters are also important in the understanding of how, when, and why metallic behaviour develops.

Recently, Teo *et al.* reported an interesting series of bimetallic Ag-Au clusters containing 25 ($Ag_{12}Au_{13}$), 37 ($Ag_{19}Au_{18}$), and 38 ($Ag_{20}Au_{18}$) metal atoms [295a]. The most interesting structural characteristic of these clusters is that they can be considered as being built from 13-atom centered icosahedral cluster units (Fig. 4.1(a)). Thus, the 25-atom cluster can be considered as two icosahedra sharing a vertex (Fig. 4.1(b)); the 37- and 38-atom clusters as three icosahedra sharing three vertices in a cyclic manner (Fig. 4.1(c)) plus one and two capping atoms, respectively.

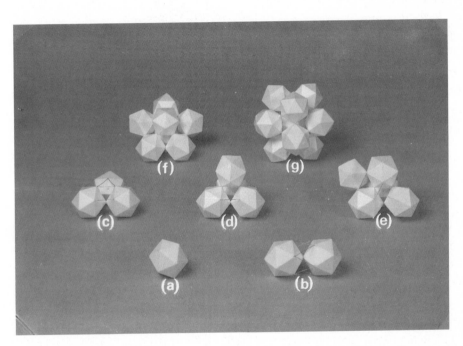

Fig. 4.1. Superclusters S_n of n-centered icosahedra (number of atoms in parentheses): a) S_1 (13); b) S_2 (25); c) S_3 (36); d) S_4 (46); e) S_5 (56); f) S_7 (76); g) S_{12} (127). Reproduced by permission from ref. [295a].

These observations, together with those of Schmid [269], led to the concept of "cluster of clusters". It is predicted that tetrahedral, trigonal-bipyramidal, and pentagonal-bipyramidal arrays of vertex-sharing icosahedra will give rise to nuclearities ($10n + 6$) of 46, 56, and 76 atom clusters, respectively, as depicted in Fig. 4.1(d)-(f). The 127-atom icosahedral cluster (Fig.4.1(g)) is formed by 12 centered icosahedra sharing 30 corners. Here the icosahedral hole created by the 12 icosahedra is filled with one additional atom. The idea of "cluster of clusters" may provide new pathways to novel high nuclearity clusters of great interest. Strong indications have also been obtained that $Ag_{20}Au_{18}Cl_{14}(PPh_3)_{12}$, whose metal core has an approximate diameter of 15 Å, possesses the free electron quasi-band structure behaviour characteristic of bulk metal [295b]. Even larger, but homonuclear, "superclusters" of ruthenium, cobalt, rhodium, platinum or gold have been reported recently and studied by scanning electron microscopy [269, 270]. Relatively little is known at present about the actual shapes and structures of the metallic particles resulting from the thermal decomposition of heterometallic cluster compounds and particularly, to what extent they keep the memory of their well-defined molecular precursors. We shall therefore not go into this here.

The main advantages expected from the use of mixed-metal cluster compounds as precursors of heterogeneous catalysts, referred to in the following as *Mixed-Metal Cluster-Derived* (MMCD) catalysts, can be detailed as follows:

(i) efficient clustering of the metals on the surface when strong metal-metal bonds are present in the molecular precursor;

(ii) low initial oxidation states of the metals and relatively facile catalyst activation by heating to a temperature usually below 475 K, where cluster decomposition would occur into reduced metal and CO, representing much less severe reduction conditions than those used for conventional metal salt precursors, thus limiting the possibility of sintering and crystallite growth and/or support alteration;

(iii) avoidance of contaminating anions such as Cl⁻ on the surface; these usually affect catalytic properties. Note, however, that phosphorus- or sulfur-containing ligands, for example, may not be innocent during the activation step of the catalyst and may also affect its properties;

(iv) at least partial retention in the mixed-metal catalyst of the unique geometries or stereochemical features of the molecular precursor;

(v) preparation of catalysts having well-dispersed and, at least initially, uniform mixed-metal phases, which would be difficult to guarantee by coimpregnation techniques;

(vi) possibility of systematic variation in particle stoichiometry by employing precursor complexes of different known compositions, stoichiometries and structures.

Critical tests are needed to evaluate the validity of this approach and they include:

(i) the comparison of catalysts derived from mixed-metal clusters with conventional catalysts containing the same metals in comparable proportions;

(ii) the comparison of catalysts $[M_xM'_y]$ derived from isostructural clusters whose cores $M_xM'_y$ are characterized by different M/M' ratios;

(iii) the comparison of catalysts derived from clusters having the same M/M' ratio but a different structure (*e.g.*, linear M-M'-M *vs.* triangulo M — M),
$$\diagdown M' \diagup$$

thus allowing an investigation of the structural influence of the molecular precursor on the properties of the MMCD catalyst.

An obvious prerequisite in this area of research is the availability of suitable molecular mixed-metal compounds, emphasizing the fundamental importance of synthetic chemistry. This may explain in part why many bimetallic systems have not yet been examined in catalysis, either because no molecular complex containing the metals of choice has yet been reported or because it can only be prepared in too low a yield to be of synthetic value [54, 130, 145, 249].

4. 2. PREPARATION OF MIXED - METAL CLUSTER - DERIVED CATALYSTS

Numerous publications and review articles have appeared which detail the preparation of heterogeneous catalysts derived from transition-metal complexes [12, 136, 169, 286]. The general methods and techniques reported can be applied to the preparation of MMCD catalysts and these will not be detailed here.

The mixed-metal cluster will be selected usually on the basis (i) of the occurrence of a specific bimetallic association expected to present interesting catalytic properties; (ii) of its availability, *i.e*, high-yield synthesis and reasonable stability under the standard conditions used; (iii) of the presence of ligands which may be easily removed during the thermal treatment phase and/or of the absence of ligands whose partial decomposition could generate catalyst poisons. The heterometallic clusters which have been used as precursors of heterogeneous bimetallic catalysts are listed at the end of Section 4, in Table 4.14, which includes references for their preparation.

Often, the inorganic support of choice, *e.g.*, alumina, silica, is previously calcinated and evacuated at elevated temperature under inert gas (*e.g.*, nitrogen, argon, helium). Typically, the mixed-metal carbonyl clusters are deposited, under an inert gas atmosphere, from an organic solution onto the inorganic support. Incipient wetness and adsorption impregnation are commonly used methods. In other cases, the solution of the cluster is sprayed over a rotating sample of the inorganic support. Dry mixing has also been used. Except in the latter case, the organic solvent will be removed under reduced pressure, at room temperature. Removal of carbon monoxide and other ligands is usually accomplished by slowly heating the material to *ca.* 570 K in a stream of dry inert gas for a certain time. Finally, it is reduced in a hydrogen stream at elevated temperature. The sequence impregnation-calcination-reduction critically determines, and is in turn influenced by, the nature and extent of the metal-support interactions. The latter have a considerable influence upon the catalytic properties of the material via, *e.g.*, the dispersion of the catalyst.

4. 3. INTERACTIONS OF MIXED - METAL CLUSTERS WITH CATALYTIC SUPPORTS

One can easily understand that the nature and extent of the interactions between the molecular cluster and the support considered will largely contribute to control the preparation of the metal particles generated after appropriate thermal treatment. Their study and understanding are therefore of great significance, particularly when associated with the evaluation of the catalytic properties of the resulting materials examined under the same conditions.

When clusters are supported, even using well-established air-free techniques [50, 281], rearrangements will often be observed, possibly including dramatic changes in nuclearity. This has been unambigously established, *e.g.*, in the case of the very stable platinum cluster $Pt_3(CO)_3(PPh_3)_4$, which upon simple impregnation onto inorganic supports under inert atmosphere, is transformed into another stable cluster, $Pt_5(CO)_6(PPh_3)_4$, which has an edge-capped tetrahedral structure (Scheme 4.1). This increase in nuclearity, from 3 to 5, was found to depend upon the nature of the support, its pre-treatment, and the solvent used [19].

SCHEME 4.1.

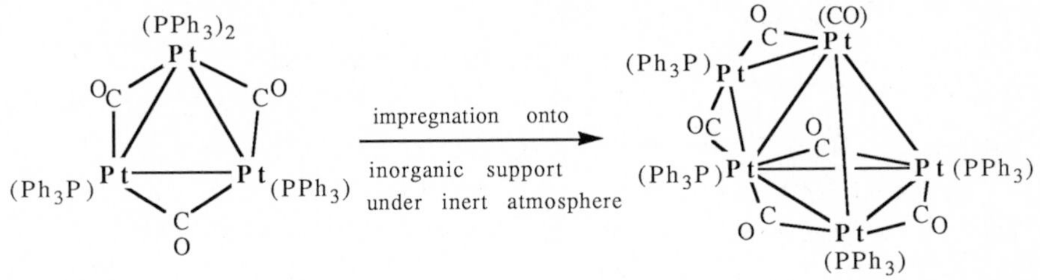

Such transformations might be even more likely with clusters of lower stability or with mixed-metal clusters, whose heterometallic bonds are intrinsically polar [49] and thus more reactive. Further alterations of the original molecular structure occur during activation or catalysis and it should not be surprising to observe transformations into the species most stable under these conditions.

Extraction of surface species is sometimes possible and anionic clusters have been extracted using solutions containing bulky cations, e.g., [Ph$_3$PNPPh$_3$]Cl. These attempts require a good knowledge of the molecular organometallic chemistry of the corresponding clusters as the reagents used may sometimes react with the cluster(s) and therefore not behave as "innocent" extractors. As expected, very different behaviours are found for different clusters but also for the same cluster on different supports. Detailed surface studies on molecular clusters adsorbed on a given support may, at best, be of only little use to understand the catalytic properties of cluster-derived catalysts prepared with the same cluster but using a different support. As already found with simpler homonuclear clusters, differences in catalytic activity of such clusters supported on, e.g., γ-Al$_2$O$_3$ with different degrees of dehydroxylation could explain some of, if not all, the difficulties encountered in preparing reproducible catalytic systems. This is even more crucial with mixed-metal clusters, if not only because of the enhanced reactivity of their heterometallic bonds. For example, the interactions between the bimetallic carbonyl clusters Fe$_2$Ru(CO)$_{12}$ (**3a**) or H$_2$FeRu$_3$(CO)$_{13}$ (**32**) and inorganic supports have been studied by infrared and Raman spectroscopy [31, 96]. After the first stage of the interaction, which is pure physisorption, their structure completely changes on alumina, particularly on the highly reactive hydrated alumina, whereas it may be partially retained on Cab-O-Sil. Thus for example, $H_2FeRu_3(CO)_{13}$ (**32**) is first physisorbed when impregnated onto a silica surface and is stable on Cab-O-Sil, unlike $Fe_3(CO)_{12}$, whereas it is fully decomposed on Al_2O_3, metal-metal bond splitting occurring [80, 96]. Findings consistent with these observations were obtained from molecular isotope exchange between the ^{12}CO ligands and gaseous ^{13}CO molecules, a higher exchange rate indicating a stronger interaction with the support [31]. Clusters with higher iron content were found to be generally more labile and the following sequence of decreasing stabilities of the clusters on either support was established: $H_2FeRu_3(CO)_{13} >> Ru_3(CO)_{12} > Fe_3(CO)_{12} > Fe_2Ru(CO)_{12}$ [136]. The subsequent decomposition of the bimetallic clusters was found to lead to species anchored onto rather uniform sites of the surface [96].

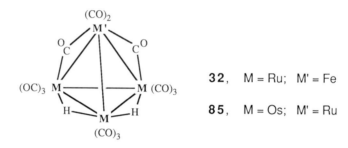

32,   M = Ru;   M' = Fe

85,   M = Os;   M' = Ru

On impregnation of a mixture of the clusters $Fe_3(CO)_{12}$ and $Ru_3(CO)_{12}$ onto Cab-O-Sil, it was found by infrared spectroscopy that interaction between iron and ruthenium does not occur in the impregnated phase, but develops during thermal decomposition [193a, 268]. The latter was found to start with $Fe_3(CO)_{12}$, as expected, whose decomposition is not influenced by the presence of ruthenium up to 400 K. However, the decomposition of $Ru_3(CO)_{12}$ was significantly modified by the presence of iron. In studies of conventional catalysts, hindrance of iron reduction in the presence of a small amount of ruthenium has been noted whereas the reduction becomes easier at higher ruthenium concentration. Addition of platinum also results in a decrease of the reduction temperature of iron [134]. Molecular bimetallic clusters are of course expected to lead to heterometallic interactions at an earlier stage than a mixture of the homonuclear clusters. This has been investigated with, *e.g.*, Fe-Ru [193, 268], Fe-Os [79] and Ru-Os [58] systems, often by comparing the behaviour of a mixture of the homonuclear clusters with that of the corresponding heterometallic clusters. This property is illustrated, for example, by the stability observed for Fe-Ru clusters impregnated on Cab-O-Sil whereas iron oxide was always formed after impregnation of a mixture of the two monometallic clusters, as a result of the reaction between surface hydroxyl groups and zerovalent iron. This observation has been related to the catalytic behavior of the corresponding particles, further characterized by, *e.g.*, Mössbauer spectroscopy and ferromagnetic resonance [193a, 193c, 268].

Adsorption of an hexane solution of the cluster $H_2RuOs_3(CO)_{13}$ (**85**) onto γ-Al_2O_3-C, previously dried at 473 K under vacuum, led to the formation of the Al_2O_3-supported anionic species $Al^+[HRuOs_3(CO)_{13}]^-$ [59] (eq. 25). When this species was heated under an equimolar H_2+CO mixture at atmospheric pressure and 373-473 K, the new surface species $Al^+[H_3RuOs_3(CO)_{12}]^-$ was formed, as shown by infrared monitoring (eq. 30). The latter species is also formed by direct deposition of $[Et_4N][H_3RuOs_3(CO)_{12}]$ [274].

$$Al^+[HRuOs_3(CO)_{13}]^- \xrightarrow[\Delta]{CO + H_2} Al^+[H_3RuOs_3(CO)_{12}]^- \qquad (30)$$

The surface bound species $[H_3RuOs_3(CO)_{12}]^-$ is stable at 473 K under H_2+CO and can be extracted intact from the alumina support. Under vacuum however, the surface cluster $Al^+[H_3RuOs_3(CO)_{12}]^-$ decomposed above 373 K, giving oxidized osmium and ruthenium carbonyl species. At 573 K, cluster decomposition occurred even in the presence of H_2+CO. Catalytic experiments showed that $Al^+[H_3RuOs_3(CO)_{12}]^-$ is active for 1-butene isomerization at 330 K and was found by infrared to be stable under the reaction conditions [59, 274], whereas it decomposed during ethylene hydrogenation [274] (see Section 3.2).

The adsorption of the cluster compounds $Co_2Rh_2(CO)_{12}$ (**8**), $Co_3Rh(CO)_{12}$ (**9**) and of the related homonuclear clusters $Co_4(CO)_{12}$ and $Rh_6(CO)_{16}$, was studied onto two typical catalyst supports, γ-alumina and Aerosil silica [5]. It was shown that $Co_2Rh_2(CO)_{12}$ is much more strongly adsorbed on γ-alumina than on silica, may be as a consequence of the greater ionicity of surface oxygen atoms in γ-alumina, making them stronger nucleophiles. An infrared study showed that initial adsorption of $Co_2Rh_2(CO)_{12}$ on γ-alumina occurred with loss of the bridging carbonyls, the remaining carbonyls being progressively lost at temperatures higher than

300 K. The resulting strong adsorption was assumed to proceed by ligand exchange with a surface anion, and it was observed that the presence of oxygen can facilitate decarbonylation. A considerable advantage found with the MMCD-catalyst was that $[Co_2Rh_2]$ had a much higher dispersion than was obtained with a conventional catalyst prepared by impregnation of the support with aqueous solutions containing equimolar concentrations of cobalt and rhodium salts. The dispersion was considerably better on $\gamma\text{-}Al_2O_3$, on which $Co_2Rh_2(CO)_{12}$ is adsorbed, than on SiO_2, on which it is only absorbed to a very limited extent, at best [5].

The preparation of highly dispersed supported metal catalysts always requires special care. At too weak a metal-support interaction, agglomeration occurs and large crystallites are obtained. At too strong an interaction, complete reduction into a metallic state becomes very difficult. This latter problem makes even more attractive the use of precursors in a low oxidation state, such as metal carbonyl compounds. Through appropriate metal-support interactions, stabilization of small metal particles can be achieved and thus a higher dispersion of the catalyst may be attained.

4. 4. OCCURRENCE OF HETEROMETALLIC INTERACTIONS IN MIXED-METAL CLUSTER - DERIVED CATALYSTS

Achieving an intimate interaction between the different metals constituting a multimetallic catalyst is one of the main objectives when using mixed-metal clusters as precursors of such catalysts. It is now well recognized that the synthetic and structural methodology of inorganic chemistry can provide suitable precursors to bimetallic catalysts, combining similar or rather disparate elements. More significantly, the catalytic properties may be uniquely related to the specific bimetallic precursor [136, 279]. This opens the possiblity of enhanced control and even of effective design of bimetallic catalytic sites. However, reports have indicated finding no evidence for bimetallic interactions influencing the activity of cluster-derived catalysts [278, 296].

The electronic structure of the dispersed metal and hence its catalytic activity is largely dependent upon phase combination and various degrees of miscibility are observed. The degree of alloy formation can be investigated, *e.g.*, by reaction activity and chemisorption. Evidence for direct Rh-Fe bonding in silica-supported catalysts has been obtained by Mössbauer and *in situ* EXAFS spectroscopic methods [163]. On the other hand, it has been observed that hydrogen will dissociatively adsorb on Ru-Cu/SiO_2

catalysts, and that hydrogen spillover occurs from Ru onto Cu, suggesting that hydrogen chemisorption cannot be used to titrate the number of surface Ru atoms in a Ru-Cu catalyst [180]. The extent to which alloy particles are formed in bimetallic systems still remains a matter of some uncertainty although recent results clearly demonstrate the advantages in using mixed-metal clusters as precursors (see below) [40, 78]. Sinfelt has reported evidence for the existence of bimetallic particles in conventional Ru-Cu and Os-Cu systems [283], although these are metal couples almost totally immiscible in the bulk. In Rh-Ag systems, however, Anderson *et al.* [4] concluded that the particle composition agreed with the expectations from the bulk behaviour where the miscibility is known to be low. Solution stabilization can result from small particle size effects.

General difficulties in the interpretation of data from bimetallic catalysts lie in the exact knowledge of the composition of the metallic particles and in the possibility of surface enrichment by the component having the lower surface energy. It has been noted, for example, that cobalt and rhodium form a continuous series of solid solutions across the entire composition range [143], so that phase separation should not occur, from a thermodynamic point of view [5]. It has therefore been deduced that catalysts prepared from $Co_2Rh_2(CO)_{12}$ will necessarily consist of metallic particles each with equal amounts of the two metals. However, a systematic and complete knowledge of the decomposition steps of carbonyl clusters impregnated on metal oxides is still lacking. With these aims, quantitative data have been obtained by temperature programmed decomposition of the series of isostructural clusters $Rh_4(CO)_{12}$, $Co_2Rh_2(CO)_{12}$ and $Co_3Rh(CO)_{12}$ [231].

The tetrahedral clusters $H_2FeRu_3(CO)_{13}$ (32), $H_2FeOs_3(CO)_{13}$ (48) and $HFeCo_3(CO)_{12}$ (49) were chemisorbed on magnesia and their thermal decomposition led to supported bimetallic particles of the same composition [78].

48

49

This important result was obtained by high-resolution analytical electron microscopy and quantitative energy dispersive analysis. For comparison, coimpregnation of two different monometallic clusters, $Fe_3(CO)_{12}$ and $Os_3(CO)_{12}$ ([Fe_3 + Os_3] or $Fe_3(CO)_{12}$ and $Ru_3(CO)_{12}$ ([Fe_3 + Ru_3] has been reported not to result in the formation of bimetallic particles [78], whereas others have found that [Fe_3 + Ru_3] leads to the formation of a bimetallic catalyst [138, 139].

Using similar techniques, it has been observed recently that the Fe-Pd clusters $FePd_2(CO)_4(Ph_2PCH_2PPh_2)_2$ (**103**) and $Fe_2Pd_2(CO)_5(NO)_2(Ph_2PCH_2P-Ph_2)_2$ (**104**), when chemisorbed on alumina and thermally decomposed, lead to bimetallic particles of composition similar to that of the metal core in **103** and **104**, respectively [40]. Interestingly, it was found that metal segregation occurs when these particles are used to catalyze the reductive carbonylation of aromatic nitro compounds [40] (see Section 4.5.4).

The possible advantages of using MMC for generating new types of bimetallic catalysts lie, in particular, in the pre-existence of heterometal-metal bonds which might favour the formation of bimetallic particles, as already shown by the examples given above, whereas the occurrence of separate metallic phases, when monometallic clusters are used, could be due to the different temperatures at which they start to decompose. This should be even more dramatic and lead to unique MMCD catalysts when bimetallic systems are considered whose metals are immiscible in the bulk.

4. 5. REACTIONS INVOLVING HETEROGENEOUS MIXED - METAL CLUSTER - DERIVED CATALYSTS

In this Section, we shall examine the different reactions for which MMCD catalysts have been employed. A summary of the various bimetallic couples which have been used for a given reaction type is presented in Table 4.12 while a complete listing of the mixed-metal clusters which have been used in heterogeneous catalysis is given in Table 4.14. Since the area covered in this review only includes transition metals, the reader interested in, *e.g.*, the use of platinum-tin complexes in hydrogenolysis and skeletal isomerization of hydrocarbons [136, 202, 318] or in the recent use of potassium carbonyl ferrates as precursors to Fischer-Tropsch catalysts [198, 199], is invited to consult the literature.

4. 5. 1. Hydrocarbon Skeletal Rearrangements

 Anderson and Mainwaring used the molecular cluster compound $Co_2Rh_2(CO)_{12}$ (**8**) as precursor of a dispersed bimetallic catalyst for the hydrogenolysis of methylcyclopentane [6]. They found a remarkable specificity for hydrogenolysis of the methyl-substituted five-membered ring to non-cyclic C_6 isomers (Scheme 4.2), with little ring expansion to cyclics and even less extensive cracking to lower molecular weight products.

SCHEME 4. 2.

Hydrogenolysis of Methylcyclopentane.

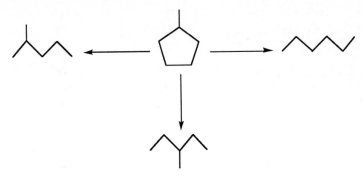

The product distribution as a function of the total methylcyclopentane conversion, shown in Fig. 4.2, remained constant with increasing catalyst use.

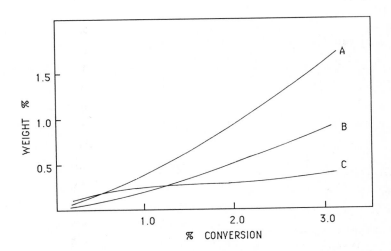

Fig. 4.2. Isomer formation as a funtion of methylcyclopentane conversion: A, 2-methylpentane; B, 3-methylpentane; C, n- hexane (adapted from ref. [6]).

A considerable dissimilarity was observed in the behaviour of this $[Co_2Rh_2]$ catalyst compared to that of conventional cobalt and rhodium catalysts. It was also found that on this MMCD catalyst, n - hexane is particularly sensitive to secondary reactions.

Also in the course of studies on the hydrogenolysis of methylcyclopentane [202], three different heteropolymetallic cobalt-platinum complexes were used for the preparation of supported bimetallic catalysts [203]. The molecular clusters displayed an interesting variation in the composition of their metal core, the Co : Pt ratio ranging from 2 : 1 in *trans*-$[Pt\{Co(CO)_4\}_2(c\text{-}C_6H_{11}NC)_2]$ **(23a)**, 2 : 2 in $Co_2Pt_2(CO)_8(PPh_3)_2$ **(28a)** to 2 : 3 in $Co_2Pt_3(CO)_9(PEt_3)_3$ **(82b)**. The $[Co_2Pt_2]$ and $[Co_2Pt_3]$ catalysts showed a higher selectivity for demethylation of methylcyclopentane ($C_6 \rightarrow C_5 + C_1$) than the $[Co_2Pt]$ catalyst. The MMCD catalysts showed a lower selectivity in isomerization of 2-methylpentane than the conventional ones in favour of an increased demethylation. Cobalt alone on alumina did not show this selectivity. These effects were tentatively attributed to the different nature of the ligands bound to the molecular precursors (phosphines *vs.* isonitriles) rather than to the change in the Co : Pt ratio. Thus, these ligands might have been difficult to completely eliminate during the thermal activation steps of the catalysts, and this could lead to modifications in the composition and/or structure of the final catalyst. It was also found that the catalysts $[Co_2Pt]$ or that derived from *trans*-$[Pt\{Fe(CO)_3NO\}_2(t\text{-}BuNC)_2]$ **(24b)** have properties similar to those of highly dispersed Pt catalysts (0.2 % Pt/Al_2O_3) when they are weakly loaded in platinum (*e.g.*, 0.3 % Pt) [136]. Interestingly, a $[Cr_2Pd_2]$ catalyst, derived from $Cr_2Pd_2Cp_2(CO)_6(PMe_3)_2$ **(90)**, was found to display the same contribution of the cyclic mechanism as a conventional Cr-Pd catalyst containing 6.6 % Pd in the isomerization of 2-methylpentane in methylcyclopentane [104].

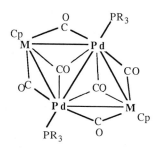

90   M = Cr ; R = Me

16   M = W ; R = Ph

The corresponding [W_2Pd_2] catalyst, prepared from the closely related tungsten-containing cluster **16**, had completely different properties and was found to selectively isomerize 2-methylpentane to 3-methylpentane [136].

Recent studies on the hydrogenolysis of cyclopentane with Re, Pt and Re/Pt catalysts supported on γ-Al_2O_3 have shown that the TOF for methane formation is over 100 times higher when using $Re_2Pt(CO)_{12}$ (**111**)

$$(OC)_5 Re-\underset{\underset{CO}{|}}{\overset{\overset{CO}{|}}{Pt}}-Re(CO)_5$$

111

as a catalyst precursor [11]. The enhancement observed with the [Re_2Pt] catalyst was related to the direct contact of the Re and Pt atoms on the surface. When using an acidic Y-zeolite as a support, hydrogenolysis of n-heptane suggested that under severe conditions (773 K, 15 atm), rhenium-platinum clusters are formed even in physical mixtures of Re/Na-Y and Pt/Na-Y [11].

The first uses of molecular noble metal-containing mixed-metal clusters as supported catalyst precursors had in fact been claimed in a patent [196]. The precursor complexes that had been prepared were given the formulations: $Ru_3Pt(CO)_{12}Py_3$ (**91**), $Ir_2Pt(CO)_7Py_2$ (**92**) and $Ir_6Pt(CO)_{15}Py_2$ (**93**). Their proposed structures may however not be those that an X-ray diffraction study would reveal so that correlations between molecular structures and catalytic properties will not be attempted. The MMCD catalysts studied, obtained by impregnation of these complexes onto inorganic oxides, have been found to be very active and selective catalysts for hydrocarbon conversion. For example, in the n-heptane conversion into, e.g., methane and the isomeric xylenes, the [Ir_6Pt] catalysts maintained a significantly higher level of conversion than a standard catalyst prepared from H_2MCl_6 (M = Ir or Pt) acid solutions, and than the [Ru_3Pt] catalyst derived from **91**. The properties of the latter catalysts were governed by the high hydrogenolysis activity of ruthenium, and would make it more useful for hydrogenation reactions. The toluene selectivities exhibited by the [Ir_6Pt] catalyst were found to be 5 to 7 % higher than those of the [Ir_2Pt] catalyst. The 30 % lower coking rate displayed by the [Ir_6Pt] compared to a conventional catalyst was found to be another advantage of the MMCD catalyst. For naphtha reforming, the [Ir$_2$Pt] and [Ir$_6$Pt] catalyst have been found to be, at least, 1.6 and 2.4 times, respectively, more active than conventional catalysts and the [Ir$_6$Pt] catalyst 1.8 times more active than the [Ir$_2$Pt] catalyst. It was suggested that unique heterometallic sites on the cluster-derived catalysts were more efficient in carrying out aromatization reactions than those present on conventional Ir-Pt catalysts [196].

A [Ru$_3$Ni] catalyst, derived from H$_3$Ru$_3$NiCp(CO)$_9$ (10), has been recently reported to transform toluene to cyclohexane via demethylation of the intermediate methylcyclohexane [67]. Cracking of the latter to n-hexane was also observed to a limited extent. Phenylacetylene is mainly hydrogenated with this [Ru$_3$Ni] catalyst but some hydrogenolysis to cyclohexane and n-hexane was also observed. The properties of this catalyst appear to be characteristic of the presence of ruthenium.

The catalytic properties of the materials obtained after decomposing the supported tungsten-iridium clusters WIr$_3$Cp(CO)$_{11}$ (94) and W$_2$Ir$_2$Cp$_2$(CO)$_{10}$ (95) were examined for n-butane hydrogenolysis [279], another structure sensitive reaction [33]. Comparisons were made with catalysts derived from Ir$_4$(CO)$_{12}$ and W$_2$Cp$_2$(CO)$_6$. The corresponding data are reported in Table 4.1.

94

9 5

Neither the decomposed W$_2$Cp$_2$(CO)$_6$/Al$_2$O$_3$ nor the alumina alone was active in the temperature range investigated (< 500 K). The [Ir$_4$] catalyst displayed a high selectivity for scission of the central carbon-carbon bond of butane, forming two molecules of ethane. The [WIr$_3$] and [2 W$_2$+Ir$_4$] catalysts also showed ethane selectivity of 70 % or more, but the [W$_2$Ir$_2$] catalyst showed less than 50 % ethane in the product stream. This different cracking pattern for the W$_2$Ir$_2$Cp$_2$(CO)$_{10}$- derived catalyst was taken as strong evidence for a residual tungsten-iridium interaction that modifies the character of the catalytic site(s).

TABLE 4.1.

n-Butane Hydrogenolysis Data for Tungsten-Iridium Catalysts (from ref. [279]).

Catalyst type	Precursor complex(es)	% ethane[a]	Turnover frequency[b]	$E_a{}^c$
$[Ir_4]$	$Ir_4(CO)_{12}$	71.0	0.59	45
$[WIr_3]$	$WIr_3Cp(CO)_{11}$ (94)	75.3	5.03	36
$[W_2Ir_2]$	$W_2Ir_2Cp_2(CO)_{10}$ (95)	49.2	0.86	32
$[2\ W_2+Ir_4]$	$W_2Cp_2(CO)_6 +$ $1/2\ Ir_4(CO)_{12}$	70.4	0.28	48

a Percent of ethane in products, the balance being methane and propane.
b Molecules butane reacted x 10^2 per iridium per s, measured at 488 K.
c Apparent activation energy (kcal/mol) over the range 473-503 K.

The activities measured were comparable with previous data for Ir/Al_2O_3 [112], but most noteworthy was that the activation energies were lower for both the $[WIr_3]$ and the $[W_2Ir_2]$ catalyst. Measurements of hydrogen and carbon monoxide chemisorption indicated very high dispersions, with particle sizes probably less than 10 Å [197]. The mixed-metal catalysts showed some loss of metal surface area after the hydrogenolysis experiments. However, electron microscopy studies revealed no discernable particles under conditions where particles larger than 20 Å were seen in conventionally prepared catalyst samples. The diminished chemisorption, particularly true of the $[W_2Ir_2]$ catalyst, may indicate the formation of carbonaceous residues during the catalytic runs [279]. These results also confirm that the catalytic properties of MMCD catalysts may be uniquely linked to the specific bimetallic precursor used.

The trinuclear clusters $Fe_3(CO)_{12}$, $Fe_2Ru(CO)_{12}$ (**3a**), $FeRu_2(CO)_{12}$ (**2a**) and $Ru_3(CO)_{12}$ were used to prepare γ-Al_2O_3 supported heterogeneous catalysts [166]. The $[FeRu_2]$ catalyst showed the maximum activity among the $[Fe_{3-x}Ru_x]$ (x = 0-3) catalyst for C_2H_4 self-homologation whereas conventional Fe-Ru catalysts, obtained by impregnation of an aqueous solution of ferric nitrate and ruthenium chloride, showed a regular decrease in activity when decreasing the ruthenium content (Fig. 4.3). Also notable was the steadily increasing selectivity ($C_3 : C_4$ ratio) with increasing Fe content for the MMCD catalysts in contrast to the conventional catalysts.

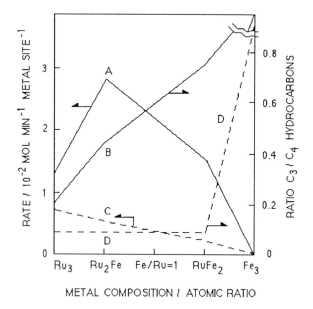

Fig. 4.3. C_2H_4 Self-homologation on Al_2O_3-supported $Ru_3(CO)_{12}$, $FeRu_2(CO)_{12}$, $Fe_2Ru(CO)_{12}$ and $Fe_3(CO)_{12}$ catalysts (A, B) and conventional Fe/Ru catalysts (C, D) at 523 K; $p\ C_2H_4$ = 33 Torr; metal = 6.0 x 10^{-5} mol; metal/ support = 1.5-3 wt % (adapted from ref. [166]).

Ethane hydrogenolysis proceeded more readily on the [Ru_3] catalyst than on the conventional Ru catalyst whereas Fe catalysts were inactive (Fig. 4.4). The [Fe_2Ru] catalyst showed more than one order of magnitude greater activity than the [$FeRu_2$] catalyst whereas the activities of the impregnation catalysts decreased with a decrease in the Ru content. Active sites of a different structure and arrangement are required in ethane hydrogenolysis and in ethylene self homologation. In both cases, however, notable differences can be found when using MMCD instead of conventional catalysts and synergic effects are observed with the [Fe_2Ru] and [$FeRu_2$] catalysts. EXAFS studies suggested that the active sites of the mixed-metal cluster catalysts consist of Fe(II) ions and small ruthenium clusters chemically bonded to the Al_2O_3 through the surface oxygen atoms [166].

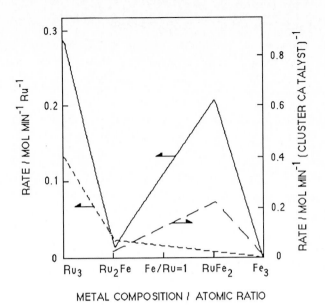

Fig. 4.4. Ethane hydrogenolysis over Al_2O_3-supported $Ru_3(CO)_{12}$, $FeRu_2(CO)_{12}$, $Fe_2Ru(CO)_{12}$ and $Fe_3(CO)_{12}$ catalysts (———,- - -) and conventional Fe/Ru catalysts (- - - - - -). Reaction temperature 443 K; $p\ C_2H_6$ = 25 Torr; $p\ H_2$ = 50 Torr (adapted from ref. [166]).

4. 5. 2. Hydrogenation Reactions

4. 5. 2. 1. Hydrogenation of Carbon-Carbon Multiple Bonds.

The cluster $Ru_3NiCp(\mu\text{-}H)_3(CO)_9$ (**10**) has been used recently as a catalyst precursor for the hydrogenation, at 80 °C, of cyclohexene and 1,3- and 1,4-cyclohexadiene to cyclohexane [67]. It is more active in the hydrogenation of benzene or toluene than the [Os_3Ni] catalyst prepared from the isostructural cluster **11** (see below).

The hydrogenation of 1,3-*cis*- and 1,3-*trans*-pentadiene has been investigated using $Os_3NiCp(\mu\text{-}H)_3(CO)_9$ (**11**) supported on Chromosorb *P*. The pretreatment times were found to play a critical role and only partial decomposition of the cluster occurred with short pre-treatment times. This resulted in both homogeneous and heterogeneous processes being operative, their relative contributions depending on the temperature and the ageing of the system. The former process was found to lead to 2-pentenes whereas the latter accounted for the non-selective total hydrogenation to pentane [68]. Cluster **11** was also suported on γ-Al_2O_3 and the derived heterogeneous catalyst [Os_3Ni] was shown to be very efficient for the hydrogenation of acetylene to ethylene and ethane [212]. A good selectivity is achieved only if the catalyst has been activated at temperatures below 453 K, otherwise quantitative conversion of acetylene to ethane occurs, even at room temperature. With this catalyst, ethylene, propylene and benzene are converted to ethane, propane and cyclohexane, respectively, at room temperature [212].

The influence of the nature of the metals and of the Os : Ni ratio was studied in the hydrogenation of benzene, acetylene, CO and CO_2 (*vide infra*), using γ-Al_2O_3 supported $Ni_2Cp_2(CO)_2$, **11**, $Os_3Ni_3Cp_3(CO)_9$ (**13**), $H_2Os_3(CO)_{10}$ and $Os_3(CO)_{12}$ as catalyst precursors. They were compared

11   **13**

with a conventional system prepared from $Ni(NO_3)_2 \cdot 6H_2O$ and $OsCl_3$. For the hydrogenation of benzene, the order of activity found was $[Os_3Ni] < [Os_3Ni_3] < [H_2Os_3] < 2.3$ moles of benzene converted per g. atom metal, with the former being more selective in the formation of cyclohexane. The hydrogenation of acetylene was studied as a function of the catalyst activation temperature. The catalyst $[Ni_2]$ was the most active (2.8 moles of C_2H_2 converted per g. atom of metal) but the activity of $[Os_3Ni]$ and $[H_2Os_3]$ increased gradually as the activation temperature increased, consistent with osmium being active in the process, although at higher activation temperatures. The systems $[Os_3Ni_3]$ and $[Os_3]$ were less active than the above ones. Selectivity for ethylene was found to decrease in the order: $[Ni_2] > [Os_3Ni] > [H_2Os_3]$ and to depend on the activation temperature [1].

4. 5. 2. 2. Hydrogenation of CO and CO_2

The hydrogenation of carbon monoxide has stimulated a considerable amount of work, mainly because of the interest in the Fischer-Tropsch synthesis [113, 175]. On the other hand, using CO_2 as an energy source is a very challenging research area, in view of the low cost and easy availability of this C_1 molecule. However, the direct incorporation of CO_2 into organic substrates is clearly the most elegant route to functionalized chemicals as it does not require the oxygen sinks, *e.g.*, hydrogen, needed when CO_2 reduction is considered. Only monometallic catalysts have been so far reported for such reactions. The hydrogenation of the carbon-oxygen multiple bonds of both CO and CO_2, up to their rupture, *i.e.*, the methanation reaction, will be examined together, as partial reduction of CO_2 is known to afford CO, which can be further reduced.

Supported bimetallic catalysts derived from the sulfido clusters $Mo_2Fe_2S_2(\eta-C_5H_5)_2(CO)_6$ (**98**) and $Mo_2Co_2S_3(\eta-C_5H_5)_2(CO)_4$ (**99**) have been found to be active for CO methanation. It has been deduced from infrared, EXAFS and Mössbauer studies that no structural change occurs upon initial adsorption of **98** whereas it is oxidized upon heating [91, 92]. When adsorbed on MgO, the [Mo_2Fe_2] catalyst was found to have a selectivity higher than 95 mol % C_2H_4 and C_2H_6 (*ca.* 1 : 2) against more than 95 mol % CH_4 when adsorbed or γ-Al_2O_3. The suggestion that the clusters are not fragmenting and re-aggregating into larger crystallites was based on the fact that the selectivities of the [Mo_2Fe_2] and [Mo_2Co_2] catalysts differ from those of Mo/Al_2O_3, MoS_2, Fe/Al_2O_3 or Co/Al_2O_3 [91]. However, recent Mössbauer and EXAFS results show only metal oxo species on the surface, even after reduction in H_2 or H_2+CO at 673 K [91]. Addition of 13 ppm H_2S to the feed stream had essentially no effect on activity or selectivity of these MMCD catalysts.

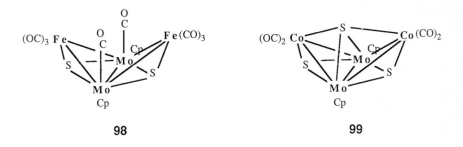

98   99

Addition of manganese to cobalt- or iron-based Fischer-Tropsch catalysts generally leads to an increased formation of light olefins whereas the catalyst activity decreases. By comparing the properties of the MMCD catalyst prepared from the heterodinuclear complex $MnCo(CO)_9$ (**96**) with those of catalysts prepared by successive or simultaneous impregnation of $Mn_2(CO)_{10}$ and $Co_2(CO)_8$ ([Mn$_2$ + Co$_2$]) and of conventional catalysts, it was found that the effect of manganese was very different in the [MnCo] catalyst (Table 4.2) [304]. Thus, the catalyst activity at 473 K and 1 atm was strongly enhanced with only little changes in selectivity and only *n*-alkanes were formed with a maximum at C_6. Furthermore, the deactivation of the catalyst was dramatically reduced. It was also concluded that manganese played an active role in the catalysis, such as increasing the rate of CO dissociation [304]. The different behaviour of the MMCD-catalyst was accounted for by two main reasons [304]:

(i) with conventional catalysts, higher temperatures are required to reduce the salts and this leads to large crystallites and inhomogeneous samples, in contrast to the situation with zerovalent precursors;

(ii) the counter ions, *e.g.*, chlorides, present when using conventional metallic salts are difficult to completely eliminate and the oxidation state of manganese is not as well known as when using zerovalent precursor complexes.

TABLE 4.2.

Properties of Manganese-Cobalt Catalysts (from ref. [304])

	Precursor Complexes		
	$Co_2(CO)_8$	$Mn_2(CO)_{10}$ + $Co_2(CO)_8$	$MnCo(CO)_9$
Catalyst	[Co$_2$]	[Mn$_2$ + Co$_2$]	[MnCo]
Mn/Co atomic ratio	0 : 1	0.93 : 1	1 : 1
% wt Co (mass)	1.92	0.97	0.69
% wt Mn (mass)	0	0.84	0.64
working time (h)	144	216	> 240
conversion (%)	13.6	48.7	31
activity at 473 K [mole CO h^{-1}(g Co)$^{-1}$]	13.6×10^{-3}	4.9×10^{-3}	11.5×10^{-3}

Further comparative studies involved manganese-iron catalysts prepared from the carbonyl cluster $[Et_4N][MnFe_2(CO)_{12}]$ (**97b**) (and not from

$[Et_4N]_2[MnFe_2(CO)_{12}]$ as in the original publication [187]) and from iron and manganese nitrates. It is interesting to note that the anion $[MnFe_2(CO)_{12}]^-$ is isoelectronic and isostructural with $Fe_3(CO)_{12}$, which was also used for comparative purposes (Table 4.3). The decrease in CO methanation and the increase in the yield of olefins and higher hydrocarbons when using the MMCD catalyst was taken as an indication that the manganese-iron interaction still occurs after the thermal decomposition of the supported cluster. This interaction was apparently absent in the catalysts prepared by supporting the nitrates of these metals. It was also observed that on MgO, the MMCD catalyst had properties similar to those of the iron-only catalysts, probably because of Mn-Fe bonds dissociation [187].

TABLE 4.3.
Comparison of Manganese-Iron Catalysts in CO Reduction at 523 K (adapted from ref. [187])

Catalyst Precursor	Support	A	B	C	D	E	F
$[Et_4N][MnFe_2(CO)_{12}]$ (97b)	Al_2O_3	0.97	2.3	0.16	12	39	96
$[Et_4N][MnFe_2(CO)_{12}]$ (97b)	SiO_2	1.33	17.4	1.05	10	28	94
$[Et_4N][MnFe_2(CO)_{12}]$ (97b)	MgO	0.74	11.4	0.37	33	33	93
$[Et_4N][MnFe_2(CO)_{12}]$ (97b)	TiO_2	0.55	5.0	0.11	5	27	84
$[Et_4N][MnFe_2(CO)_{12}]$ (97b)	ZrO_2	0.45	0.6	0.06	5	34	98
$Mn_2(CO)_{10}$	TiO_2	1.5	inactive				
$Fe_3(CO)_{12}$	TiO_2	2.10	0.1	0.12	32	34	85
$[Mn(NO_3)_2+Fe(NO_3)_3]$	Al_2O_3	0.97	0.03	0.01	26	35	94
$[Mn(NO_3)_2+Fe(NO_3)_3]$	SiO_2	1.35	0.73	0.20	33	38	79
$[Mn(NO_3)_2+Fe(NO_3)_3]$	TiO_2	0.55	inactive				
$[Mn(NO_3)_2+Fe(NO_3)_3]$	ZrO_2	10.0	0.98	0.89	29	43	20

A- Fe content (wt. %);
B- activity (mol CO x 10^5 g. atom Fe^{-1} s^{-1}) (20 h after starting the experiment);
C- CO conversion (%);
D- methane content (wt. %);
E- content of $C_2 + C_3$ hydrocarbons (wt. %);
F- olefin portion in $C_2 + C_3$ fraction (%).

Independently, Turney et al. [53] compared the properties of silica-supported catalysts derived from $K[MnFe_2(CO)_{12}]$ (97a) with those of catalysts prepared by aqueous impregnation of metal nitrate solutions. They found the former to be more selective for the CO/H_2 conversion into C_2-C_4 olefins. Using $[Et_4N][MnFe_2(CO)_{12}]$ (97b) as precursor showed that potassium had little effect under these conditions and no oxygenates were detected. At elevated temperatures, aromatics (mainly toluene and xylenes) were found with the MMCD catalysts, with a maximum of *ca.* 20 % at 673 K. In contrast, conventional catalysts gave almost exclusively aliphatic products. The temperature dependence of selectivity for the catalyst derived from **97a** on silica is shown in Fig. 4.5. The authors noted that not only do the individual C_2, C_3 and C_4 olefin/paraffin ratios increase with temperature but also the selectivities for C_1 and C_2 (both C_2H_4 and C_2H_6).

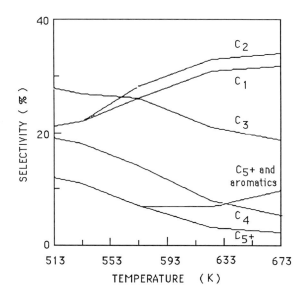

Fig. 4.5. Variation in selectivity with temperature. Catalyst prepared from 0.5 % $K[MnFe_2(CO)_{12}]$ on silica (adapted from ref. [53]).

The degree of support loading was found to markedly affect the product selectivity. Thus, as shown in Fig. 4.6, a 0.5 % Fe loading, for a catalyst derived from **97b** on silica, results in a selectivity for olefins C_2-C_4 of 64 %, whereas a 10 % loading decreases it to 43 %. The latter catalyst affords an overall olefin selectivity higher than 75 %. Studies by transmission electron microscopy and X-ray diffraction on the used catalysts at low loadings indicated a relatively high dispersion, but at 10 % loading, the presence of micron-sized iron crystallites was evident. This appears to be a consequence of the relatively weak interaction with the silica support [53]. With the exclusion of methane, products from both conventional and MMCD catalysts followed approximately a Schultz-Flory distribution. It was suggested that under the particular reaction conditions used, the MMCD catalyst produced the higher C_2-C_4 olefin yield because the secondary reactions of ethene are minimized.

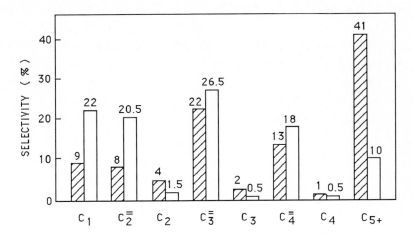

Fig. 4.6. Variation of selectivity with metal loading in catalysts derived from [Et$_4$N][MnFe$_2$(CO)$_{12}$] (**97b**) at 543 K, ▨ = 10% Fe, ☐ = 0.5 % Fe (adapted from ref. [53]).

More recently, carbon-supported Mn-Fe and K-Mn-Fe catalysts were employed for the selective synthesis of C_2-C_4 olefins from CO and H_2 [305c]. In order to study the effect of varying the Mn : Fe ratio and that of potassium, the following mixed-metal clusters were used as precursors: K[MnFe(CO)$_9$] (**112a**), [Et$_4$N][MnFe(CO)$_9$] (**112b**), Mn$_2$Fe(CO)$_{14}$ (**113**) and [Et$_4$N][MnFe$_2$(CO)$_{12}$] (**97b**) (and not [Et$_4$N][MnFe$_2$(CO)$_{13}$] as in the original publication [305c]). This resulted in the preparation of highly dispersed Mn-Fe catalysts which have selectivities to C_2-C_4 olefins as high as 85-90 wt %, with the balance being methane. The highest selectivities to light olefins were obtained with MMCD catalyst having Mn : Fe ratios of 0.5, and using potassium as a counter ion in the carbonyl clusters further enhanced both selectivity and activity, although its presence increases CO_2 production. Conventional catalysts prepared from nitrate salts, having the same Mn : Fe ratio did not exhibit these interesting catalytic properties. In view of the specific behaviour of the catalyst derived from **97b**, it was suggested that the Fe$_2$MnO$_4$ mixed-metal spinel could dominate in the unpromoted catalyst after a low temperature reduction step. It was shown that the high olefin : paraffin ratios and activities achieved with specific Mn-Fe and K-Mn-Fe MMCD catalysts are not due to differences in conversion and temperature and are retained after long periods on-stream [305d]. This represents significant improvements over previous oxide-supported catalysts.

The surface species resulting from the interactions of Fe_2Ru-$(CO)_{12}$ (**3a**), $FeRu_2(CO)_{12}$ (**2a**) and $Ru_3(CO)_{12}$ with γ–Al_2O_3 were suggested to be $HFe_2Ru(CO)_{10}(OAl\lessgtr)$, $HFeRu_2(CO)_{10}(OAl\lessgtr)$ and $HRu_3(CO)_{10}(OAl\lessgtr)$, respectively [9]. The temperature desorption spectra of $[Fe_3]$ or $[Ru_3]$ or of $[Fe_3 + Ru_3]$ exhibited negligible or very small CH_4 peaks compared with large CO, H_2 and CO_2 peaks, whereas a large CH_4 peak was observed for $[Fe_2Ru]$ or $[FeRu_2]$ at *ca.* 610 K. In CO hydrogenation, the $[Fe_2Ru]$ catalyst was found to be *ca.* 50 times more active and *ca.* 6.2 times more selective towards C_2-C_5 hydrocarbons than the $[Fe_3]$ catalyst. A good selectivity was retained on further replacement of Fe atoms by Ru atoms. In contrast, conventional Fe-Ru catalysts, prepared by impregnation of an aqueous solution of $Fe(NO_3)_3$ and $RuCl_3$, produced predominantly CH_4 under identical conditions [166].

The activity and selectivity for the CO/H_2 reaction of highly dispersed catalysts prepared from $Fe_2Ru(CO)_{12}$ (**3a**) and $H_2FeRu_3(CO)_{13}$ (**32**) have also been compared to those of catalysts prepared from a mixture of the homonuclear clusters $Fe_3(CO)_{12}$ and $Ru_3(CO)_{12}$ [137, 267]. Such a $[Fe_3+Ru_3]$ mixture has been shown to lead to the formation of a bimetallic catalyst, by comparison of its properties with those of a $[Fe_2Ru]$ MMCD catalyst [138, 139]. The mechanism of the CO/H_2 reaction was discussed in terms of the different types of surface carbons, *i.e.*, those involved in methane formation, in olefin formation and in catalyst deactivation [193b, 193d]. All the catalysts decomposed in helium, consisting of smaller particles, had higher activity than those decomposed in hydrogen. The catalyst $[2Fe_3+Ru_3]$ showed the same activity as $[Fe_2Ru]$, in agreement with their composition, whereas the catalyst $[FeRu_3]$, with the highest amount of Ru, was the most active among them [267].

Baird *et al.* [110, 111] have studied the hydrogenation of carbon monoxide on catalysts derived from the clusters $H_2Ru_4(CO)_{13}$, $H_2FeRu_3(CO)_{13}$ (**32**) and $[PPN][Ru_3Co(CO)_{13}]$ (**70b**). The authors used γ-alumina, silica and Na-Y zeolite as supports. Significant differences in activity were found but the main product of CO hydrogenation was in all cases methane. The formation of CO_2 suggested the occurrence of the water gas shift reaction $(H_2O + CO \longrightarrow H_2+CO_2)$ and the possibility, at low CO/H_2 ratio (1 : 4), to have some CO_2 methanation. The results in Table 4.4 show that the best yields under these conditions were obtained with alumina as a support. The reasons for the anomalously low activities of the $[Ru_3Co]$ catalyst on silica and zeolite were not clear and may arise from the decomposition of the PPN cation and coating of the metal surface during catalyst activation.

TABLE 4.4.

Comparative CO Methanation at 623 K, CO/H_2 = 1 : 2 (from ref. [110])

Catalyst precursor	Methane yields (%, ± 5 %)		
	$\gamma\text{-Al}_2O_3$	SiO_2	Na-Y zeolite
$H_2Ru_4(CO)_{12}$	56	35	38
$H_2FeRu_3(CO)_{13}$ (32)	42	43	40
[PPN][Ru$_3$Co(CO)$_{13}$] (70b)	60	inactive	12

A detailed study of the properties of carbon-supported catalysts prepared from $Fe_3(CO)_{12}$, $Ru_3(CO)_{12}$, $FeRu_2(CO)_{12}$ (2a) , $Fe_2Ru(CO)_{12}$ (3a) and $H_2FeRu_3(CO)_{13}$ (32) was reported [173]. Highly dispersed Fe-Ru bimetallic crystallites were obtained on an amorphous carbon black when using mixed-metal carbonyl precursors. If not exposed to air, only a low-temperature pre-treatment at 473 K was required to reduce essentially all the iron in the [Fe$_3$] catalyst and to activate it. In contrast, air exposure did not affect the reduction behaviour of the ruthenium-containing catalysts. The presence of ruthenium facilitates reduction of the iron and a low-temperature reduction at 473 K is sufficient to reduce and activate these clusters and this could be taken as evidence for Fe-Ru contact. In addition, reduction of these clusters on carbon appears to be more facile and complete than on many oxide supports. In agreement with previous studies of bulk Fe-Ru alloys, surface enrichment in iron seems to occur in these small crystallites after reduction although these particles do seem to retain an overall homogeneous composition, as expected from the bulk phase diagram. Turnover frequencies (TOF) for CH_4 and total hydrocarbon formation increased in parallel with the increase in surface ruthenium concentration whereas the TOF for CO_2 smoothly decreased. This trend is consistent with other data showing that ruthenium is more active than iron for CO hydrogenation but is less active than iron in forming CO_2. The catalytic activity for CO hydrogenation and the product selectivity to hydrocarbons are listed in Tables 4.5 and 4.6, respectively. The increased air sensitivity of supported metal carbonyl clusters compared to the unsupported clusters was shown by the much lower activity for the [Fe$_3$] catalyst following a brief air exposure and a low temperature pretreatment. The activities of the Fe-Ru catalysts were much higher than those of the iron carbonyl catalyst when treated at low temperature and were equal to or higher than activities of the same Fe-Ru catalysts after a high temperature pretreatment. There was a gradual change in selectivity over the Fe-Ru catalysts when their intermetallic composition was varied (methane increased with Ru content), but the [Fe₃+Ru₃] catalyst was found to lie somewhat off all the curves provided by the MMCD catalysts, indicating some inhomogeneity among the metal crystallites of the former system. This variation between Fe-Ru catalysts with nearly identical metal loading and overall composition demonstrates the advantage of utilizing stoichiometric mixed-metal clusters to facilitate metal-metal contact and uniformity among well-dispersed metal crystallites [173].

TABLE 4.5.

Catalytic Activity in CO Hydrogenation for Iron-Ruthenium Systems (from ref. [173])

Catalyst precursors	Pretreat-ment[a]	Metal Loadings			Activity[b] $[mole \cdot s^{-1}(mole\ M_t metal)^{-1}] \times 1000$			
		Fe (wt %)	Ru (wt %)	Fe : Ru mole ratio	CO to hydro-carbons	CH₄	CO₂	% Conversion of CO to hydrocarbons
Fe(NO₃)₃	H	5.0	-	-	0.34	0.138	0.153	3.1
Fe₃(CO)₁₂	L	4.4	-	-	0.0045	0.0045	0.0056	0.04
	H				1.44	0.430	0.770	6.0
Fe₃(CO)₁₂ + Ru₃(CO)₁₂	L	4.2	2.0	3.76	1.74	0.494	1.591	7.8
	H				1.34	0.318	1.43	6.0
Fe₂Ru(CO)₁₂ (3a)	L	0.97	0.77	2.27	2.52	1.82	1.23	2.9
	H				2.79	1.56	1.20	3.2
FeRu₂(CO)₁₂ (2a)	L	0.70	2.35	0.536	6.56	4.83	0.822	5.8
	H				5.03	3.72	0.623	4.4
H₂FeRu₃(CO)₁₃(32)	L	0.47	2.6	0.32	6.36	5.95	0.903	5.2
	H				5.51	4.75	0.496	4.9
Ru₃(CO)₁₂	L	-	6.3	0	4.00	3.96	0.192	6.5
	H				3.62	3.54	0.160	5.9

[a] L : low temperature, H : high temperature.
[b] P = 1 atm, H₂ : CO = 3, T = 548 K, M_t total number of metal atoms.

In comparison, the sequentially impregnated [Fe₃+Ru₃] catalyst produced a higher olefin : paraffin ratio than any of the bimetallic Fe-Ru clusters examined. This is similar to the selectivity of silica-supported Fe-Ru catalysts prepared by sequential impregnation of Ru and Fe salts [306]. Again, this shows that varying the Fe-Ru precursor and the impregnation technique can affect the catalytic selectivity, probably owing to a variation in the contact between the two metal components, which is optimized with the bimetallic clusters.

TABLE 4.6
Product Selectivity^a in CO Conversion (from ref. [173])

Catalyst Precursors	% CO to hydrocarbon conversion	$\frac{Fe_t}{Ru_t}$	Hydrocarbons (mole %)							C_2-C_3 Olefin/paraffin ratio
			C_1	$C_{2=}$	C_2	$C_{3=}$	C_3	C_4	C_{5+}	
Fe(NO$_3$)$_3$	3.1	-	61	3	27	-	7	1	1	0.09
Fe$_3$(CO)$_{12}$	6.0	-	56	5	19	3	4	7	5	0.36
Fe$_3$(CO)$_{12}$ + Ru$_3$(CO)$_{12}$	6.0	3.76	50	10	9	16	-	7	7	2.9
Fe$_2$Ru(CO)$_{12}$ (3a)	3.2	2.27	76	3	16	-	-	3	1	0.19
FeRu$_2$(CO)$_{12}$ (2a)	4.4	0.55	88	1	8	-	2	1	1	0.08
H$_2$FeRu$_3$(CO)$_{13}$(32)	4.9	0.32	95	-	4	-	0.6	0.6	0.3	-
Ru$_3$(CO)$_{12}$	5.9	-	99	-	1	-	-	-	-	-

^{a} Measured after high temperature pretreatment ; T = 548 K, P = 1 atm, H$_2$: CO = 3

Initial scanning transmission electron microscopy studies showed that the metal crystallites (20-60 Å) form rafts within the small pores of the high surface area carbon used, indicating that the previously reported estimations of particle size using spheres or cube-shaped models may not apply here. This morphology may be responsible for the stability of the kinetic behaviour observed under the reaction conditions. This investigation has clearly shown that mixed-metal clusters can be used to produce active, bimetallic crystallites and that carbon is an excellent support to provide high dispersions and facilitate reduction of all the metal components [173].

After impregnation of silica with a dichloromethane solution of H$_2$FeOs$_3$(CO)$_{13}$ (48), infrared and Raman spectroscopies have indicated that the cluster was physisorbed on the silica surface [79]. Thermal activation of the physisorbed species under argon at ca. 400 K cleaved the heteronuclear Fe-Os metal-metal bonds, leading to "Fe metal" and trinuclear osmium species which, at ca. 473 K, would be broken down into mononuclear osmium species. Metallic particles were formed (ca. 16 Å) under argon at temperatures greater than 523 K. Their Fischer-Tropsch properties have been investigated and compared with those of catalysts derived from $Fe_3(CO)_{12}$ and $Os_3(CO)_{12}$ on silica (Table 4.7). As expected from the breakdown of the mixed-metal cluster, the MMCD catalyst exhibited activity and selectivity patterns intermediate between those of the [Fe_3] and [Os_3] homonuclear systems[79].

TABLE 4.7.

Fischer-Tropsch Activity of a Fe-Os System on Silica (from ref. [79]).

Catalyst Precursor	Activity [a]	Selectivity (%)			
		C_1	C_2	C_3	C_4
$Fe_3(CO)_{12}$	0.52	0.42	0.22	0.24	0.12
$Os_3(CO)_{12}$	0.30	0.89	0.06	0.04	0.01
$H_2FeOs_3(CO)_{13}$ (48)	0.65	0.61	0.14	0.18	0.07

[a] Mol C_1-C_4/ h/ metal atom at 523 K.

When γ-Al_2O_3 was used as a support [57], infrared measurements indicated that the surface chemistry of $H_2FeOs_3(CO)_{13}$ (48) was similar to that of $H_2RuOs_3(CO)_{13}$ (85) [59, 274]. As deduced from infrared spectroscopy, $H_2Os_3Rh(CO)_{10}(acac)$ (100) was initially physisorbed on the γ-Al_2O_3 support and after treatment in $CO+H_2$ and catalysis, break-up of the supported cluster had occurred, first giving mononuclear rhodium complexes and triosmium clusters [57]. This is in fact not too surprising in view of the well-established solution chemistry of cluster 100 [109], which has been shown to react very rapidly with CO (eq. 31):

$$H_2Os_3Rh(CO)_{10}(acac) + 3\ CO \longrightarrow Rh(acac)(CO)_2 + H_2Os_3(CO)_{11} \qquad (31)$$

At higher temperatures, rhodium crystallites and mononuclear osmium complexes in an oxidized state were formed on the support and it was inferred that the small particles observed by electron microscopy (15-30 Å for the [$RhOs_3$] catalyst) contained almost entirely rhodium. The alumina-supported catalysts prepared from $H_2FeOs_3(CO)_{13}$ (48), $H_2Os_3Rh(CO)_{10}(acac)$ (100) and $Rh_4(CO)_{12}$ were evaluated [57]. A comparative study involving rhodium was of interest because this metal exhibits various selectivities depending on the dispersion, the nature of the support and its oxidation state [174]. Each catalyst was found to be active for conversion of $CO+H_2$, the major product observed in each experiment was methane and the hydrocarbon products were formed in approximately a Schulz-Flory-Anderson distribution (Table 4.8). Small yields of dimethyl ether, formed from methanol, were produced. The performance of the [FeOs$_3$] catalyst was different from that observed with catalysts derived from Fe$_3$(CO)$_{12}$ deposited on γ-Al$_2$O$_3$ [89], which have a high selectivity for light olefins. The different behaviour of the [FeOs$_3$] catalyst may by attributed to its lower iron loading than in the experiments with the [Fe$_3$] catalyst and also to the stability of the intermediate surface cluster Al$^+$[H$_3$FeOs$_3$(CO)$_{12}$]$^-$. The [FeOs$_3$] catalyst was found to be two orders of magnitude less active at 543 K than the [Os$_3$Rh] catalyst, but showed a high selectivity for ether formation [57]. It apparently consisted of small iron oxide particles (20-30 Å) and mononuclear osmium complexes. The selectivity of this catalyst for dimethyl ether formation increased markedly with increasing time on stream (Fig. 4.7 and Table 4.8). The performance of the [Os$_3$Rh] catalyst was comparable to that of the [Rh$_4$] catalyst. It was concluded that virtually all the catalytic activity of the [Os$_3$Rh] catalyst was due to rhodium, osmium, at most, increasing slightly the selectivity for oxygenate formation (Fig. 4.7).

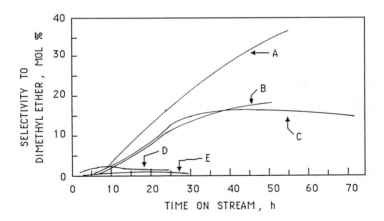

Fig. 4.7. Change in selectivity for formation of dimethyl ether from an equimolar mixture of CO and H$_2$ at 10 atm. The catalysts are specified in Table 4.8; A: [FeOs$_3$], 543 K; B: [Rh$_4$], 473 K; C: [Os$_3$Rh], 473 K; D: [Os$_3$Rh], 473 K; E: [Rh$_4$], 543 K (adapted from ref. [57]).

TABLE 4. 8.

Catalyst Activities and Selectivities in CO Hydrogenation[a] (after ref. [57])

Catalyst precursor	Metal loading (wt %)[b]			Reaction temp. (°C)	Time on steam (h)	Conversion (%)	10^4 x rate (CO molecules/ metal atom·s)	Product composition (mol %)						
	Fe	Rh	Os					CH4	C2	C3	C4	C5	C6	Me2O
Rh4(CO)12	-	0.36	-	270	3	2.6	2.13	87.1	5.6	3.6	2.4	0.79	0.31	0.33
					5.5	2.4	1.97	87.4	5.4	3.4	2.4	0.80	0.33	0.40
					28.5	1.7	1.39	89.1	4.6	2.5	2.2	0.62	0.39	0.71
Rh4(CO)12	-	0.36	-	200	2.5	0.065	0.053	87.7	4.0	5.1	2.4	0.8	-	-
					7	0.089	0.073	83.0	4.5	6.2	3.1	1.7	0.9	0.7
					24	0.12	0.098	70.5	4.8	5.8	4.4	2.7	0.7	11.1
					31	0.12	0.098	69.4	4.6	5.3	3.8	2.1	0.7	14.0
					48	0.11	0.090	67.7	4.3	4.9	3.8	2.3	0.8	16.4
H2Os3Rh(CO)10(acac)[c](100)	-	0.35	1.97	270	3.5	1.5	0.26	68.8	8.9	10.1	5.8	2.6	1.4	1.5
					7	1.4	0.25	69.1	8.6	9.8	5.7	2.4	2.0	2.4
					24	1.0	0.18	72.1	8.3	9.0	5.4	2.4	1.3	1.5
H2Os3Rh(CO)10(acac)[d](100)	-	0.35	1.97	200	2	0.099	0.026	62.1	8.3	22.7	4.9	2.0	-	-
					4	0.079	0.021	73.8	7.0	13.4	4.2	1.5	-	-
					6	0.073	0.019	75.7	7.5	11.3	3.4	1.1	-	1.0
					30	0.078	0.020	67.2	4.9	7.0	4.1	2.1	-	14.8
					72	0.070	0.018	67.7	3.7	6.2	4.8	2.5	0.7	14.5
H2FeOs3(CO)13 (48)	1.17	-	1.49	270	11	0.033	0.0033	67.4	17.9	5.0	3.2	1.9	0.7	4.2
					24	0.032	0.0032	62.7	15.9	1.6	1.6	1.3	0.5	16.4
					55	0.036	0.0036	49.0	11.2	1.0	0.8	1.0	0.6	36.4

[a] Reactor pressure = 10 atm

[b] Metal loadings were determined from uptake of catalyst precursor

[c] The Os content measured after 24 h on stream was 0.90 %.

[d] The Os content measured after 24 h on stream was 0.36 %.

Similarly, an alumina-supported catalyst was prepared from $H_2RuOs_3(CO)_{13}$ (**85**) and compared with a catalyst obtained by combination of $H_4Ru_4(CO)_{12}$ and $H_2Os_3(CO)_{10}$ (catalyst [Ru$_4$+Os$_3$]) and with the corresponding [Ru$_4$] monometallic system [58]. Each catalyst had an initial Ru content of 0.3 wt % and was found to be slightly active in the formation of hydrocarbons and dimethyl ether from CO+H$_2$. The method of preparation of the catalyst was found to play a critical role. With about the same ruthenium content, the [Ru$_4$] catalyst had twice the activity of the [RuOs$_3$] catalyst prepared in the same way. The low intrinsic activity of osmium for CO hydrogenation is known but it was surprising that it would actually suppress the activity of ruthenium. However, osmium increased the selectivity for oxygenate formation, reduced the chain-growth probability and inhibited catalyst deactivation. The electron micrographs of a [RuOs$_3$] catalyst after use provided evidence of aggregated metal; most of the visible particles were 20-30 Å in diameter. This was supported by XPS data, indicative of zerovalent ruthenium but of no zerovalent osmium. In the used catalyst, oxidized ruthenium was found and osmium was predominantly in the form of Os(II) carbonyl complexes. The hypothesis was that methanol formation involved cationic ruthenium centers and that hydrocarbon formation involved zerovalent ruthenium. Since there was no evidence for bimetallic aggregates in the used catalyts, it was assumed that there was no significant advantage in using the mixed-metal cluster as a precursor, consistent with a marked influence of osmium on the ruthenium in the [Ru$_4$ + Os$_3$] catalyst and with a rapid break-up of the heterometallic cluster, probably during the catalyst pretreatment [59].

Carbon-supported iron-cobalt and potassium-iron-cobalt carbonyl cluster-derived catalysts have also been studied in CO hydrogenation [305b]. The mixed-metal precursor complexes were K[FeCo(CO)$_8$] (**114**), HFeCo$_3$(CO)$_{12}$ (**49**), K[FeCo$_3$(CO)$_{12}$] (**59d**), K[Fe$_3$Co(CO)$_{13}$] (**60a**) and [Et$_4$N]-[Fe$_3$Co(CO)$_{13}$] (**60b**). These well-dispersed catalysts can be very active for CO hydrogenation and their behaviour was found to be intermediate between that of iron catalysts, which have a much lower TOF but a much higher olefin : paraffin ratio, and of cobalt catalysts, which are very active but produce only paraffins. Addition of potassium was found to markedly decrease the catalytic activity but greatly enhanced olefin selectivity [305b].

Mixed cobalt-rhodium clusters have also been used as precursors of heterogeneous catalysts in CO+H$_2$ reactions. Typical results are summarized in Fig. 4.8. The proportion of ethanol in the oxygenated products was found to increase with increased cobalt content, while carbon efficiencies of the oxygenated products decreased [160]. The yields of C_1-C_4 hydrocarbons increased at the expense of the oxygenated products. Catalysts derived from $Co_4(CO)_{12}$ provided preferentially methane and higher hydrocarbons, similar to the Fischer-Tropsch catalysts consisting of iron, cobalt and ruthenium. In another study, it was also found that this [Co_4] catalyst was very active and selective, all the CO reacting to form CH_4 which was the sole hydrocarbon [110]. This was consistent with the formation of small crystallites on decomposition of $Co_4(CO)_{12}$. Interestingly, CO_2 was also formed in amounts equal to 60-80 % of the amounts of methane. Cobalt metal, which catalyses the dissociative chemisorption of CO, may promote the concentration of surface hydrocarbons in a CO+H_2 reaction. This could result in the enhancement of the C_2-oxygenated products such as ethanol.

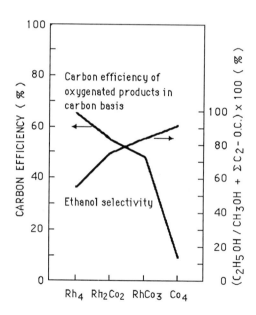

Fig. 4.8. Carbon efficiencies of the oxygenated products in a CO+H_2 reaction and ethanol selectivity over a bimetallic Co-Rh cluster-derived catalyst impregnated on ZrO_2 (adapted from ref. [160]).

A comparison of the catalytic properties for CO reduction was made between alumina- supported catalysts derived from $HMoOs_3Cp(CO)_{12}$ (**101**), $HWOs_3Cp(CO)_{12}$ (**102**) and $Os_3(CO)_{12}$ [278]. No observable difference was found and this was explained by the fragmentation of the surface bound clusters during thermal activation or by a poisoning effect of the group 6 metal atom by carbon originating from the cyclopentadienyl ligands.

101 , M = Mo

102 , M = W

The tungsten-iridium clusters $WIr_3Cp(CO)_{11}$ (**94**) and $W_2Ir_2Cp_2(CO)_{10}$ (**95**) and the corresponding homonuclear complexes $Ir_4(CO)_{12}$ and $W_2Cp_2(CO)_6$ were adsorbed onto γ-Al_2O_3 from cyclohexane solutions [279]. The supported compounds were then subjected to temperature-programmed decomposition in flowing hydrogen (Fig. 4.9). The results for $Ir_4(CO)_{12}/Al_2O_3$ showed that most of the coordinated CO was hydrogenated to CH_4 rather than released intact. This feature is true also for the supported bimetallic clusters but to a decreasing extent with an increasing W/Ir ratio. A general trend toward increasing temperatures for methane evolution with increasing tungsten content of the precursor compounds was noticed. Significantly, however, the methane profiles for the $[WIr_3]$ and $[W_2Ir_2]$ catalysts are different, and neither is the composite of the profiles for the $[Ir_4]$ and $[W_2]$ systems. In contrast, the profile for $[Ir_4+W_2]/Al_2O_3$ (W/Ir = 1 : 1) appeared as the composite of the separate profiles. This indicates that different sites or a different distribution of sites for CH_4 production are created during the decomposition of the two bimetallic clusters. The cyclopentadienyl ligands of the mixed clusters were hydrogenated and cracked above 573 K. However, quantitative data indicate that an upper limit of 10% of the CH_4 produced can be due to C_5H_5 cracking [279].

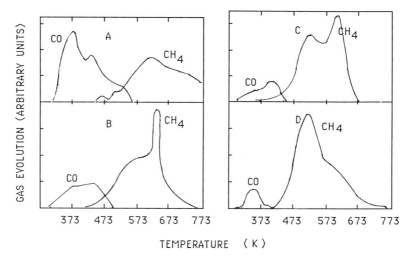

Fig. 4.9. Profiles for CO and CH$_4$ evolution during the temperature-programmed decomposition of alumina-supported tungsten-iridium complexes in flowing H$_2$; A: [W$_2$]; B: [W$_2$Ir$_2$]; C: [WIr$_3$]; D: [Ir$_4$] (adapted from ref.[279]).

When studying the influence of the precursor complexes Ni$_2$Cp$_2$-(CO)$_2$, Os$_3$NiCp(μ-H)$_3$(CO)$_9$ (11), Os$_3$Ni$_3$Cp$_3$(CO)$_9$ (13), H$_2$Os$_3$(CO)$_{10}$ or Os$_3$(CO)$_{12}$ on the hydrogenating properties of the corresponding γ-Al$_2$O$_3$-supported derived catalyst, it was found that [Os$_3$Ni] was the best catalyst in CO methanation. It gave good conversions (0.83 mole of CO converted per g. atom of metals) and high selectivity in CH$_4$ (96-100 %) at temperatures above 523 K, with small amounts of CO$_2$ and C$_2$ hydrocarbons as by-products. Oxygenates can be formed at lower temperatures [212]. For comparison, the conventional catalyst prepared from OsCl$_3$ and Ni(NO$_3$)$_2$. 6 H$_2$O gave lower conversions but good selectivity. For CO$_2$ methanation, [Os$_3$-Ni] was found to give yields superior to 90 % at temperatures between 523 and 623 K [1, 212]. It is more effective than the [H$_2$Os$_3$], [Os$_3$] or [Ni$_2$] catalysts in terms of conversion and selectivity (0.58 mole CO$_2$ per g. atom of metals converted with 96.7 % selectivity in CH$_4$ when [Os$_3$Ni] was activated at 573 K for 4 h and with a reaction temperature of 562 K).

4. 5. 2. 3. Hydrogenation of Ketones.

A heterogeneous hydrogenation catalyst allowing the reduction of acetone to propane was prepared from the cluster Os$_3$NiCp(μ-H)$_3$(CO)$_9$ (11) supported on Chromosorb-*P* and thermally treated under H$_2$ [63]. A multistep reaction pattern was proposed, which involves first hydrogenation of acetone to isopropanol, dehydration of the latter on the support with formation of propylene, followed by hydrogenation to propane.

4. 5. 3. Hydroformylation of Olefins

The bimetallic iron-rhodium carbonyl clusters $[Me_4N]_2[FeRh_4(CO)_{15}]$ (**115**), $Fe_3Rh_2C(CO)_{14}$ (**116**), $[TMBA][FeRh_5(CO)_{16}]$ (**117**) (TMBA = NMe_3CH_2Ph) and $[TMBA]_2[Fe_2Rh_4(CO)_{16}]$ (**118**) have been employed as molecular precursors for the preparation of new SiO_2-supported bimetallic catalysts [162c]. A promoter effect of iron on olefin hydroformylation was observed and Mössbauer data suggest that Fe atoms are mostly in the state of Fe^{3+} ions even after H_2 reduction at 673 K. Bimetallic $Rh-Fe^{3+}$ ensembles appear to be generated on these MMCD catalysts. They are highly active for migratory CO insertion as judged by the rates of the hydroformylation of ethylene and propene which are dramatically enhanced when compared with those of the monometallic catalysts or of physically mixed Fe-Rh catalysts prepared by conventional methods [162b, 162c].

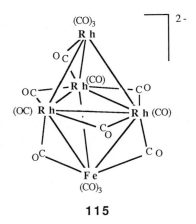

115

Carbon-supported ruthenium-cobalt catalysts were studied for the selective hydroformylation of ethylene and propene, using the tetranuclear clusters $HRuCo_3(CO)_{12}$ (**53**), $H_3Ru_3Co(CO)_{12}$ (**51**) or $[Et_4N][Ru_3Co(CO)_{13}]$ (**70a**) as precursors [162d]. Higher rates and selectivities to normal n-alcohol production were observed with the Co-rich MMCD catalysts ($[HRuCo_3]$ > $[H_3Ru_3Co]$ ~ $[Ru_3Co]$) and they were much higher than those with the $[Et_4N][Ru_3Co(CO)_{11}]$- or $Co_4(CO)_{12}$-derived catalysts [162d].

Olefin hydroformylation has also been studied on bimetallic cobalt-rhodium catalysts, using well-known molecular precursors. Thus, the structurally related tetranuclear clusters $Co_4(CO)_{12}$, $Co_3Rh(CO)_{12}$ (9), $Co_2Rh_2(CO)_{12}$ (8) and $Rh_4(CO)_{12}$ were deposited on ZnO and evaluated as catalytic precursors for the vapor-phase hydroformylation of ethylene and propene [161]. The activities and selectivities of the corresponding catalysts were studied as a function of the composition of the clusters. The hydroformylation rate decreased in the order: [Rh$_4$] (100%) > [Co$_2$Rh$_2$] (60) > [Co$_3$Rh] (42) > [Co$_4$] (5.2) (The figures in parentheses are relative rates of aldehyde formation at 433 K). Precursors with larger cobalt contents produced higher proportions of linear aldehyde (Table 4.9) and the [Co$_4$] catalyst exhibited the highest selectivity for linear aldehyde (>90%) but had the lowest specific activity. The yield of aldehyde per Rh atom in the mixed catalysts was comparable to that observed with [Rh$_4$] under similar conditions.

TABLE 4.9.

Selectivity in Propene Hydroformylation over ZnO-Supported Cobalt-Rhodium Catalysts (adapted from ref. [161a])

Precursor Complexes	Catalyst [a]	Normal isomer[b] selectivity (mole %)
$Rh_4(CO)_{12}$	[Rh$_4$]	58 ± 2
$Co_2Rh_2(CO)_{12}$ (8)	Co$_2$Rh$_2$]	70 ± 3
$Co_3Rh(CO)_{12}$ (9)	[Co$_3$Rh]	80 ± 1
$Co_4(CO)_{12}$	[Co$_4$]	90 ± 2
1 $Co_4(CO)_{12}$ + 1 $Rh_4(CO)_{12}$	[Co$_4$+Rh$_4$]	62 ± 3
3 $Co_4(CO)_{12}$ + 1 $Rh_4(CO)_{12}$	[3 Co$_4$+Rh$_4$]	65 ± 2

[a] The carbonyl clusters (0.15 mmol) were deposited on ZnO (20 g) from hexane solutions. Pyrolysis at 433 K under evacuation.
[b] [n-C_3H_7CHO/ (n-C_3H_7CHO + iso-C_3H_7CHO)] x 100 (mole %).

The n /iso selectivity of the [Co$_4$+Rh$_4$] and [3 Co$_4$+Rh$_4$] catalysts is comparatively lower than that of the [Co$_2$Rh$_2$] or [Co$_3$Rh] catalysts. This may reflect the advantage resulting from a more intimate bimetallic interaction already at the molecular precursor level. The supported bimetallic Co-Rh particles on ZnO might behave as "highly dispersed alloys" having compositions similar to those of the corresponding precursor clusters, as Anderson et al. have previously suggested for similar systems supported on Al_2O_3 [5].

Furthermore, it was suggested that in dispersed Co-Rh crystallites, cobalt atoms might behave as electron donor ligands, possibly owing to the lower ionization potential of cobalt than rhodium, influencing the rhodium sites in olefin hydroformylation reactions and enhancing the n /iso selectivity [161a]. Other results on carbon-supported $[Co_2Rh_2]$ and $[Co_3Rh]$ catalysts, prepared from clusters **8** or **9**, respectively, have provided further evidence for the higher activity of the MMCD catalysts for the gas phase hydroformylation of ethylene and propene [162b]. The n /iso selectivity was also enhanced by increasing the Co content in the bimetallic catalyst and this was suggested to originate from the possibility of retaining cobalt in a reduced state on carbon [162b].

4. 5. 4. Reductive Carbonylation of Nitro Derivatives

The preparation of aromatic isocyanates by carbonylation of the corresponding nitro derivatives, as shown in eq. 32:

$$Ar-NO_2 + 3\ CO \longrightarrow Ar-N = C = O + 2\ CO_2 \tag{32}$$

has been investigated recently using MMCD catalysts. There is a considerable academic and industrial interest for this reaction, which represents a way to aromatic isocyanates avoiding phosgene, classically used to convert amino derivatives into this important class of chemicals. It is also hoped that the use of MMCD catalysts will improve our knowledge of the mechanism(s) of this complex, redox reaction. Conventional bimetallic palladium-based catalysts have been extensively studied [13, 144, 219]. The recent development of high-yield syntheses of molecular bimetallic clusters containing palladium [37] has thus allowed comparative studies to be performed between conventional and cluster-derived catalysts. The results shown in Table 4.10 indicate that the $[Mo_2Pd_2]$ catalyst prepared from the planar cluster $Mo_2Pd_2Cp_2(CO)_6(PPh_3)_2$ (**14**) gives rise to higher selectivity in phenylisocyanate than conventional catalysts prepared by mixing the individual components [38, 179]. In a control experiment, it was shown that addition of PPh_3 to the latter did not modify their properties, indicating that residual phosphorus, possibly present, was not responsible for the better performances of the MMCD catalyst [178]. In all these experiments, comparable metal loadings and the same support (γ-Al_2O_3) were used throughout.

TABLE 4.10.

Compared Properties of Conventional and MMCD Catalysts for the Conversion of $PhNO_2$ in PhNCO (from ref. [38])

Catalyst Precursors[a]	Conversion (%)	Selectivity (%)
Pd/Al_2O_3[b]	0	0
MoO_3	100	19
$Pd/Al_2O_3 + MoO_3$	100	49
$Pd(OAc)_2 + [NH_4]_2[Mo_2O_7]$	100	67
$Mo_2Pd_2Cp_2(CO)_6(PPh_3)_2$ (14)	100	71-80

[a] Support is γ-Al_2O_3. Reaction conditions: the solvent used was o-dichlorobenzene (100 mL); initial CO pressure: 200 atm; T= 513 K; addition of pyridine (1 g); molar ratios $[PhNO_2]$/ Pd and $[PhNO_2]$/ Mo: ca. 140.
[b] A low activity was noted after a long induction period.

A more recent example concerns the use of neutral iron-palladium clusters for a similar reaction. Thus, $FePd_2(CO)_4(Ph_2PCH_2PPh_2)_2$ (103) and $Fe_2Pd_2(CO)_5(NO)_2(Ph_2PCH_2PPh_2)_2$ (104) have been evaluated as precursors of silica-supported heterogeneous catalysts transforming o-nitrophenol to benzoxazol-2-one [40, 42] (eq. 33).

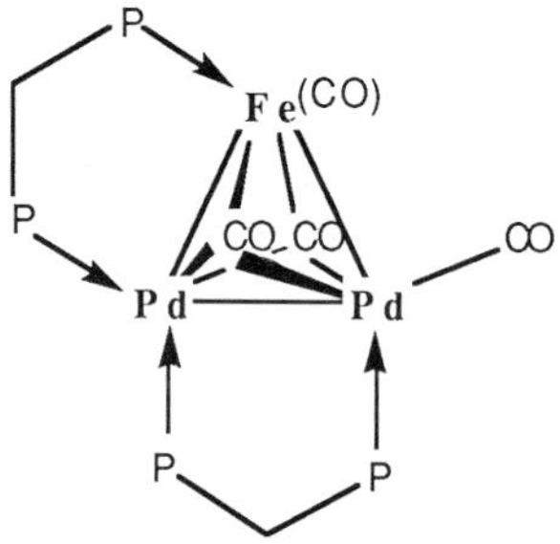

103

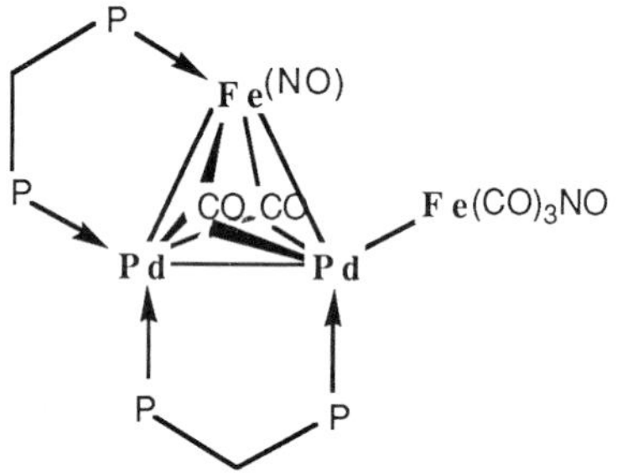

104

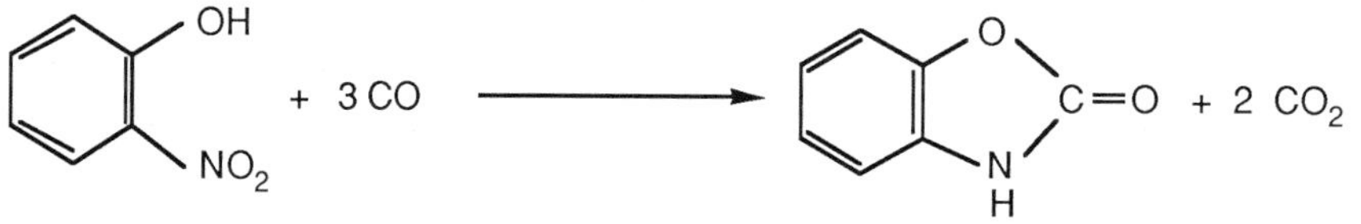

(33)

The results, given in Table 4.11, clearly indicate that both MMCD catalysts have significantly improved properties over the corresponding conventional catalysts under comparable conditions. In the case of these Fe/Pd catalysts, it was observed that using silica instead of alumina as a support gave much better results.

TABLE 4.11.
Compared Properties of Conventional and MMCD Catalyts for the Carbonylation of o-Nitrophenol to Benzoxazol-2-one (from ref. [40, 42])

Catalyst Precursors[a]	Conversion (%)	Selectivity (%)
$Fe_2(C_2O_4)_3$ + $Pd(OAc)_2$	40	90
$FePd_2(CO)_4(DPPM)_2$ (**103**)	98.5	95
$Fe_2Pd_2(CO)_5(NO)_2(DPPM)_2$ (**104**)	99.3	96

[a] The support is SiO_2. Reaction conditions: the solvent used was o-dichlorobenzene (100 mL); initial CO pressure: 200 atm; T= 473 K; 4 h; addition of pyridine (1 g); o-nitrophenol: 0.1 mole; molar ratio [$ArNO_2$]/Pd: ca. 100.

Recycling of these catalysts provides and important and useful test for their stability under catalytic conditions since either their activity or their selectivity will be affected by, e.g., leaching of one the metals. It has been verified, for example, that recycling of the [Mo_2Pd_2] catalyst derived from **14** leads to similar conversion and selectivity [38]. Preliminary analytical electron microscopy studies of the Fe/Pd catalysts indicate that metal segregation occurs in the MMCD catalysts during reaction and that highly dispersed palladium metal is formed [40]. Obviously, these catalytic systems have been investigated under comparable experimental conditions: this allows the conclusion to be drawn that under the conditions studied, appropriate MMCD catalysts offer a clear advantage over conventional catalysts for the reductive carbonylation of nitro derivatives.

4. 5. 4. Hydrodesulfurization

Environmental problems have emphasized the role of catalysts capable of removing sulfur and hydrodesulfurization (HDS) has become an increasingly important reaction in both coal and petroleum refining [92]. The availability of new molecular sulfido bimetallic clusters [37] now allows the properties of the corresponding MMCD catalysts to be compared with those of the classical MoS_2-promoted catalysts. When studying [Mo_2Fe_2] and [Mo_2Co_2] catalysts derived from the butterfly clusters $Mo_2Fe_2S_2Cp_2(CO)_6$ (**98**) and $Mo_2Co_2S_3Cp_2(CO)_4$ (**99**), Curtis, Schwank et al. have found that the cluster-derived catalysts have activities for S-removal from thiophene comparable to commercial HDS catalysts while showing decreased H_2 consumption (less butane formation). By comparison to conventionally prepared CoMoS catalysts, it appears that the sulfided clusters are more efficient in producing the active site for HDS [92].

4. 6. CONCLUSION

Supported bimetallic catalysts derived from well-defined molecular compounds, particularly from mixed-metal carbonyl clusters, constitute still a new class of catalytic materials. They have already found application for the various catalytic reactions presented in Table 4.12. They are becoming increasingly important as new heterometallic couples and supports are found to allow a fine tuning of the properties of highly dispersed catalysts [286]. A list of the heterogeneous catalysts derived from molecular mixed-metal clusters is given in Table 4.14. Many questions remain unanswered yet and much remains to be learned about the nature and the structure of the heterometallic species and of the sites involved in the working catalysts. The following statement, made in the context of homonuclear osmium clusters, is even more true when considering mixed-metal clusters, whose chemistry is richer and more subtle: " Without doubt unequivocal characterization of supported cluster compounds is a difficult, if not impossible, task" [168]. It has been noted that the study of support interactions is an area where cluster-derived catalysts may provide information not readily available from conventional catalysts [168]. Future research will certainly address these problems. When considering the main advantages, listed in Section 4.1, to be anticipated for MMCD over conventional catalysts, it is clear that very significant results have already been obtained. We have seen that comparative studies have been carried out between MMCD and conventional catalysts as well as between related MMCD catalysts. When the molecular precursors are in a low oxydation state, preferably carbonyl complexes, the comparison is most meaningful since it allows the minimum number of parameters to be changed at the same time. On the other hand, comparing the properties of a MMCD catalysts with those of a conventional catalyst prepared from inorganic salts of the metals, *e.g.*, chlorides or nitrates, is perhaps more informative from an economical point of view if one wishes to replace "old" but currently used catalysts with new ones. Studies have provided evidence, in many instances, that residual interactions between the different metals which were bonded to each other in the precursor organometallic cluster can distinctly modify the character of the catalytic site(s), the latter being often more homogeneous in the MMCD catalytic systems [173, 279]. In this context, it is noteworthy that recent analytical electron microscopy studies on $[Ni_{38}Pt_6(CO)_{48}H]^{5-}$ (Fig. 1.3) have allowed observation at low beam intensities of uniform size particles, corresponding to the dimensions of the metallic skeleton of the cluster (radius 10.9 Å). Elemental analysis on these particles, using electron energy loss spectroscopy, provided results very close to that expected for a $Ni_{38}Pt_6$ core. At higher electron beam intensities, particle agglomeration was observed together with formation of graphitic carbon and a face-centred cubic Ni/Pt alloy [148]. It will also be very important to address the question of metal segregation, possibly occurring not only during the preparation of the MMCD catalyst but also under catalytic conditions [40]. Furthermore, the study of bimetallic systems involving metals not miscible in the bulk but for which molecular clusters are known, *e.g.*, ruthenium and copper, should reveal particularly important information concerning specific properties associated with MMCD catalysts.

Despite all the results available, much remains to be done in performing the critical tests allowing a clear-cut evaluation of MMCD catalysts. This is largely due to the number of parameters involved in such studies, *e.g.*, nature of the support, of the impregnation method used, of the thermal decomposition and activation procedures of the catalysts, which make comparisons between results from different research groups very difficult and sometimes meaningless. It also appears that the study of structure-sensitive reactions [32, 33], *i.e.*, whose mechanism and kinetics depend on the surface structure and the crystalline size of the catalyst, is very promising and should be increasingly applied to MMCD catalysts where the latter might turn out to be most valuable. We also anticipate that "heterometal-sensitive reactions", *i.e.*, reactions which require at least two different metals to proceed with high activity and selectivity, should be particularly worthwhile to study as their results will be very informative, even if only qualitatively, about the specificity of MMCD catalysts. Studying cluster-derived catalysts should, in general, help our understanding of fundamental aspects of catalytic chemistry.

TABLE 4. 12.

Cluster-Derived Bimetallic Couples used in Heterogeneous Catalysis

Catalyzed Reaction	Bimetallic Couples
Hydrocarbon Skeletal Rearrangements	Cr-Pd W-Ir, W-Pd Re-Pt Fe-Ru, Fe-Pt Ru-Ni, Ru-Pt Co-Rh, Co-Pt Rh-Ir, Rh-Pt Ir-Pt
Hydrogenation / Isomerization of Carbon-Carbon Multiple Bonds	Ru-Os, Ru-Ni Os-Ni
Hydrogenation of CO and CO_2	Mo-Fe, Mo-Os, Mo-Co W-Os Mn-Co, Mn-Fe Fe-Ru, Fe-Os, Fe-Co Ru-Co, Ru-Os Os-Rh, Os-Ni Co-Rh
Hydrogenation of Ketones	Os-Ni
Hydroformylation of Olefins	Fe-Rh Ru-Co Co-Rh
Reductive Carbonylation of Nitro Derivatives	Mo-Pd Fe-Pd
Hydrodesulfurization	Mo-Fe, Mo-Co

This area of modern research has not yet reached the stage where conclusions can be presented about systematic comparisons of catalysts derived from clusters having the same M/M' ratio but a different structure. Those concerning the comparison of catalysts $[M_xM'_y]$ derived from isostructural clusters having a different M/M' ratio will become increasingly available as more and more molecular precursors can be synthesized in high yields and as more scientists performing heterogeneous catalysis are working on such systems. It may also be anticipated that the increasing number of molecular mixed-metal clusters containing three or four different metals will stimulate their use as precursors of a new generation of polymetallic catalysts. These will be expected to contain, at least initially, three or four metals, respectively, in a well defined stoichiometry. Just to give an idea about the considerable potential of such systems, it is appropriate to remember that, even restricted to platinum group metals, little has been done to characterise the 20 possible ternary systems, in marked contrast with the detailed investigations concerning binary alloys [195]. Among the possible ternary systems of the platinum group metals, presented in Table 4.13, only those in italics appear to have given rise to published information [195].

TABLE 4.13.
Possible Ternary Systems of the Platinum Group Metals[a]

Ir-Os-Pd	Ir-Pd-Rh	Pd-Pt-Os	Pd-Ru-Os
Ir-Os-Pt	Ir-Pd-Ru	*Pd-Pt-Rh*	*Pt-Rh-Ru*
Ir-Os-Rh	*Ir-Pt-Rh*	Pd-Pt-Ru	*Pt-Rh-Os*
Ir-Os-Ru	*Ir-Pt-Ru*	Pd-Rh-Ru	Pt-Ru-Os
Ir-Pd-Pt	Ir-Rh-Ru	Pd-Rh-Os	Rh-Ru-Os

[a] The metals are listed alphabetically.

However, it remains a challenge to the synthetic chemists to develop high yield syntheses for all these multimetallic clusters but there is little doubt about their future success!

5.   ACKNOWLEDGEMENTS

We are grateful to Profs./Drs. R. D. Adams, A. Choplin, M. D. Curtis, B. C. Gates, G. L. Geoffroy, L. Guczi, M. Hidai, I. T. Horvath, M. Ichikawa, H. D. Kaesz, G. Maire, V. Ponec, E. Sappa, B. K. Teo, H. Vahrenkamp and K. Weiss for sharing results prior to publication and/or for valuable comments. Our own work in this area was supported by the CNRS, the Université Louis Pasteur de Strasbourg, ATOCHEM and NATO.

TABLE 4.14.

Utilization of Molecular Mixed-Metal Clusters in Heterogeneous Catalysis.

Bimetallic Couple[a]	Precursor Cluster	Ref. Synth.	Notation of the MMCD Catalyst[b]	Support	Catalyzed Reactions	Ref. Catal.
Cr-Pd	$Cr_2Pd_2Cp_2(CO)_6(PMe_3)_2$ **(90)**	22	$[Cr_2Pd_2]$	γ-Al_2O_3	Hydrocarbon Rearrangements	104
Mo-Fe	$Mo_2Fe_2S_2Cp_2(CO)_6$ **(98)**	41, 312	$[Mo_2Fe_2]$	γ-Al_2O_3	Hydrodesulfurization	91
				γ-Al_2O_3, MgO	CO Hydrogenation	91
Mo-Os	$HMoOs_3Cp(CO)_{12}$ **(101)**	86	$[MoOs_3]$	γ-Al_2O_3	CO Hydrogenation	278
Mo-Co	$Mo_2Co_2S_3Cp_2(CO)_4$ **(99)**	93	$[Mo_2Co_2]$	γ-Al_2O_3	Hydrodesulfurization	91
					CO Hydrogenation	91
Mo-Pd	$Mo_2Pd_2Cp_2(CO)_6(PPh_3)_2$ **(14)**	22	$[Mo_2Pd_2]$	γ-Al_2O_3	$ArNO_2$ Carbonylation	38, 179
W-Os	$HWOs_3Cp(CO)_{12}$ **(102)**	85, 86	$[WOs_3]$	γ-Al_2O_3	CO Hydrogenation	278
W-Ir	$WIr_3Cp(CO)_{11}$ **(94)**	279	$[WIr_3]$	γ-Al_2O_3	n-Butane Hydrogenolysis	279
	$W_2Ir_2Cp_2(CO)_{10}$ **(95)**	279	$[W_2Ir_2]$	γ-Al_2O_3	n-Butane Hydrogenolysis	279
W-Pd	$W_2Pd_2Cp_2(CO)_6(PPh_3)_2$ **(16)**	22	$[W_2Pd_2]$	γ-Al_2O_3	Hydrocarbon Rearrangements	104, 136
Mn-Fe	$C[MnFe(CO)_9]$ (C = K, Et_4N) **(112)**	258	$[MnFe]$	carbon	CO Hydrogenation	305c,d
	$C[MnFe_2(CO)_{12}]$ (C = K, Et_4N) **(97)**	3	$[MnFe_2]$	SiO_2	CO Hydrogenation	53
	$[Et_4N][MnFe_2(CO)_{12}]$ **(97b)**	3	$[MnFe_2]$	SiO_2, Al_2O_3	CO Hydrogenation	187
	$Mn_2Fe(CO)_{14}$ **(113)**	115b,c	$[Mn_2Fe]$	carbon	CO Hydrogenation	305c,d

Metals	Complex	Ref.	Species	Support	Reaction	Ref.
Mn-Co	MnCo(CO)$_9$ (96)	172, 200, 258, 314	[MnCo]	Al$_2$O$_3$	CO Hydrogenation	304
Re-Pt	Re$_2$Pt(CO)$_{12}$ (111)	302	[Re$_2$Pt]	γ-Al$_2$O$_3$	Hydrocarbon Rearrangements	11
Fe-Ru	FeRu$_2$(CO)$_{12}$ (2a)	316	[FeRu$_2$]	γ-Al$_2$O$_3$	C$_2$H$_4$ self Homologation	166
				γ-Al$_2$O$_3$	Hydrocarbon Rearrangements	166
				carbon	CO Hydrogenation	166
				SiO$_2$	CO Hydrogenation	173
	Fe$_2$Ru(CO)$_{12}$ (3a)	316	[Fe$_2$Ru]	SiO$_2$	CO Hydrogenation	137, 166, 267
				carbon	CO Hydrogenation	173
				γ-Al$_2$O$_3$	Hydrocarbon Rearrangements	166
				γ-Al$_2$O$_3$	C$_2$H$_4$ self Homologation	166
				carbon	CO Hydrogenation	166
	H$_2$FeRu$_3$(CO)$_{13}$ (32)	129	[FeRu$_3$]	SiO$_2$	CO Hydrogenation	137, 267
				carbon	CO Hydrogenation	137, 267
				γ-Al$_2$O$_3$, SiO$_2$, Na-Y zeolite	CO, CO$_2$ Hydrogenation	110, 111
Fe-Os	H$_2$FeOs$_3$(CO)$_{13}$ (48)	84, 129a, 234	[FeOs$_3$]	SiO$_2$	CO Hydrogenation	79
				γ-Al$_2$O$_3$	CO Hydrogenation	57
Fe-Co	KFeCo(CO)$_8$ (114)	258	[FeCo]	carbon	CO Hydrogenation	305b
	HFeCo$_3$(CO)$_{12}$ (49)	76	[FeCo$_3$]	carbon	CO Hydrogenation	305b
	K[FeCo$_3$(CO)$_{12}$] (59d)	76	[KFeCo$_3$]	carbon	CO Hydrogenation	305b

Fe-Co	[Fe3Co]	C[Fe3Co(CO)13] (C = K, Et4N) (60)	290	carbon	CO Hydrogenation	305b
Fe-Rh	[FeRh4]	[Me4N]2[FeRh4(CO)15] (115)	74a	SiO2	Olefin Hydroformylation	162c
	[Fe3Rh2]	Fe3Rh2C(CO)14 (116)	74b	SiO2	Olefin Hydroformylation	162c
	[FeRh5]	[TMBA][FeRh5(CO)16] (117)	74a	SiO2	Olefin Hydroformylation	162c
	[Fe2Rh4]	[TMBA]2[Fe2Rh4(CO)16] (118)	74a	SiO2	Olefin Hydroformylation	162c
Fe-Pd	[FePd2]	FePd2(CO)4(DPPM)2 (103)	42	SiO2	ArNO2 Carbonylation	42, 43
	[Fe2Pd2]	Fe2Pd2(CO)5(NO)2(DPPM)2 (104)	43, 44	SiO2	ArNO2 Carbonylation	40, 43
Fe-Pt	[Fe2Pt]	Fe2Pt(CO)6(NO)2(t-BuNC)2 (24b)	15	γ-Al2O3	Hydrocarbon Rearrangements	319
Ru-Os	[RuOs3]	H2RuOs3(CO)13 (85)	60, 129a, 274	γ-Al2O3	Alkene Isomerization	274
					CO Hydrogenation	58
Ru-Co	[HRuCo3]	HRuCo3(CO)12 (53)	48, 152	carbon	Olefin Hydroformylation	162d
	[H3Ru3Co]	H3Ru3Co(CO)12 (51)	131	carbon	Olefin Hydroformylation	162d
	[Ru3Co]	[Et4N][Ru3Co(CO)13] (70a)	290	carbon	Olefin Hydroformylation	162d
	[Ru3Co]	[PPN][Ru3Co(CO)13] (70b)	290	SiO2, γ-Al2O3, Na-Y zeolite	CO, CO2 Hydrogenation	110, 111
Ru-Ni	[Ru3Ni]	H3Ru3NiCp(CO)9 (10)	70, 192, 263	Chromosorb P	Hydrocarbon Rearrangements	67
					Hydrogenation of alkenes, alkynes	67
Ru-Pt	[Ru3Pt]	Ru3Pt(CO)12Py3 (91)	196	oxides, carbon	Hydrocarbon Rearrangements	196
Os-Rh	[Os3Rh]	H2Os3Rh(CO)10(acac) (100)	109	γ-Al2O3	CO Hydrogenation	57
Os-Ni	[NiOs3Ni]	H3Os3NiCp(CO)9 (11)	70, 192, 280	γ-Al2O3	Hydrogenation of alkenes, alkynes	1, 212
					CO and CO2 Hydrogenation	1, 212
				Chromosorb P	Hydrogenation of 1,3-pentadienes	68
				Chromosorb P	Acetone Reduction	63

			Core composition[b]			
Os-Ni	Os$_3$Ni$_3$Cp$_3$(CO)$_9$ (13)	262	[Os$_3$Ni$_3$]	γ-Al$_2$O$_3$	Hydrogenation of benzene, alkynes	1
					CO Hydrogenation	1
Co-Rh	Co$_2$Rh$_2$(CO)$_{12}$ (8)	209	[Co$_2$Rh$_2$]	SiO$_2$	Hydrocarbon Rearrangements	6
				ZnO	Olefin Hydroformylation	161
				ZrO$_2$	CO Hydrogenation	160
	Co$_3$Rh(CO)$_{12}$ (9)	209	[Co$_3$Rh]	ZnO	Olefin Hydroformylation	161
Co-Pt	Co$_2$Pt(CO)$_8$(c-C$_6$H$_{11}$NC)$_2$ (23a)	15	[Co$_2$Pt]	γ-Al$_2$O$_3$	Hydrocarbon Rearrangements	203
	Co$_2$Pt$_2$(CO)$_8$(PPh$_3$)$_2$ (28a)	39	[Co$_2$Pt$_2$]	γ-Al$_2$O$_3$	Hydrocarbon Rearrangements	203
	Co$_2$Pt$_3$(CO)$_9$(PEt$_3$)$_3$ (82b)	16	[Co$_3$Pt$_2$]	γ-Al$_2$O$_3$	Hydrocarbon Rearrangements	203
Ir-Pt	Ir$_2$Pt(CO)$_7$Py$_2$ (92)	196	[Ir$_2$Pt]	oxides, carbon	Hydrocarbon Rearrangements	196
	Ir$_6$Pt(CO)$_{15}$Py$_2$ (93)	196	[Ir$_6$Pt]	oxides, carbon	Hydrocarbon Rearrangements	196

a The metals are listed with increasing number of their group.
b This notation is meant to indicate the core composition of the molecular precursor but has no implication as far as the particle size, shape or composition of the catalyst are concerned.

6. REFERENCES

1.   Albanesi, G.; Bernardi, R.; Moggi, P.; Predieri, G.; Sappa, E. *Gazz. Chim. Ital.* **1986**, *116*, 385.
2.   Alexiev, V. D.; Binsted, N.; Evans, J.; Greaves, G. N.; Price, R. J. *J. Chem. Soc., Chem. Commun.* **1987**, 395, and references cited therein.
3.   Anders, U.; Graham, W. A. G. *J. Chem. Soc., Chem. Commun.* **1966**, 291.
4.   Anderson, J. H., Jr.; Conn, P. J.; Brandenberger, S. G. *J. Catal.* **1970**, *16*, 326, 404.
5.   Anderson, J. R.; Elmes, P. S.; Howe, R. F.; Mainwaring, D. E. *J. Catal.* **1977**, 50, 508.
6.   Anderson, J. R.; Mainwaring, D. E. *J. Catal.* **1974**, *35*, 162.
7.   Anderson, J. R.; Mainwaring, D. E. *Ind. Eng. Chem. Prod. Res. Dev.* **1978**, *17*, 202.
8.   (a) Arndt, L. W.; Delord, T.; Darensbourg, M. Y. *J. Am. Chem. Soc.* **1984**, *106*, 456. (b) Arndt, L. W.; Darensbourg, M. Y.; Delort, T.; Trzcinska Bancroft, B. *J. Am. Chem. Soc.* **1986**, *108*, 2617.
9.   Asakura, K.; Yamada, M.; Iwasawa, Y.; Kuroda, H. unpublished results quoted in ref. 166.
10.   (a) Attali, S.; Dahan, F.; Mathieu, R. *Organometallics* **1986**, *5*, 1376. (b) Attali, S.; Mathieu, R. *J. Organomet. Chem.* **1985**, *291*, 205. (c) Alami, M. K.; Dahan, F.; Mathieu, R. *J. Chem. Soc., Dalton Trans.* **1987**, 1983.
11.   Augustine, S. M.; Nacheff, M. S.; Tsang, C. M.; Butt, J. B.; Sachtler, W. M. H. *Proceedings, Congress on Catalysis;* San Diego (USA), **1987**.
12.   Bailey, D. C.; Langer, S. H. *Chem. Rev.* **1981**, *81*, 109.
13.   Balabanov, G. P.; Dergunov, Yu. I.; Khosdurdyev, Kh. O.; Manov-Yuvenskii, V. I.; Nefedov, B. K.; Rysikhin, A. I. *U.S. Patent* 4 207 212; *Chem. Abstr.* **1980**, *93*, 150873.
14.   Barbier, J.-P. Thèse de Docteur Ingénieur, Université Louis Pasteur, Strasbourg, **1978**.
15.   Barbier, J.-P.; Braunstein, P. *J. Chem. Res., Synop.* **1978**, 412; *J. Chem. Res., Miniprint* **1978**, 5029.
16.   Barbier, J.-P.; Braunstein, P.; Fischer, J.; Ricard, L. *Inorg. Chim. Acta* **1978**, *31*, L361.
17.   Bardi, R.; Piazzesi, A. M.; Cavinato, G.; Cavolli, P.; Toniolo, L. *J. Organomet. Chem.* **1982**, *224*, 407.
18.   Basset, J.-M.; Choplin, A. *J. Mol. Catal.* **1983**, *21*, 95   and references cited therein.
19.   Bender, R.; Braunstein, P. *J. Chem. Soc., Chem. Commun.* **1983**, 334.
20.   Bender, R.; Braunstein, P.; Dusausoy, Y.; Protas, J. *Angew. Chem., Int. Ed. Engl.* **1978**, *17*, 596.

21. Bender, R.; Braunstein, P.; Dusausoy, Y.; Protas, J. *J. Organomet. Chem.* **1979**, 172, C51.

22. Bender, R.; Braunstein, P.; Jud, J.-M.; Dusausoy, Y. *Inorg. Chem.* **1983**, 22, 3394.

23. Bender, R.; Braunstein, P.; Jud, J.-M.; Dusausoy, Y. *Inorg. Chem.* **1984**, 23, 4489.

24. Benson, B. C.; Jackson, R.; Joshi, K. K.; Thompson, D. T. *Chem. Commun.* **1968**, 1506.

25. Beurich, H.; Vahrenkamp, H. *Angew. Chem., Int. Ed. Engl.* **1978**, 17, 863.

26. Beurich, H.; Vahrenkamp, H. *Angew. Chem., Int. Ed. Engl.* **1981**, 20, 98.

27. Beurich, H.; Vahrenkamp, H. *Chem. Ber.* **1982**, 115, 2385, 2409.

28. Bhaduri, S.; Sharma, K. R. *J. Chem. Soc., Dalton. Trans.* **1984**, 2309.

29. Bhaduri, S.; Sharma, K. R.; Clegg, W.; Sheldrick, G. M.; Stalke, D. *J. Chem. Soc., Dalton Trans.* **1984**, 2851.

30. Binsted, N.; Cook, S. L.; Evans, J.; Greaves, G. N. *J. Chem. Soc., Chem. Commun.* **1985**, 1103.

31. (a) Böszörményi, I.; Dobos, S.; Guczi, L.; Marko, L.; Lazar, K.; Reiff, W. M.; Schay, Z.; Takacs, L.; Vizi-Orosz, A. In *Proceedings, 8th International Congress on Catalysis;* Berlin (Ger.); Verlag Chemie, Weinheim, **1984**, Vol. 5, p 183. (b) Böszörményi, I.; Guczi, L. *Inorg. Chim. Acta* **1986**, 112, 5.

32. Boudart, M. *Adv. Catal.* **1969**, 20, 153.

33. Boudart, M. *J. Mol. Catal.* **1985**, 30, 27; *J. Mol. Catal. Review Issue* **1986**, 29.

34. Bradford, C. W.; van Bronswijk, W.; Clark, R. J. H.; Nyholm, R. S. *J. Chem. Soc. A* **1970**, 2889.

35. Bradley, J. S. *J. Am. Chem. Soc.* **1979**, 101, 7419.

36. Bradley, J. S.; Pruett, R. L.; Hill, E.; Ansell, G. B.; Leonowicz, M. E.; Modrick, M. A. *Organometallics* **1982**, 1, 748.

37. Braunstein, P. *Nouv. J. Chim.* **1986**, 10, 365, and references cited therein.

38. Braunstein, P.; Bender, R.; Kervennal, J. *Organometallics* **1982**, 1, 1236.

39. Braunstein, P.; Dehand, J.; Nennig, J.-F. *J. Organomet. Chem.* **1975**, 92, 117.

40. Braunstein, P.; Devenish, R.; Gallezot, P.; Heaton, B. T.; Humphreys, C. J. Kervennal, J.; Mulley, S.; Ries, M. unpublished results.

41. Braunstein, P.; Jud, J.-M.; Tiripicchio, A.; Tiripicchio Camellini, M.; Sappa, E. *Angew. Chem., Int. Ed. Engl.* **1982**, 21, 307.

42. Braunstein, P.; Kervennal, J.; Richert, J.-L. *Angew. Chem., Int. Ed. Engl.* **1985**, 24, 768.

43. Braunstein, P.; Kervennal, J.; Richert, J.-L.; Ries, M. *French Patent* 2 558 074, *U.S. Patent* 4 609 639 (**1985**), *Chem. Abstr.* **1986**, *104*, 19 681 (to Atochem)

44. Braunstein, P.; de Méric de Bellefon, C.; Ries, M.; Fischer, J. *Organometallics* **1988**, *7*, in the press.

45. Braunstein, P.; Rosé, J. *Gold Bulletin* **1985**, *18*, 17, and references cited therein.

46. Braunstein, P.; Rosé, J.; Bars, O. *J. Organomet. Chem.* **1983**, *252*, C101.

47. Braunstein, P.; Rosé, J.; Dedieu, A.; Dusausoy, Y.; Mangeot, J.-P.; Tiripicchio, A.; Tiripicchio Camellini, M. *J. Chem. Soc., Dalton Trans.* **1986**, 225.

48. Braunstein, P.; Rosé, J.; Dusausoy, Y.; Mangeot, J.-P. *C.R. Hebd. Acad. Sci., Ser. II* **1982**, *294*, 967.

49. Braunstein, P.; Schubert, U.; Burgard, M. *Inorg. Chem.* **1984**, *23*, 4057 and references cited therein.

50. Brenner, A.; Burwell, R. J. *J. Catal.* **1978**, *52*, 353.

51. Brown, E. S. *U.S. Patent* 3 989 799 (**1976**), *Chem. Abstr.* **1977**, *86*, 96 646; *U.S. Patent* 3 929 969 (**1975**), *Chem. Abstr.* **1976**, *84*, 127 210.

52. Brown, T. H. *J. Mol. Catal.* **1981**, *92*, 41.

53. Bruce, L.; Hope, G.; Turney, T. W. *React. Kinet. Catal. Lett.* **1982**, *20*, 175.

54. Bruce, M. I. *J. Organomet. Chem.* **1985**, *283*, 339; *J. Organomet. Chem. Libr.* **1985**,*17*, 399.

55. Bruce, M. I.; Shaw, G.; Stone, F. G. A. *J. Chem. Soc., Dalton Trans.* **1972**, 1082.

56. Bruce, M. I.; Shaw, G.; Stone, F. G. A. *J. Chem. Soc., Dalton Trans.* **1972**, 1781.

57. Budge, J. R.; Lücke, B. F.; Gates, B. C.; Toran, J. *J. Catal.* **1985**, *91*, 272.

58. Budge, J. R.; Lücke, B. F.; Scott, J. P.; Gates, B. C. In *Proceedings, 8th International Congress on Catalysis;* Berlin (Ger.); Verlag Chemie, Weinheim, **1984**, Vol. 5, p 89.

59. Budge, J. R.; Scott, J. P.; Gates, B. C. *J. Chem. Soc., Chem. Commun.* **1983**, 342.

60. Burkhardt, E. W.; Geoffroy, G. L. *J. Organomet. Chem.* **1980**, *198*, 179.

61. Burlitch, J. M.; Hayes, S. E.; Lemley, J. T. *Organometallics* **1985**, *4*, 167.

62. Carlton, L.; Lindsell, W. E.; McCullough, K. J.; Preston, P. N. *J. Chem. Soc., Dalton Trans.* **1984**, 1693.

63. Castiglioni, M.; Giordano, R.; Sappa, E. *J. Mol. Catal.* **1986**, *37*, 287.

64. Castiglioni, M.; Giordano, R.; Sappa, E. *J. Organomet. Chem.* **1983**, *258*, 217.

65. Castiglioni, M.; Giordano, R.; Sappa, E. *J. Organomet. Chem.* **1984**, *275*, 119.

66. Castiglioni, M.; Giordano, R.; Sappa, E. *J. Organomet. Chem.* **1987**, *319*, 167.

67. (a) Castiglioni, M.; Giordano, R.; Sappa, E. *J. Mol. Catal.* **1987**, *40*, 65. *Ibid.* **1987**, *42*, 307.

68. Castiglioni, M.; Giordano, R.; Sappa, E.; Predieri, G.; Tiripicchio, A. *J. Organomet. Chem.* **1984**, *270*, C7.

69. Castiglioni, M.; Giordano, R.; Sappa, E.; Tiripicchio, A.; Tiripicchio Camellini, M. *J. Chem. Soc., Dalton Trans.* **1986**, 23.

70. Castiglioni, M.; Sappa, E.; Valle, M.; Lanfranchi, M.; Tiripicchio, A. *J. Organomet. Chem.* **1983**, *241*, 99.

71. Catton, G. A.; Jones, G. F. C.; Mays, M. J.; Howell, J. A. S. *Inorg. Chim. Acta* **1976**, *20*, L41.

72. Ceriotti, A.; Demartin, F.; Longoni, G.; Manassero, M.; Marchionna, M.; Piva, G.; Sansoni, M. *Angew. Chem., Int. Ed. Engl.* **1985**, *24*, 697.

73. (a) Ceriotti, A.; Garlaschelli, L.; Longoni, G.; Malastesta, M. C.; Strumolo, D.; Fumagalli, A.; Martinengo, S. *J. Mol. Catal.* **1984**, *24*, 309. (b) *Ibid.* **1984**, *24*, 323.

74. (a) Ceriotti, A.; Longoni, G.; Pergola, R. D.; Heaton, B. T.; Smith, D. O. *J. Chem. Soc., Dalton Trans.* **1983**, 1433. (b) Hriljac, J. A.; Holt, E. M.; Shriver, D. F. *Inorg. Chem.* **1987**, *26*, 2943. (c) Tachikawa, M.; Geerts, R. L.; Muetterties, E. L. *J. Organomet. Chem.* **1981**, *213*, 11.

75. (a) Chini, P. *Gazz. Chim. Ital.* **1979**, *109*, 225. (b) *J. Organomet. Chem.* **1980**, *200*, 37.

76. Chini, P.; Colli, L.; Peraldo, M. *Gazz. Chim. Ital.* **1960**, *90*, 1005.

77. Choplin, A.; Huang, L.; Basset, J.-M.; Mathieu, R.; Siriwardane, U.; Shore, S. G. *Organometallics* **1986**, *5*, 1547   and references cited therein.

78. Choplin, A.; Huang, L.; Theolier, A.; Gallezot, P.; Basset, J.-M.; Siriwardane, U.; Shore, S. G.; Mathieu, R. *J. Am. Chem. Soc.* **1986**, *108*, 4224.

79. Choplin, A.; Leconte, M.; Basset, J.-M.; Shore, S. G.; Hsu, W.- L. *J. Mol. Catal.* **1983**, *21*, 389.

80. Choplin, A.; Theolier, A.; D'Ornelas, L.; Defour, P.; Huang, L.; Basset, J.-M. unpublished results.

81. Choukroun, R.; Gervais, D.; Jaud, J.; Kalck, P.; Senocq, F. *Organometallics*, **1986**, *5*, 67.

82. (a) Choukroun, R.; Iraqi, A.; Gervais, D. *J. Organomet. Chem.* **1986**, *311*, C60. (b) Choukroun, R.; Iraqi, A.; Gervais, D.; Daran, J.-C.; Jeannin, Y. *Organometallics*, **1987**, *6*, 1197.

128

83. Christensen, P. H.; Morup, S.; Clausen, B. S.; Topsoe, H. In *Proceedings, 8th International Congress on Catalysis;* Berlin (Ger.); Verlag Chemie, Weinheim, **1984**, Vol. 2, p 545.

84. Churchill, M. R.; Bueno, C.; Hsu, W. L.; Plotkin, J. S.; Shore, S. G. *Inorg. Chem.* **1982**, *21*, 1958.

85. Churchill, M. R.; Hollander, F. J. *Inorg. Chem.* **1979**, *18*, 843.

86. Churchill, M. R.; Hollander, F. J.; Shapley, J. R.; Foose, D. S. *J. Chem. Soc., Chem. Commun.* **1978**, 534.

87. Ciapetta, F. G.; Plank, C. J. In *Catalysis*; Emmet, P. H., Ed.; Reinhold: New-York, **1954**, Vol. 1, p 315.

88. Collier, G.; Hunt, D. J.; Jackson, S. D.; Moyes, R. B.; Pickering, I. A.; Wells, P. B.; Simpson, A. F.; Whyman, R. *J. Catal.* **1983**, *80*, 154.

89. Commereuc, D.; Chauvin, Y.; Hugues, F.; Basset, J.-M.; Olivier, D. *J. Chem. Soc., Chem. Commun.* **1980**, 154.

90. Costa, L. C. *Catal. Rev. Sci. Eng.* **1983**, *25*, 325.

91. Curtis, M. D. personal communication.

92. Curtis, M. D.; Schwank, J.; Thompson, L.; Williams, P. D.; Baralt, O. *Prepr., Fuel Division, Am. Chem. Soc.* **1986**, *31(3)*, 44.

93. Curtis, M. D.; Williams, P. D. *Inorg. Chem.* **1983**, *22*, 2661.

94. Dagan, L.; Weixu, Z.; Zhengshi, C.; Yanwen, S.; Xiuru, Z.; Zhongheng, W. *Huaxue Xuebao* **1986**, *44*, 990.

95. Dehand, J.; Nennig, J.-F. *Inorg. Nucl. Chem. Lett.* **1974**, *10*, 875.

96. Dobos, S.; Böszörményi, I.; Mink, J.; Guczi, L. *Inorg. Chim. Acta* **1986**, *120*, 135, 145.

97. Dombek, B. D. *Adv. Catal.* **1983**, *32*, 325.

98. Dombek, B. D. *Organometallics* **1985**, *4*, 1707.

99. Doyle, G. *Eur. Patent* 30 434 (**1981**), *Chem. Abstr.* **1982**, *96*, 006 151 (to Exxon Research and Engineering Co) .

100. (a) Doyle, G. *J. Mol. Catal.* **1981**, *13*, 237. (b) *Ibid.* **1983**, *18*, 251.

101. Dubois, R. A.; Garrou, P. E.; Hartwell, G. E.; Hunter, D. L. *Organometallics* **1984**, *3*, 95.

102. Engler, B.; Vruger, C.; Keim, W.; Sekutowski, J. C. *Erdöl Kohle, Erdgas Petrochem. Brennst. Chem.* **1978**, *3*, 87.

103. Ertl, G. In *Studies in Surface Science and Catalysis, Vol. 29: Metal Clusters in Catalysis*; Gates, B. C.; Guczi, L.; Knözinger, H., Eds.; Elsevier Science Publishers: Amsterdam **1986**, p 577 - 604.

104. Esteban Puges, P.; Garin, F.; Girard, P.; Bernhardt, P.; Maire, G. In *Proceedings, Iberoamerican Congress on Catalysis;* Lisbon (Portugal) **1984**, Vol. 2, p 1111.

105. Evans, J. *Chem. Soc. Rev.* **1981**, *10*, 159.

106. Evans, J.; Jingxing, G. *J. Chem. Soc., Chem. Commun.* **1985**, 39.

107. Evans, J.; Street, A. C.; Webster, M. *Organometallics* **1987**, *6*, 794.

108. Farrugia, L. J.; Green, M.; Hankey, D. R.; Orpen, A. G.; Stone, F. G. A. *J. Chem. Soc., Chem. Commun.* **1983**, 310.

109. Farrugia, L. J.; Howard, J. A. K.; Mitrprachachon, P.; Stone, F. G. A.; Woodward, P. *J. Chem. Soc., Dalton Trans.* **1981**, 171.

110. Ferkul, H. E.; Berlie, J. M.; Stanton, D. J.; McCowan, J. D.; Baird, M. C. *Can. J. Chem.* **1983**, *61*, 1306.

111. Ferkul, H. E.; Stanton, D. J.; McCowan, J. D.; Baird, M. C. *J. Chem. Soc., Chem. Commun.* **1982**, 955.

112. Foger, K.; Anderson, J. R. *J. Catal.* **1979**, *59*, 325.

113. Ford, P. C., Ed.; *Catalytic Activation of Carbon Monoxide*; *A C S Symposium Series N°152*, Am. Chem. Soc.: Washington, D. C. **1981**.

114. Ford, P. C.; Rinker R. C.; Unkermann, C.; Laine, R. M.; Landis, V.; Moya, S. A. *J. Am. Chem. Soc.* **1978**, *100*, 4595.

115. (a) Forster, A.; Johnson, B. F. G.; Lewis, J.; Matheson, T. W.; Robinson, B. H.; Jackson, W. G. *J. Chem. Soc., Chem. Commun.* **1974**, 1042.
     (b) Evans, G. O.; Wozniak, W. T.; Sheline, R. K. *Inorg. Chem.* **1970**, *9*, 979. (c) Schubert, E. H.; Sheline, R. K. *Z. Naturforsch.* **1965**, *20b*, 1306.

116. Fox, J. R.; Gladfelter, W. L.; Geoffroy, G. L. *Inorg. Chem.* **1980**, *19*, 2574.

117. Fox, J. R.; Gladfelter, W. L.; Geoffroy, G. L.; Tavanaiepour, I.; Abdel-Mequid, S.; Day, V. W. *Inorg. Chem.* **1981**, *20*, 3230.

118. Freeman, M. J.; Green, M.; Orpen, A. G.; Salter, I. D.; Stone, F. G. A. *J. Chem. Soc., Chem. Commun.* **1983**, 1332.

119. Freeman, M. J.; Orpen, A. G.; Salter, I. D. *J. Chem. Soc., Dalton Trans.* **1987**, 379.

120. Fumagalli, A.; Martinengo, S.; Chini, P.; Albinati, A.; Bruckner, S.; Heaton, B. T. *J. Chem. Soc., Chem. Commun.* **1978**, 195.

121. Fumagalli, A.; Martinengo, S.; Chini, P.; Galli, D.; Heaton, B. T.; Pergola, R. D. *Inorg. Chem.* **1984**, *23*, 2947.

122. Fusi, A.; Ugo, R.; Psaro, R.; Braunstein, P.; Dehand, J. *Phil. Trans. R. Soc. Lond.* A **1982**, *308*, 125; *J. Mol. Catal.* **1982**, *16*, 217.

123. Garlaschelli, L.; Longoni, G.; Marchionna, M. In *Proceedings, XIX Congresso Nazionale di Chimica Inorganica, VI Congresso Nazionale di Catalysi*; Santa Margherita di Pula-Cagliari (Italy), October **1986**, p 491.

124. Gates, B. C. *NSF-CNRS International Seminar on the Relationship Between Metal Cluster Compounds, Surface Science and Catalysis*; Asilomar (USA), November **1979**.

125. Gates, B. C. In *Studies in Surface Science and Catalysis, Vol. 29: Metal Clusters in Catalysis*; Gates, B. C.; Guczi, L.; Knözinger, H., Eds.; Elsevier Science Publishers: Amsterdam **1986**, p 415- 425.

130

126. Gates, B. C.; Guczi, L.; Knözinger, H., Eds., *Studies in Surface Science and Catalysis, Vol. 29: Metal Clusters in Catalysis*; Elsevier Science Publishers: Amsterdam **1986**.

127. Gates, B. C.; Lieto, J. *CHEMTECH* **1980**, *10*, 195, 248.

128. (a) Gauthier-Lafaye, J.; Perron, R.; Colleuille, Y. *L' Actualité Chimique*, **1983**, *9*, 11. (b) Gauthier-Lafaye, J.; Perron, R. *Methanol et Carbonylation*, Rhône-Poulenc Recherches, Ed. Technip: Paris, **1986**, and references cited therein.

129. (a) Geoffroy, G. L.; Fox, J. R.; Burkhardt, E.; Foley, H. C.; Harley, A. D.; Rosen, R. *Inorg. Synth.* **1982**, *21*, 57.
(b) Geoffroy, G. L.; Gladfelter, W. L. *J. Am. Chem. Soc.* **1977**, *99*, 7565.

130. Gladfelter, W. L.; Geoffroy, G. L. *Adv. Organometal. Chem.* **1980**, *18*, 207.

131. Gladfelter, W. L.; Geoffroy, G. L.; Calabrese, J. C. *Inorg. Chem.* **1980**, *19*, 2569.

132. Golodov, V. A. *J. Res. Inst. Catal., Hokkaido Univ.* **1981**, *29*, 49.

133. Good, M. L.; Akbarnejad, M.; Donner, J. T. *Prepr., Div. Pet. Chem., Am. Chem. Soc.* **1980**, *25*, 763.

134. Guczi, L. *Catal. Rev. Sci. Eng.* **1981**, *23*, 329.

135. Guczi, L. In *Studies in Surface Science and Catalysis, Vol. 29: Metal Clusters in Catalysis*; Gates, B. C.; Guczi, L.; Knözinger, H., Eds., Elsevier Science Publishers: Amsterdam **1986**, p 209-230.

136. Guczi, L. In *Studies in Surface Science and Catalysis, Vol. 29: Metal Clusters in Catalysis*; Gates, B. C.; Guczi, L.; Knözinger, H., Eds.; Elsevier Science Publishers: Amsterdam **1986**, p 547-574.

137. Guczi, L.; Schay, Z.; Lazar, K.; Vizi, A.; Marko, L. *Surf. Sci.* **1981**, *106*, 516.

138. Guczi, L.; Schay, Z.; Lazar, K.; Vizi, A.; Marko, L. *2nd International Symposium on Small Particles and Metallic Clusters*; Lausanne (Switzerland), September 8-12, **1980**.

139. Guczi, L.; Schay, Z.; Matusek, K.; Bogyay, I.; Steffler, G. In *Proceedings, 7th International Congress on Catalysis*; Tokyo (Japan); Kodansha, Tokyo and Elsevier, Amsterdam, **1981**, Part A, p 211.

140. Guglielminotti, E.; Osella, D.; Stanghellini, P. L. *J. Organomet. Chem.* **1985**, *281*, 291.

141. Haelg, P.; Consiglio, G.; Pino, P. In *Proceedings, 2nd International Symposium on Homogeneous Catalysis*, Düsseldorf (Ger.) 1-3 Sept **1980**, p 22.

142. Haelg, P.; Consiglio, G.; Pino, P. *J. Organomet. Chem.* **1985**, *296*, 281.

143. Hansen, M. *Constitution of Binary Alloys*; McGraw-Hill: New-York, **1958**; Elliot, R. P. *Constitution of Binary Alloys, First supplement*; McGraw-Hill: New-York, **1965**.

144. Hardy, W. B.; Bennett, R. P. *Tetrahedron Lett.* **1967**, 961.

145. Hartley, F. R. *Supported Metal Complexes, Catalysis by Metal Complexes*; Reidel Publ. Co.: Dordrecht (The Netherlands), **1985**.

146. Hartley, F. R.; Vezey, P. N. *Adv. Organomet. Chem.* **1977**, *15*, 189.

147. Hartwell, G. E.; Garrou, E. P. *U.S. Patent* 4 144 191 (**1979**), *Chem. Abstr.* **1979**, *90*, 203 478 (to Dow Chemical Co.).

148. Heaton, B. T.; Ingallina, P.; Devenish, R.; Humphreys, C. J.; Ceriotti, A.; Longoni, G.; Marchionna, M. *J. Chem. Soc., Chem. Commun.* **1987**, 765.

149. Hemmerich, R.; Keim, W.; Röper, M. *J. Chem. Soc., Chem. Commun.* **1983**, 428.

150. Hidai, M.; Fukuoka, A.; Koyasu, Y.; Uchida, Y. *J. Chem. Soc., Chem. Commun.* **1984**, 516.

151. Hidai, M.; Fukuoka, A.; Koyasu, Y.; Uchida, Y. *J. Mol. Catal.* **1986**, *35*, 29.

152. Hidai, M.; Orisaku, M.; Ue, M.; Koyasu, Y.; Kodama, T.; Uchida, Y. *Organometallics* **1983**, *2*, 292.

153. Hidai, M.; Orisaku, M.; Ue, M.; Uchida, Y.; Yasufuku, K.; Yamazaki, H. *Chem. Lett.* **1981**, 143.

154. Hieber, W.; Teller, U. *Z. Anorg. Allgem. Chem.* **1942**, *249*, 43.

155. Higginson, G. W. *Chem. Eng.* (N.Y.) **1974**, *81*, 98.

156. Horvath, I. T.; Zsolnai, L.; Huttner, G. *Organometallics* **1986**, *5*, 180.

157. (a) Horvath, I. T. *Organometallics* **1986**, *5*, 2333. (b) Horvath, I. T.; Bor, G.; Garland, M.; Pino, P. *Organometallics* **1986**, *5*, 1441. (c) Horvath, I. T.; Garland, M.; Bor, G.; Pino, P. submitted for publication.

158. Hoskins, B. F.; Steen, R. J.; Turney, T. W. *Inorg. Chim. Acta* **1983**, *77*, L69.

159. Huttner, G.; Schneider, J.; Müller, H.-D.; Mohr, G.; von Seyerl, J.; Wohlfahrt, L. *Angew. Chem., Int. Ed. Engl.* **1979**, *18*, 76.

160. Ichikawa, M. *CHEMTECH* **1982**, 674.

161. (a) Ichikawa, M. *J. Catal.* **1979**, *59*, 67. (b) *Ibid.* **1979**, *56*, 127.

162. (a) Ichikawa, M. In *Tailored Metal Catalysts*; Iwasawa, Y., Ed.; Reidel Publ. Co.: Dordrecht (The Netherlands), **1986**, p 183. (b) Ichikawa, M. In *Homogeneous and Heterogeneous Catalysis, Proceedings, 5th International Symposium on Relations between Homogeneous and Heterogeneous Catalysis*; Novosibirsk (USSR), July 15-19, **1986**; Yermakov, Yu.; Likholobov, V., Eds.; VNU Science Press: Utrecht (The Netherlands). (c) Fukuoka, A.; Ichikawa, M.; Hriljac, J. A.; Shriver, D. F. *Inorg. Chem.* **1987**, *26*, 3643. (d) Fukuoka, A.; Matsuzaka, H.; Hidai, M.; Ichikawa, M. *Chem. Lett.* **1987**, 941.

163. Ichikawa, M.; Fukushima, T.; Yokoyama, T.; Kosugi, N.; Kuroda, H. *J. Phys. Chem.* **1986**, *90*, 1222.

164. Iiskola, E.; Pakkanen, T.A. unpublished results.

165. Innes, W. B. In *Catalysis*; Emmett, P. H., Ed., Reihold: New-York, **1954**, Vol. 1, p 245.

166. Iwasawa, Y.; Yamada, M. *J. Chem. Soc., Chem. Commun.* **1985**, 675.

167. Jacobs, P. A. In *Studies in Surface Science and Catalysis, Vol. 29: Metal Clusters in Catalysis*; Gates, B. C.; Guczi, L.; Knözinger, H., Eds.; Elsevier Science Publishers: Amsterdam **1986**, p 357- 414.

168. Jackson, S. D.; Wells, P. B. *Platinum Metals Rev.* **1986**, *30*, 14.

169. Jackson, S. D.; Wells, P. B.; Whyman, R.; Worthington, P. In *Catalysis* (Specialist Periodical Report), Royal Society of Chemistry, **1981**, Vol. 4.

170. Johnson, B. F. G. *Transition Metal Cluster Compounds*; Wiley-Interscience: Chichester, **1980** and references cited therein.

171. Johnson, B. F. G.; Kaner, D. A.; Lewis, J.; Raithby, P. R. *J. Organomet. Chem.* **1981**, *215*, C33.

172. Joshi, K. K.; Pauson, P. L. *Z. Naturforsch.* **1962**, *176*, 565.

173. Kaminsky, M.; Yoon, K. J.; Geoffroy, G. L.; Vannice, M. A. *J. Catal.* **1985**, *91*, 338.

174. Katzer, J. R.; Sleight, A. W.; Gajardo, P.; Michel, J. B.; Gleason, E. F.; McMillan, S. *Faraday Discuss. Chem. Soc.* **1981**,*72*, 121 and references cited therein.

175. Keim, W., Ed., *Catalysis in C_1 Chemistry*; Reidel Publ. Co., Dordrecht (The Netherlands) **1983**.

176. Keim, W.; Anstock, M.; Röper, M.; Shlupp, J. *C_1 Mol. Chem.* **1984**, *1*, 21.

177. Keim, W.; Berger, M.; Schlupp, J. *J. Catal.* **1980**, *61*, 359.

178. Kervennal, J. private communication.

179. Kervennal, J.; Cognion, J.-M.; Braunstein, P. *French Patent* 2 515 640; *U.S. Patent* 4 478 757 (**1982**), *Chem. Abstr.* **1983**, *99*, 139 487 (to PCUK Produits Chimiques Ugine Kuhlmann).

180. King, T. S.; Wu, X.; Gerstein, B. C. *J. Am. Chem. Soc.* **1986**, *108*, 6056.

181 Klabunde, K. J.; Imizu, Y. *J. Am. Chem. Soc.* **1984**, *106*, 2721.

182. Knifton, J. F. *J. Chem. Soc., Chem. Commun.* **1983**, 729.

183. Knight, J.; Mays, M. J. *J. Chem. Soc. A* **1970**, 711.

184. Knözinger, H. In *Studies in Surface Science and Catalysis, Vol. 29: Metal Clusters in Catalysis*; Gates, B. C.; Guczi, L.; Knözinger, H., Eds.; Elsevier Science Publishers: Amsterdam **1986**, p 123- 207 and 259-264.

185. Knox, S. A. R.; Koepke, J. W.; Andrews, M. A.; Kaesz, H. D. *J. Am. Chem. Soc.* **1975**, *97*, 3942.

186. Kuznetsov, V. L.; Bell, A. T.; Yermakov, Y. I. *J. Catal.* **1980**, *65*, 374.

187. Kuznetsov, V. L.; Danilyuk, A. F.; Kolosova, I. E.; Yermakov, Yu. I. *React. Kinet. Catal. Lett.* **1982**, *21*, 249.

188. Labroue, D.; Poilblanc, R. *J. Mol. Catal.* **1977**, *2*, 329.

189. Labroue, D.; Qeau, R.; Poilblanc, R. *J. Organomet. Chem.* **1982**, *233*, 359.

190. Laine, R. M. *J. Mol. Catal.* **1982**, *14*, 137.

191. Lauher, J. *J. Catal.* **1980**, *66*, 237.

192. Lavigne, G.; Papageorgiou, F.; Bergounhou, C.; Bonnet, J. J. *Inorg. Chem.* **1983**, *22*, 2485.

193. (a) Lazar, K.; Matusek, K.; Mink, J.; Guczi, L.; Vizi-Orosz, A.; Marko, L.; Reiff, W. M. *J. Catal.* **1984**, *87*, 163. (b) Lazar, K.; Reiff, W. M.; Guczi, L. *Hyperfine Interactions* **1986**, *28*, 871. (c) Lazar, K.; Reiff, W. M.; Mörke, W.; Guczi, L. *J. Catal.* **1986**, *100*, 118. (d) Lazar, K.; Schay, Z.; Guczi, L. *J. Mol. Catal.* **1982**, *17*, 205.

194. Lieto, J.; Wolf, M.; Matrana, B. A.; Prochazka, M.; Tesche, B.; Knözinger, H.; Gates, B. C. *J. Phys. Chem.* **1985**, *89*, 991.

195. McGill, I. R. *Platinum Metals Rev.*, **1987**, *31*, 74.

196. McVicker, G. B. *U.S. Patent* 4 217 249 (**1980**), *Chem. Abstr.* **1981**, *94*, 106198; *U.S. Patent* 4 302 400 (**1981**) (to Exxon Research and Engineering Co.).

197. McVicker, G. B.; Baker, R. T. K.; Garten, R. L.; Kugler, E. L. *J. Catal.* **1980**, *65*, 207.

198. McVicker, G. B.; Vannice, M. A. *U.S. Patent* 4 154 751 (**1979**), *Chem. Abstr.* **1979**, *91*, 76 694.

199. McVicker, G. B.; Vannice, M. A. *J. Catal.* **1980**, *63*, 25.

200. Madach, T.; Vahrenkamp, H. *Chem. Ber.* **1980**, *113*, 2675.

201. Mahé, C.; Patin, H.; Le Marouille, J.-Y.; Benoit, A. *Organometallics* **1983**, *2*, 1051.

202. Maire, G. In *Studies in Surface Science and Catalysis, Vol. 29: Metal Clusters in Catalysis*; Gates, B. C.; Guczi, L.; Knözinger, H., Eds.; Elsevier Science Publishers: Amsterdam **1986**, p 509 - 530.

203. Maire, G.; Zahraa, O.; Garin, F.; Crouzet, C.; Aeiyach, S.; Legaré, P.; Braunstein, P. *J. Chim. Phys.* **1981**, *78*, 951.

204. Mani, D.; Vahrenkamp, H. *J. Mol. Catal.* **1985**, *29*, 305.

205. Mani, D.; Vahrenkamp, H. *Chem. Ber.* **1986**, *119*, 3639.

206. Marchionna, M.; Longoni, G. *J. Mol. Catal.* **1986**, *35*, 107.

207. Marko, L.; Vizi-Orosz, A. In *Studies in Surface Science and Catalysis, Vol. 29: Metal Clusters in Catalysis*; Gates, B. C.; Guczi, L.; Knözinger, H., Eds.; Elsevier Science Publishers: Amsterdam **1986**, p 89 - 120.

208. Marrakchi, H.; Haimeur, M.; Escalant, P.; Lieto, J.; Aune, J.-P. *Nouv. J. Chim.* **1986**, *10*, 159.

209. Martinengo, S.; Chini, P.; Albano, V. G.; Cariati, F.; Salvatori, T. *J. Organomet. Chem.* **1973**, *59*, 379.

210. Meier, P. F.; Pennella, F.; Klabunde, K. J.; Imizu, Y. *J. Catal.* **1986**, *101*, 545.

211. Michelin Lausarot, P.; Vaglio, G. A.; Valle, M. *J. Organomet. Chem.* **1984**, *275*, 233.

212. Moggi, P.; Albanesi, G.; Predieri, G.; Sappa, E. *J. Organomet. Chem.* **1983**, *252*, C89.

213. Moroz, B. L.; Semikolenov, V. A.; Likholobov, V. A.; Yermakov, Y. I. *J. Chem. Soc., Chem. Commun.* **1982**, 1286.

214. Moss, J. R.; Graham, W. A. G. *J. Organomet. Chem.* **1970**, *23*, C23.

215. Muetterties, E. L.; Krause, M. J. *Angew. Chem., Int. Ed. Engl.* **1983**, *22*, 135.

216. Muetterties, E. L.; Rhodin, T. N.; Band, E.; Brucker, C. F.; Pretzer, W. R. *Chem. Rev.* **1979**, *79*, 91.

217. Müller, M.; Vahrenkamp, H. *Chem. Ber.* **1983**, *116*, 2311, 2322.

218. Müller, M.; Vahrenkamp, H. *Chem. Ber.* **1984**, *117*, 1039.

219. Niyazov, A. N.; Nefedov, B. K.; Khoshdurdyev, Kh. O.; Manov-Yuvenskii, V. I. *Dokl. Akad. Nauk. SSSR.* **1981**, *258*, 1120 and references cited therein.

220. (a) Ojima, I. *J. Mol. Catal.* **1986**, *37*, 25. (b) Ojima, I.; Okabe, M.; Kato, K.; Kwon, H. B.; Horvath, I. T. *J. Am. Chem. Soc.* **1988**, *110*, 150.

221. Ollis, D. F. *J. Catal.* **1971**, *23*, 131.

222. Ozin, G. A.; Andrews, M. P. In *Studies in Surface Science and Catalysis, Vol. 29: Metal Clusters in Catalysis*; Gates, B. C.; Guczi, L.; Knözinger, H., Eds.; Elsevier Science Publishers: Amsterdam **1986**, p 265- 356.

223. Park, J. T.; Shapley, J. R.; Churchill, M. R.; Bueno, C. *J. Am. Chem. Soc.* **1983**, *105*, 6182.

224. Parkyns, N. D. In *Proceedings, 3rd International Congress on Catalysis;* Sachtler, W. H. M.; Schuit, G. C. A.; Zwietering, P., Eds.; North-Holland, Amsterdam, **1965**, p 194.

225. Pettifer, R. F. In *Studies in Surface Science and Catalysis, Vol. 29: Metal Clusters in Catalysis*; Gates, B. C.; Guczi, L.; Knözinger, H., Eds.; Elsevier Science Publishers: Amsterdam **1986**, p 231- 258.

226. Phillips, J.; Dumesic, J. A. *Appl. Catal.* **1984**, *9*, 1.

227. Piacenta, F.; Matteoli, U.; Bianchi, M.; Frediani, P.; Menchi, G. In *Proceedings, XIX Congresso Nazionale di Chimica Inorganica; VI Congresso Nazionale di Catalisi*; Santa Margherita di Pula - Cagliari; October **1986**, p 389.

228. Pierantozzi, R.; McQuade, K. J.; Gates, B. C. In *Proceedings, 7th International Congress on Catalysis*; Tokyo (Japan), Kodansha, Tokyo and Elsevier, Amsterdam, **1981**, Part B, p 941.

229. Pierantozzi, R.; McQuade, K. J.; Gates, B. C.; Wolf, M.; Knözinger, H.; Ruhmann, W. *J. Am. Chem. Soc.* **1979**, *101*, 5436.

230. Pierantozzi, R.; Valagene, E. G.; Nordquist, A. F.; Dyer, P. N. *J. Mol. Catal.* **1983**, *21*, 189.

231. Pince, R.; Queau, R.; Labroue, D. *J. Organomet. Chem.* **1986**, *306*, 251.

232. Pittman, C. U., Jr.; Honnick, W.; Absi-Halabi, M.; Richmond, M. G.; Bender, R.; Braunstein, P. *J. Mol. Catal.* **1985**, *32*, 177.

233. (a) Pittman, C. U., Jr.; Richmond, M. G.; Absi-Halabi, M.; Beurich, H.; Richter, F.; Vahrenkamp, H. *Angew. Chem., Int. Ed. Engl.* **1982**, *21*, 786. (b) Pittman, C. U., Jr.; Ryan, R. C.; Wilson, W. D.; Wileman, G.; Absi-Halabi, M. *Prepr., Div. Pet. Chem., Am. Chem. Soc.* **1980**, *25*, 714.

234. Plotkin, J. S.; Alway, D. G.; Weisenberger, C. R.; Shore, S. G. *J. Am. Chem. Soc.* **1980**, *102*, 6156.

235. Pruett, R. L. *Ann. N. Y. Acad. Sci.* **1977**, *295*, 239.

236. Pruett, R. L.; Bradley, J. S. *U.S. Patent* 4 301 086 (**1982**); *Chem. Abstr.* **1982**, *96*, 35 540; *U.S. Patent* 4 342 838 (**1982**); *Chem. Abstr.* **1982**, *97*, 184 362 (to Exxon Research and Engineering Co.).

237. Pruett, R. L.; Walther, W. E. *U.S. Patent* 3 833 634 (**1974**); *Chem. Abstr.* **1973**, *79*, 78 088 (to Union Carbide Co.).

238. (a) Psaro, R.; Ugo, R. In *Studies in Surface Science and Catalysis, Vol. 29: Metal Clusters in Catalysis*; Gates, B. C.; Guczi, L.; Knözinger, H., Eds.; Elsevier Science Publishers: Amsterdam **1986**, p 427- 496. (b) Lamb, H. H.; Gates, B. C.; Knözinger, H. *Angew. Chem., Int. Ed. Engl.* to be published.

239. Pursiainen, J.; Karjalainen, K.; Pakkanen, T. A. *J. Organomet. Chem.* **1986**, *314*, 227.

240. Pursiainen, J.; Pakkanen, T. A. *J. Chem. Soc., Chem. Commun.* **1984**, 252.

241. Pursiainen, J.; Pakkanen, T. A.; Heaton, B. T.; Seregni, C.; Goodfellow, R. W. *J. Chem. Soc., Dalton Trans.* **1986**, 681.

242. Pursiainen, J.; Pakkanen, T. A.; Jääskeläinen, J. *J. Organomet. Chem.* **1985**, *290*, 85.

243. Pursiainen, J.; Pakkanen, T. A.; Smolander, K. *J. Chem. Soc., Dalton Trans.* **1987**, 781.

244. Richmond, M. G.; Absi-Halabi, M.; Pittman, C. U., Jr. *J. Mol. Catal.* **1984**, *22*, 367.

245. Richter, F.; Beurich, H.; Vahrenkamp, H. *J. Organomet. Chem.* **1979**, *166*, C5.

246. Richter, F.; Vahrenkamp, H. *Angew. Chem., Int. Ed. Engl.* **1978**, *17*, 864.

247. Richter, F.; Vahrenkamp, H. *Angew. Chem., Int. Ed. Engl.* **1979**, *18*, 531.

248. Richter, F.; Vahrenkamp, H. *Angew. Chem., Int. Ed. Engl.* **1980**, *19*, 65.

249. Roberts, D. A.; Geoffroy, G. L. In *Comprehensive Organometallic Chemistry*; Wilkinson, G.; Stone, F. G. A.; Abel, E. W., Eds.; Pergamon Press, Oxford, **1982**, Chapter 40.

250. Robertson, J.; Webb, G. *Proc. R. Soc. London Ser. A* **1974**, *341*, 383.
251. Roland, E.; Bernhardt, W.; Vahrenkamp, H. *Chem. Ber.* **1986,** *119*, 2566.
252. Roland, E.; Vahrenkamp, H. *Angew. Chem., Int. Ed. Engl.* **1981**, *20*, 679.
253. Roland, E.; Vahrenkamp, H. *J. Mol. Catal.* **1983**, *21*, 233.
254. Roland, E.; Vahrenkamp, H. *Organometallics* **1983**, *2*, 183.
255. Roland, E.; Vahrenkamp, H. *Chem. Ber.* **1984**, *117*, 1039.
256. Rosé, J. Thèse de Doctorat d'Etat, Université Louis Pasteur, Strasbourg, **1985**.
257. Röper, M.; Schieren, M.; Fumagalli, A. *J. Mol. Catal.* **1986**, *34*, 173.
258. Ruff, J. K. *Inorg. Chem.* **1968**, *7*, 1818.
259. Ryndin, Yu. A.; Kuznetsov, B. N.; Kovalchuk, V. I.; Miner, G.; Mizel, V.; Yermakov, Yu. I. *Suported Metal Catalysts for the Conversion of Hydrocarbons*; Institute of Catalysis, Novosibirsk (USSR), **1978**, p 231.
260. Salter, I. D. *J. Organomet. Chem.* **1985**, *295*, C17.
261. Salter, I. D.; Stone, F. G. A. *J. Organomet. Chem.* **1984**, *260*, C71.
262. Sappa, E.; Lanfranchi, M.; Tiripicchio, A.; Tiripicchio Camellini, M. *J. Chem. Soc., Chem Commun.* **1981**, 995.
263. Sappa, E.; Manotti Lanfredi, A. M.; Tiripicchio, A. *J. Organomet. Chem.* **1981**, *221*, 93.
264. Sappa, E.; Tiripicchio, A.; Braunstein, P. *Coord. Chem. Rev.* **1985**, *65*, 219.
265. Sappa, E.; Valle, M.; Predieri, G.; Tiripicchio, A. *Inorg. Chim. Acta* **1984**, *88*, L23.
266. Satterfield, C. N. *Heterogeneous Catalysis in Practice*; McGraw-Hill: New-York, **1980**.
267. Schay, Z.; Guczi, L. *Acta Chim. Acad. Sci. Hung.* **1982**, *111*, 607.
268. Schay, Z.; Lazar, K.; Mink, J.; Guczi, L. *J. Catal.* **1984**, *87*, 179.
269. Schmid, G. *Nachr. Chem. Tech. Lab.* **1987**, *34*, 249.
270. Schmid, G.; Klein, N. *Angew. Chem., Int. Ed. Engl.* **1986**, *25*, 922.
271. Schneider, J.; Huttner, G. *Chem. Ber.* **1983**, *116*, 917.
272. Schrauzer, G. N.; Bastian, B. N.; Fosselius, G. A. *J. Am. Chem. Soc.* **1966**, *88*, 4890.
273. Schrauzer, G. N.; Ho, R. K. Y.; Schlesinger, G. *Tetrahedron Lett.* **1970**, 543.
274. Scott, J. P.; Budge, J. R.; Rheingold, A. ; Gates, B. C. *J. Am. Chem. Soc.*, **1987**, *109*, 7736.
275. Scrivanti, A.; Berton, A.; Toniolo, L.; Botteghi, C. *J. Organomet. Chem.* **1986**, *314,* 369.
276. Senocq, F.; Randrianalimanana, C.; Thorez, A.; Kalck, P.; Choukroun, R.; Gervais, D. *J. Chem. Soc., Chem. Commun.* **1984**, 1376.

277. Senocq, F.; Randrianalimanana, C.; Thorez, A.; Kalck, P.; Choukroun, R.; Gervais, D. *J. Mol. Catal.* **1986**, *35*, 213.

278. Shapley, J. R.; Hardwick, S. J.; Foose, D. S.; Stucky, G. D. *Prepr., Div. Pet. Chem., Am. Chem. Soc.* **1980**, *25*, 780.

279. Shapley, J. R.; Hardwick, S. J.; Foose, D. S.; Stucky, G. D.; Churchill, M. R.; Bueno, C.; Hutchinson, J. P. *J. Am. Chem. Soc.* **1981**, *103*, 7383.

280. Shore, S. G.; Hsu, W.-L.; Weisenberger, C. R.; Caste, M. L.; Churchill, M. R.; Bueno, C. *Organometallics* **1982**, *1*, 1405.

281. Shriver, D. F. *The Manipulation of Air Sensitive Compounds*; McGraw-Hill Book Company, **1969**.

282. Simpson, A. F.; Whyman, R. *J. Organomet. Chem.* **1981**, *213*, 157.

283. Sinfelt, J. H. *J. Catal.* **1973**, *29*, 308.

284. Sinfelt, J. H. *Acc. Chem. Res.* **1977**, *10*, 15, and references cited therein.

285. Sinfelt, J. H. *J. Phys. Chem.* **1986**, *90*, 4711.

286. Sinfelt, J. H. *Bimetallic Catalysts, Discoveries, Concepts and Applications*; Wiley: New York, **1983**.

287. Sinfelt, J. H.; Via, G. H. *J. Catal.* **1979**, *56*, 1, and references cited therein.

288. Smith, G. C.; Chojnacki, T. P.; Dasgupta, S. R.; Iwatate, K.; Watters, K. L. *Inorg. Chem.* **1975**, *14*, 1419.

289. (a) Spindler, F. H.; Bor, G.; Dietler, U. K.; Pino, P. *J. Organomet. Chem.* **1981**, *213*, 303. (b) Spindler, F. H. Dissertation E. T. H. (Zürich) N° 7012, **1982**. (c) Pino, P.; von Bézard, D. *Ger. Offen.* 2 807 251 (**1978**), *Chem. Abstr.* **1979**, *90*, 6235.

290. Steinhardt, P. C.; Gladfelter, W. L.; Harley, A. D.; Fox, J. R.; Geoffroy, G. L. *Inorg. Chem.* **1980**, *19*, 332.

291. Steinmetz, G. R.; Larkins, T. H. *Organometallics* **1983**, *2*, 1879.

292. Stone, F. G. A. *Angew. Chem., Int. Ed. Engl.* **1984**, *23*, 89.

293. Tanaka, K.; Watters, K. L.; Howe, R. F. *J. Catal.* **1982**, *75*, 23.

294. Teo, B. K.; Keating, K. *J. Am. Chem. Soc.* **1984**, *106*, 2224.

295. (a) Teo, B. K.; Hong, M. C.; Zhang, H.; Huang, D. B. *Angew. Chem., Int. Ed. Engl.* **1987**, *26*, 897. (b) Teo, B. K.; Keating, K.; Kao, Y.-H. *J. Am. Chem. Soc.* **1987**, *109*, 3494. (c) Tzou, M. S.; Teo, B. K.; Sachtler, W. M. H. *Langmuir* **1986**, *2*, 773, and references cited therein.

296. Thomas, T.; Mistalski, G.; Hucul, D. A.; Brenner, A. *179th National Meeting Am. Chem. Soc.*, Houston, TX, March 1980; Am. Chem. Soc.: Washington, D. C. **1980**, Abstr. COLL 11.

297. Thompson, D. T. *Platinum. Met. Rev.* **1975**, *19*, 88.

298. Tooley, P. A.; Arndt, L. W.; Darensbourg, M. Y. *J. Am. Chem. Soc.* **1985**, *107*, 2422.

138

299. Turney, T. W.; Bruce, L.; Hope, G. *React. Kinet. Catal. Lett.* **1980**, *20*, 175.

300. Uchida, Y. *Rep. Asahi Glass Found. Ind., Technol.* **1982**, *40*, 49.

301. Ungermann, C.; Landis, V.; Moya, S. A.; Cohen, H.; Walker, H.; Pearson, R. G.; Rinker, R. G.; Ford, P. C. *J. Am. Chem. Soc.* **1979**, *101*, 5922.

302. Urbancic, M. A.; Wilson, S. R., Shapley, J. R. *Inorg. Chem.* **1984**, *23*, 2954.

303. Vahrenkamp, H. *Adv. Organomet. Chem.* **1983**, *22*, 169.

304. Vanhove, D.; Makambo, L.; Blanchard, M. *J. Chem. Res., Synop.* **1980**, 335; *J. Chem. Res., Miniprint* **1980**, 4121.

305. (a) Vannice, M. A. *J. Catal.* **1975**, *37*, 449. (b) Chen, A. A.; Kaminsky, M.; Geoffroy, G. L.; Vannice, M. A. *J. Phys. Chem.* **1986**, *90*, 4810. (c) Venter, J.; Kaminsky, M.; Geoffroy, G. L.; Vannice, M. A. *J. Catal.* **1987**, *103*, 450. (d) *Ibid.* **1987**, *103*, 155.

306. Vannice, M. A.; Lam, Y.; Garten, R. L. *Hydrocarbon Synthesis*, In *ACS Symposium Series N° 178*, Am. Chem. Soc.: Washington, D. C. **1979**, p 26.

307. Venäläinen, T.; Iiskola, E.; Pursiainen, J.; Pakkanen, T. A.; Pakkanen, T. T. *J. Mol. Catal.* **1986**, *34*, 293.

308. Venäläinen, T.; Pakkanen, T. T. *J. Organomet. Chem.* **1984**, *266*, 269.

309. Vogt, W.; Koch, J.; Glaser, H. *Ger. DE* 4 186 112 (**1980**); *Chem. Abstr.* **1978**, *89*, 118 482 (to Hoechst A. G.).

310. Walker, W. E.; Brown, E. S.; Pruett, R. L. (a) *U.S. Patent* 3 878 292 (**1975**); *Chem. Abstr.* **1975**, *83*, 45 426. (b) *U.S. Patent* 3 878 290 (**1975**); *Chem. Abstr.* **1975**, *83*, 45 427. (c) *U.S. Patent* 3 878 214 (**1975**); *Chem. Abstr.* **1975**, *83*, 45 428 (to Union Carbide Co.).

311. (a) Weiss, K.; Krauss, H.-L. private communication; Weiss, K.; Guthmann, W.; Maisuls, S. *Angew. Chem.* **1988**, *100*, 268.
(b) Wenjuan, X.; Fengying, J.; Sukun, Y.; Yuanqi, Y.; Dagang, L. *Cuihua Xuebao* **1985**, *6*, 172.

312. Williams, P. D.; Curtis, M. D.; Duffy, D. N.; Butler, W. M. *Organometallics*, **1983**, *2*, 165.

313. Wolf, M.; Knözinger, H.; Tesche, B. *J. Mol. Catal.* **1984**, *25*, 273.

314. Wozniak, W. T.; Sheline, R. K. *J. Inorg. Nucl. Chem.* **1973**, *35*, 1199.

315. Yates, D. J. C.; Taylor, W. F.; Sinfelt, J. H. *J. Am. Chem. Soc.* **1964**, *86*, 2996.

316. Yawney, D. B. M.; Stone, F. G. A. *J. Chem. Soc. A.* **1969**, 502.

317. Yermakov, Yu. I. *J. Mol. Catal.* **1983**, *21*, 35.

318. Yermakov, Yu. I.; Kuznetsov, B. N.; Zakharov, V. A. *Catalysis by Supported Complexes*; Elsevier Scientific: Amsterdam, **1981**.

319. Zahraa, O. *Thèse de Doctorat d' Etat*, Université Louis Pasteur, Strasbourg, **1980**; Zahraa, O.; Maire, G. unpublished results.

320. Zwart, J.; Snel, R. *J. Mol. Catal.* **1985**, *30*, 305.

CONFORMATIONAL ANALYSIS FOR LIGANDS BOUND TO THE CHIRAL AUXILIARY [(C$_5$H$_5$)Fe(CO)(PPh$_3$)]

Brent K. Blackburn, Stephen G. Davies and Mark Whittaker

CONFORMATIONAL ANALYSIS FOR LIGANDS BOUND TO THE CHIRAL AUXILIARY [(C5H5)Fe(CO)(PPh3)] [†]

BRENT K. BLACKBURN, STEPHEN G. DAVIES   AND MARK WHITTAKER

The Dyson Perrins Laboratory, University of Oxford, South Parks Road, Oxford, OX1 3QY, UK

[†]The descriptor η^5 for the cyclopentadienyl ligand is omitted throughout for clarity.

LIST OF ABBREVIATIONS

PMO   Perturbation Molecular Orbital
LUMO   Lowest Unoccupied Molecular Orbital
HOMO   Highest Occupied Molecular Orbital
SHOMO   Second Highest Occupied Molecular Orbital
Me   methyl
Et   ethyl
^{i}Pr   *iso*-propyl
^{t}Bu   *tert*-butyl
Ph   phenyl

1. INTRODUCTION

There are many examples in organotransition metal chemistry of complexes which incorporate the moiety [(C5H5)Fe(CO)(PPh3)], because of their ease of synthesis and general stability. The first complex, reported in 1963, was the acetyl [(C5H5)Fe(CO)(PPh3)COMe].[1] Since this time syntheses of many different types of such complexes have been demonstrated, including alkyls, acyls, acetylides, vinylidenes, vinyls, alkoxyvinyls, carbenes, alkoxycarbenes, olefins, *etc.* More recently it has been reported that many reactions of ligands attached to [(C5H5)Fe(CO)(PPh3)] occur with remarkably high degrees of stereoselectivity. Such stereocontrol indicates that the auxiliary [(C5H5)Fe(CO)(PPh3)] is able to control effectively both the three dimensional space around attached ligands, and the orientation of such ligands within that space. The need to understand these effects, therefore, provides the impetus for a detailed conformational analysis of ligands attached to the auxiliary [(C5H5)Fe(CO)(PPh3)].

This review surveys our current understanding of the conformational properties of ligands attached to [(C5H5)Fe(CO)(PPh3)]. The analysis has been developed from examination of the many X-ray crystal structures of such complexes now available, molecular graphics and theoretical modelling studies and consideration of the stereocontrol observed in reactions of the various ligands.

2. THE CHIRAL AUXILIARY [(C5H5)Fe(CO)(PPh3)]

2.1. Geometry:

Complexes [(C5H5)Fe(CO)(PPh3)R] derived from complexation of an organic fragment R, *e.g.* alkyl, acyl, *etc.*, to the iron auxiliary [(C5H5)Fe(CO)(PPh3)] adopt a geometry close to octahedral.[2,3] The triphenylphosphine, carbon monoxide and organic group R occupy three adjacent sites of the pseudo-octahedral structure, with each lying approximately orthogonal to the plane formed by the other two and the metal. The remaining three coordination sites available on the iron are jointly occupied by the cyclopentadienyl ligand. The centroid of the cyclopentadienyl ligand to metal line thus subtends an angle of approximately 125° to each of the bonds between the iron and the three other ligands (Figure 1).

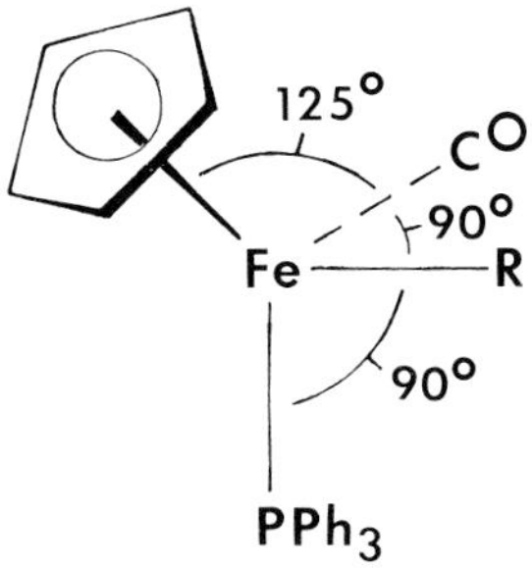

Figure 1: Idealised octahedral geometry of the complexes [(C5H5)Fe(CO)(PPh3)R].

2.2. Chirality at Iron:

Complexes of the type [(C5H5)Fe(CO)(PPh3)R] contain an iron atom bonded to four different ligands. Such complexes are, therefore, chiral since they do not possess an alternating axis of symmetry, and are consequently nonsuperimposable on their mirror images. Small scale resolutions for several complexes in this series have been achieved (R = CO2menthyl,[4] CH2Omenthyl,[5] CH2CO2menthyl,[5] and COCH2CH2Omenthyl[6]) and the absolute configuration at iron has been established.[6,7] Furthermore, the complex [(C5H5)Fe(CO)(PPh3)COMe] is currently commercially available in optically pure form as either the (+)- or (-)-enantiomer.[8]

The Cahn-Ingold-Prelog system[9] has been modified for application to transition metal complexes.[10] Ligands are assigned apparent molecular weight values defined as the sum of the atomic weights of all the contiguous atoms directly bonded to the metal (formal metal to atom X double bonds give an apparent atomic weight of double the real atomic weight of X). The Cahn-Ingold-Prelog rules are then applied as normal using these apparent atomic weights. For the complexes [(C5H5)Fe(CO)-(PPh3)R] , therefore, the order of ligand priority is cyclopentadienyl (apparent atomic weight 60), phosphorus (atomic weight 31), carbon monoxide (apparent atomic weight 24) with the organic fragment R having the lowest priority. Figure 2 shows the R and S configurations of [(C5H5)Fe(CO)(PPh3)R] as Newman projections looking from the *alpha* carbon of ligand R to the metal. For rapid assignment of configuration, when looking at this Newman projection with the iron to phosphorus bond positioned vertically downwards, the carbon monoxide ligand is on the right in the R-configuration and on the left in the S-configuration.

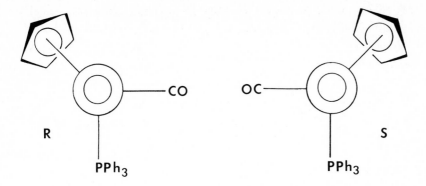

Figure 2: The R and S configurations of [(C5H5)Fe(CO)(PPh3)R] viewed as Newman projections from the *alpha* carbon to iron.

2.3. Triphenylphosphine Rotors:

The phenyl rings of triphenylphosphine adopt the structure of a rotor in order to minimise mutual steric interactions. For triphenylphosphine there are two degenerate configurations of the rotor, one with a clockwise arrangement of the phenyl groups the other with an anticlockwise arrangement (Figure 3).[11]

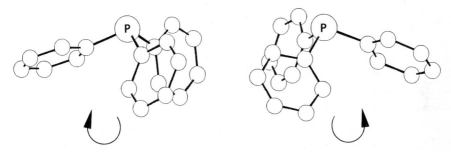

Figure 3: Rotor configurations for triphenylphosphine.

When triphenylphosphine is ligated to a chiral metal centre as in the complexes [(C5H5)Fe(CO)(PPh3)R] the two possible rotor configurations are no longer degenerate. In fact, for complexes [(C5H5)Fe(CO)(PPh3)R] the triphenylphosphine adopts the clockwise rotor configuration when attached to complexes with the R-configuration at iron, and conversely the anticlockwise rotor configuration when attached to complexes with the S-configuration at iron. Only in exceptional circumstances (*vide infra*) is the triphenylphosphine forced to adopt an anticlockwise rotor configuration while attached to an iron of R-configuration and *vice versa*. The origins of the marked general preference for one particular rotor appear to lie in steric interactions between the triphenylphosphine and the other ligands, in particular the cyclopentadienyl ligand. Furthermore, generally the most stable conformation for the triphenylphosphine in these complexes with respect to rotation about the iron phosphorus bond has the carbon monoxide ligand staggered between two phosphorus C_{ipso} bonds. Both the preferred conformation and relative configuration of the rotor are shown in Figure 4, which illustrates a computer simulation[12] of the structure of the complex [(C5H5)Fe(CO)(PPh3)CH3]. The same effects will be apparent in the X-ray crystal structures which appear later in this review.

3. CONFORMATIONAL ANALYSIS FOR [(C5H5)Fe(CO)(PPh3)CH2R]

3.1. [(C5H5)Fe(CO)(PPh3)CH3]:

For the chiral iron auxiliary [(C5H5)Fe(CO)(PPh3)] the ligand with the dominant steric effect is the triphenylphosphine.[2] The next most sterically demanding ligand is the cyclopentadienyl ligand, but its effect is tempered by virtue of it occupying three rather than one of the coordination sites. Carbon monoxide is by far the smallest ligand. For the complex [(C5H5)Fe(CO)(PPh3)CH3] the most stable conformation for the methyl group will be the one where eclipsing interactions between the methyl hydrogens and the ligands on iron are minimised. Unlike normal hydrocarbons where, due to the tetrahedral geometry at carbon, a completely staggered arrangement is possible,[13] the pseudo-octahedral nature of the chiral auxiliary [(C5H5)Fe(CO)(PPh3)] excludes perfect staggering of the methyl group. Calculation,[12] taking into account only van der Waals interactions, of the conformational energy profile for rotation about the iron to methyl carbon bond, revealed three degenerate minima. These each correspond to the most stable conformation (Figure 5) having one methyl hydrogen *anti* periplanar to the carbon monoxide ligand and, therefore, with the other two methyl hydrogens straddling the carbon monoxide ligand. The energy barriers between the degenerate minima are, as expected, small. This correlates well with the experimentally measured barrier (5.4 Kcal mol-1),[14] the theoretically estimated barriers (2.9 Kcal mol-1)[3] for [(C5H5)Fe-(CO)2CH3] and the observation that in the [1]H n.m.r. spectrum of [(C5H5)Fe(CO)(PPh3)CH3] the methyl protons are equivalent indicating essentially free rotation of the methyl group.

3.2. [(C5H5)Fe(CO)(PPh3)CH2R] (R = alkyl):
3.2.1. General Conformational Model:

The preferred conformations for CH2R (R = alkyl) attached to the chiral auxiliary will be determined by steric interactions. Given the relative sizes of the other ligands

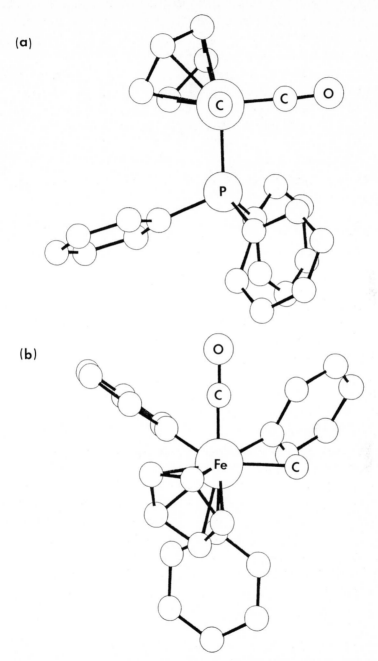

Figure 4. Molecular sructure for [(C₅H₅)Fe(CO)(PPh₃)Me]. The structure is arranged so as to view along the (a) *alpha* carbon to iron bond and (b) iron to phosphorus bond. The protons are removed for clarity.

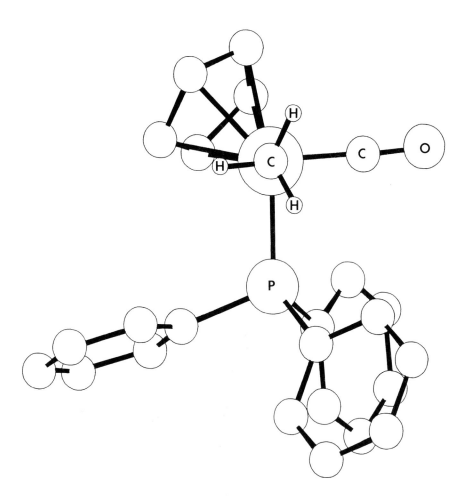

Figure 5. The theoretically determined most stable conformation for [(C5H5)Fe(CO)(PPh3)Me]. The diagram shows a Newman projection along the *alpha* carbon to iron bond showing a staggered arrangement. Selected protons are removed for clarity.

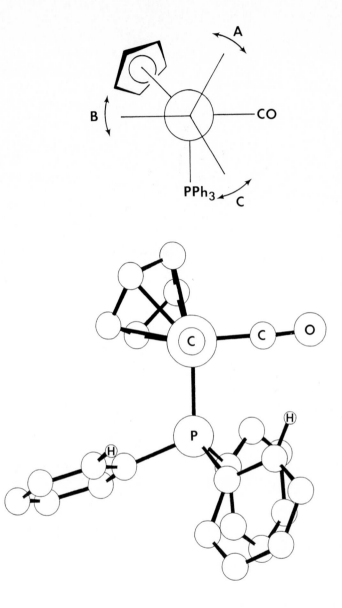

Figure 6. (a) General conformational model for $[(C_5H_5)Fe(CO)(PPh_3)CH_2R]$. (b) The diagram shows the *ortho* hydrogen atoms on the phenyl rings partially occupying zones B and C.

attached to the iron, PPh3 >> C5H5 >> CO, and a requirement to minimise eclipsing interactions then, as evident from consideration of the methyl complex above, zone A (Figure 6), between the small and medium ligands, will be the least sterically encumbered space. Zone C between the carbon monoxide and triphenylphosphine ligands will be virtually inaccessible to all but the smallest groups (*e.g.* hydrogen), particularly due to the pseudo-octahedral nature of the complexes which places these two ligands in close proximity. In addition, an *ortho* hydrogen on one of the phenyl groups partially occupies zone C. The relatively large angle between the centroid of the cyclopentadienyl to iron and phosphorus to iron bonds should make zone B, between the medium and large ligands, accessible although, as with zone C, one of the phenyl groups will tend to block partially this zone as well.

On the basis of the above elementary analysis it is predicted that for the complexes [(C5H5)Fe(CO)(PPh3)CH2R] (R = alkyl) three stable conformations will exist (Figure 7) with the order of stability being I >> II >> III, where the alkyl group occupies zones A, B and C respectively.[2]

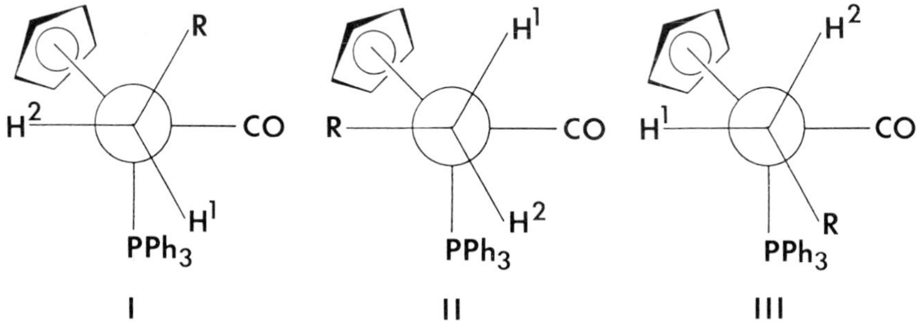

Figure 7: Stable conformations for the complexes [(C5H5)Fe(CO)(PPh3)CH2R] (R = alkyl).

Information about preferred conformations in the solid state is available from X-ray crystal structure data. In solution conformational preferences may be deduced from proton n.m.r. data.[15] In the ^{1}H n.m.r. spectrum of the complexes [(C5H5)Fe(CO)(PPh3)CH2R] a hydrogen in zone C, because it is positioned over the centre of a phenyl ring, will experience an upfield chemical shift relative to a hydrogen in zone A (see Figure 5): A hydrogen in zone B, because it is positioned close to the edge of a phenyl ring will experience a downfield chemical shift relative to a hydrogen in zone A.[16] Consistent with the Karplus relationship between torsional angle and three bond coupling constants,[17] a hydrogen in zone B will exhibit only a small $^3J_{PH}$ coupling, while hydrogens in zones A or C will show large $^3J_{PH}$ couplings. Additional confirmatory evidence of solution conformational preferences can, for certain complexes, be obtained from long range coupling constants and nuclear Overhauser effects.[18]

3.2.2. Conformer Populations:

The above criteria can be used to establish the preferred conformation in solution. They are unlikely to give, however, either any indication as to whether more than one conformation is populated or the relative energies of different conformations. Population of a single conformation for the complexes $[(C_5H_5)Fe(CO)(PPh_3)CH_2R]$ may be established by the lack of variation of $^3J_{PH}$ coupling constants with temperature. If more than one conformation were populated then significant variations in the $^3J_{PH}$ couplings would be apparent. For example, if the complex $[(C_5H_5)Fe(CO)(PPh_3)CH_2R]$ existed mainly as conformation I in equilibrium with conformation II then increasing the temperature should not affect the phosphorus to H^1 coupling, but should significantly increase the phosphorus to H^2 coupling. The same arguments may be applied to any two or more conformations in equilibrium. No significant variation in either of the $^3J_{PH}$ couplings with temperature will occur if only one conformation is significantly populated at ambient temperatures or, if two or more degenerate conformations are populated. These two possiblities may be distinquished by examination of the coupling constant values as described above, since in the latter case they would reflect average values.[15]

3.2.3. $[(C_5H_5)Fe(CO)(PPh_3)CH_2CH_3]$:

The X-ray crystal structure of the ethyl complex $[(C_5H_5)Fe(CO)(PPh_3)-CH_2CH_3]$[15] is shown in Figure 8. The pseudo-octahedral nature of the complex, and the clockwise screw of the triphenylphosphine rotor relative to the R configuration at iron are apparent and selected bond angles and torsional angles are given. The conformation adopted by the ethyl group in the solid state is as predicted by the conformational analysis, that is with the methyl group occupying zone A between the cyclopentadienyl and carbon monoxide ligands. This conformation places H^2 approximately *anti* periplanar to the carbon monoxide, while H^1 is situated in zone C directly over one of the phenyl rings of the triphenylphosphine ligand.

The complex $[(C_5H_5)Fe(CO)(PPh_3)CH_2CH_3]$ also adopts the same conformation in solution.[15] In the 300MHz ^1H n.m.r. spectrum H^2 appears at δ 1.87 with $^3J_{PH} = 2.0Hz$, whereas H^1 appears at δ 1.07 with $^3J_{PH} = 12.1Hz$. These observed coupling constants correlate with the expected dihedral angles for conformation I, and the upfield chemical shift of H^1 relative to H^2 is consistent with H^1 being deshielded by the proximate phenyl group. Furthermore, the methyl protons exhibit a long range four bond coupling of 2.1Hz to phosphorus. Such long range coupling is observed

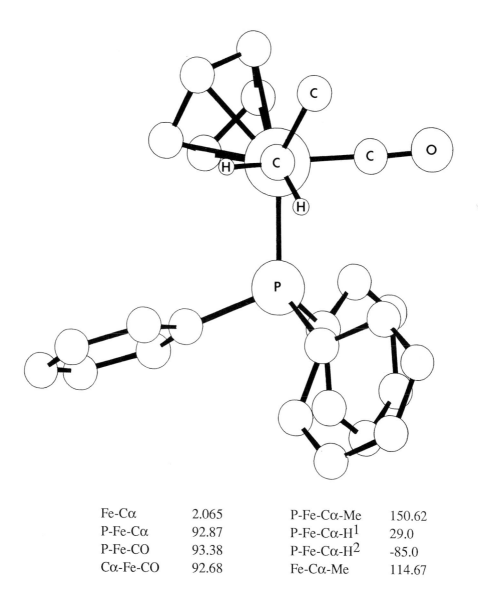

Fe-Cα	2.065	P-Fe-Cα-Me	150.62
P-Fe-Cα	92.87	P-Fe-Cα-H^1	29.0
P-Fe-CO	93.38	P-Fe-Cα-H^2	-85.0
Cα-Fe-CO	92.68	Fe-Cα-Me	114.67

Figure 8. X-ray crystal structure and selected bond lengths (Å), angles and torsional angles (º) for [(C$_5$H$_5$)Fe(CO)(PPh$_3$)CH$_2$Me]. The diagram shows the Newman projection along the *alpha* carbon to iron bond. Selected protons are removed for clarity.

153

between nuclei in a 'W' arrangement,[19] which for the complex [(C5H5)Fe(CO)-(PPh3)CH2CH3] is only achieved in conformations where the methyl is in zone A.

The magnitude and lack of variation with temperature of the $^3J_{PH}$ to H^1 and H^2 and $^4J_{PH}$ to CH3 coupling constants demonstrates that conformation I is the only conformation populated in solution at ambient temperatures for [(C5H5)Fe(CO)-(PPh3)CH2CH3].[15] The population of more than one conformation would result in significant variations of at least one of these couplings with temperature.

3.2.4. [(C5H5)Fe(CO)(PPh3)CH2R] (R = Et, iPr, tBu):

Each of the complexes [(C5H5)Fe(CO)(PPh3)CH2R] (R = Et, iPr, and tBu) also exists in the single stable conformation with the *alpha* substituent, R, in zone A. Figure 9 shows this conformation for the three complexes together with the observed chemical shifts and $^3J_{PH}$ coupling constants for the methylene protons, which in each case are only consistent with conformation I. The lack of variation of these coupling constants with temperature excludes the population of any other conformation. For [(C5H5)Fe(CO)(PPh3)CH2tBu] the observation of a negative nuclear Overhauser enhancement between H^1 and the cyclopentadienyl protons dictates that H^1, H^2 and the circle swept by the cyclopentadienyl protons must subtend an obtuse angle.[15,18] Such a situation only occurs in conformations close to I.

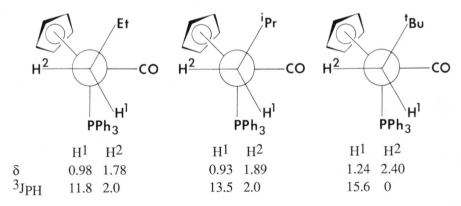

	H^1	H^2		H^1	H^2		H^1	H^2
δ	0.98	1.78		0.93	1.89		1.24	2.40
$^3J_{PH}$	11.8	2.0		13.5	2.0		15.6	0

Figure 9: Most stable conformations of [(C5H5)Fe(CO)(PPh3)CH2R] (R = Et, iPr and tBu). Chemical shift and coupling constant (Hz) data for H^1 and H^2 are shown.

3.2.5. Calculated Conformational Energy Profiles:

Using the X-ray crystal structure of [(C5H5)Fe(CO)(PPh3)CH2CH3] it is possible to derive the conformational energy profile for this complex by a number of methods. The calculated[12] conformational energy profile for rotation about the iron to *alpha*

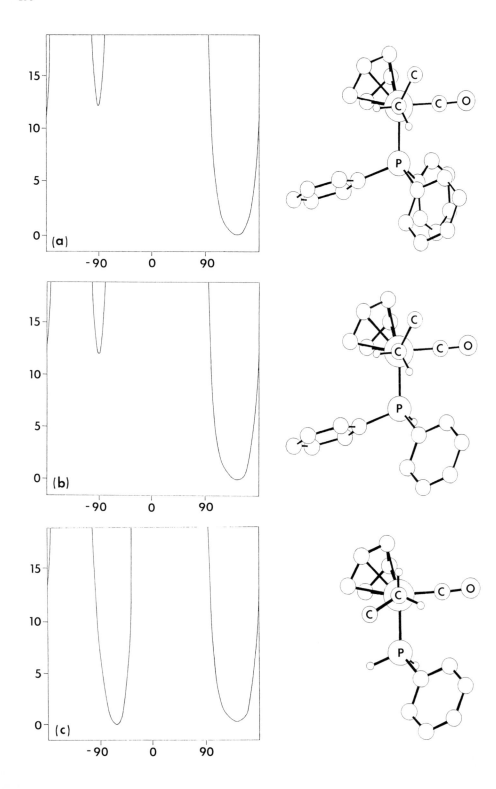

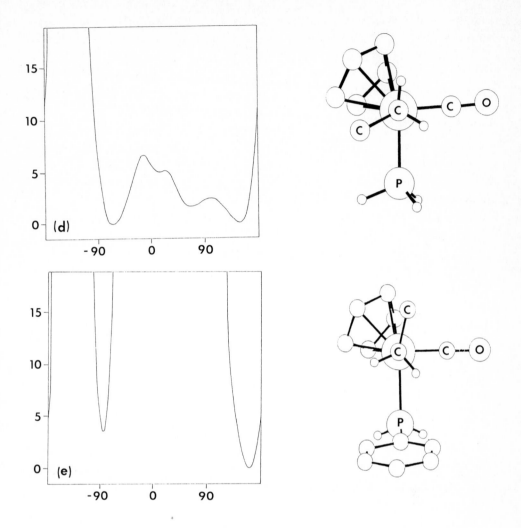

Figure 10. Conformational energy profile for (a) [(C$_5$H$_5$)Fe(CO)(PPh$_3$)Et], (b) [(C$_5$H$_5$)Fe(CO)(PHPh$_2$)Et], (c) [(C$_5$H$_5$)Fe(CO)(PH$_2$Ph)Et], (d) [(C$_5$H$_5$)Fe(CO)(PH$_3$)Me], (e) [(C$_5$H$_5$)Fe(CO)(PH$_2$Ph)Et]:[20] Energy (Kcal mol^{-1}) *vs* P-Fe-Cα-Me torsional angle ($^{\circ}$). The calculated most stable conformation is shown with each energy profile viewing along the *alpha* carbon to iron bond. Selected protons are removed for clarity.

carbon bond, derived taking into account only van der Waals interactions, is shown in Figure 10a. As predicted from the elementary analysis above the stability order for the methyl group occupying the three zones is calculated to be A >> B >> C. Essentially the same conformational energy profile is obtained when the triphenylphosphine in the complex [(C$_5$H$_5$)Fe(CO)(PPh$_3$)CH$_2$CH$_3$] is modelled by PPh$_2$H obtained by replacing, for the purposes of the calculation, the phenyl group distal (Figure 10b) to the *alpha* carbon by hydrogen. However, modelling of the triphenylphosphine by PPhH$_2$ with only the phenyl group proximate to the *alpha* carbon being retained is not satisfactory (Figure 10c), the conformations I and II in this case are calculated as being close to degenerate. Nonetheless, a reasonable model for triphenylphosphine is provided by PPhH$_2$ orientated such that the phosphorus to C$_{ipso}$ bond eclipses the iron to *alpha* carbon bond with the plane of the phenyl parallel to the plane formed by the *alpha* carbon, iron and carbon monoxide.[2,20] Using this latter model both the calculation shown here[12] (Figure 10e) and molecular orbital calculations of the extended Huckel type[2] give essentially the same energy profile, which was similar to that obtained for [(C$_5$H$_5$)Fe(CO)-(PPh$_3$)CH$_2$CH$_3$] (Figure 10a). On the other hand modelling the sterically demanding PPh$_3$ with the small PH$_3$ proved completely unsatisfactory (Figure 10d).

The above results indicate that in the complexes [(C$_5$H$_5$)Fe(CO)(PPh$_3$)CH$_2$R] one phenyl group blocks zone C while a second phenyl group effectively blocks zone B leaving zone A energetically the most accessible.

The calculated[12] conformational energy profiles for the complexes [(C$_5$H$_5$)Fe-(CO)(PPh$_3$)CH$_3$], [(C$_5$H$_5$)Fe(CO)(PPh$_3$)CH$_2$Me] and [(C$_5$H$_5$)Fe(CO)(PPh$_3$)-CH$_2^t$Bu] are also shown in Figure 11. For [(C$_5$H$_5$)Fe(CO)(PPh$_3$)CH$_3$] three degenerate minima are obtained connected by relatively low energy barriers. For the complex [(C$_5$H$_5$)Fe(CO)(PPh$_3$)CH$_2^t$Bu] only the single conformation with the t-butyl group in zone A is energetically feasible.

3.3. [(C$_5$H$_5$)Fe(CO)(PPh$_3$)CH$_2$R] (R = aryl, vinyl):

For the complexes [(C$_5$H$_5$)Fe(CO)(PPh$_3$)CH$_2$R] (R = alkyl) described in Section 3.2. above the *alpha* substituent is essentially spherically symmetrical. For the complexes where R is aryl or vinyl, however, the *alpha* substituent can be either effectively smaller than a methyl group (face on) or larger than a t-butyl group (edge on) depending on the orientation.[21] For this reason when calculating the conformational energy profile for the complexes [(C$_5$H$_5$)Fe(CO)(PPh$_3$)CH$_2$R], where R is aryl or vinyl, it is essential to allow the substituent R to relax its orientation about the *alpha* carbon to *beta* carbon bond during the energy calculation for rotation about the iron to *alpha* carbon bond.

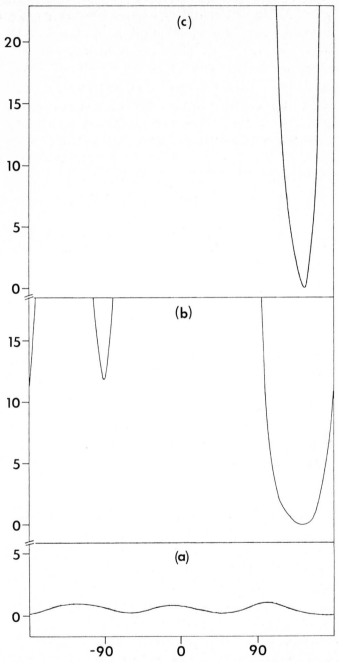

Figure 11. Conformational energy profile for (a) [(C$_5$H$_5$)Fe(CO)(PPh$_3$)Me]: Energy (Kcal mol^{-1}) *vs* P-Fe-Cα-H torsional angle (^{0}), (b) [(C$_5$H$_5$)Fe(CO)(PPh$_3$)CH$_2$Me]: Energy (Kcal mol^{-1}) *vs* P-Fe-Cα-Me torsional angle (^{0}), (c) [(C$_5$H$_5$)Fe(CO)(Ph$_3$) CH$_2$^{t}Bu]: Energy (Kcal mol^{-1}) *vs* P-Fe-Cα-C torsional angle (^{0}).

3.3.1. [(C5H5)Fe(CO)(PPh3)CH2(mesityl)]:

The calculated energy profile for [(C5H5)Fe(CO)(PPh3)CH2(mesityl)] indicates that only the conformation IV having the *alpha* carbon to *ipso* carbon bond eclipsing the centroid of the cyclopentadienyl to iron bond, with the plane of the mesityl ring face on to the cyclopentadienyl ring (Figure 12), is energetically feasible.[2] This conformation, IV, places the two bulky *ortho* methyl groups in the sterically least encumbered zones A and B. Any deviation from conformation IV towards I or II would introduce severe steric interactions between an *ortho* methyl group and either the carbon monoxide or triphenylphosphine ligands (Figure 12).

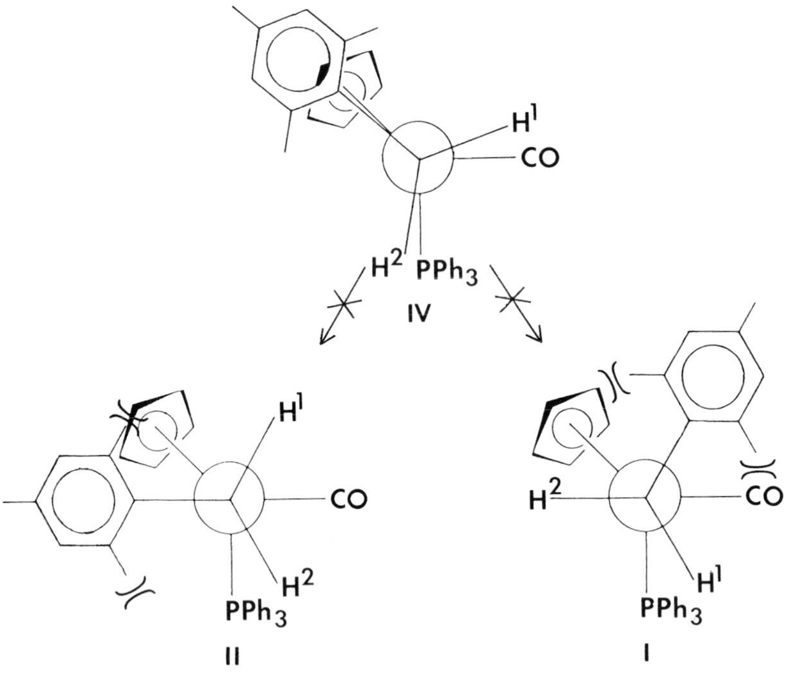

Figure 12: Most stable conformation for [(C5H5)Fe(CO)(PPh3)CH2(mesityl)].

Consistent with the above calculations, n.m.r. results show that H^1 has a small $^3J_{PH}$ (1.7 Hz) coupling constant with an upfield chemical shift (δ 1.75) relative to H^2 (δ 2.25) which shows a large $^3J_{PH}$ (12.0Hz) coupling constant.[15a] These coupling constants do not vary with temperature indicating that conformation IV is the only one populated. As expected from examination of molecular models and from energy calculations, rotation about the *alpha* carbon to *ipso* carbon bond in conformation IV is very hindered having an experimentally measured activation energy of 12.95 Kcal mol^{-1}.[15a]

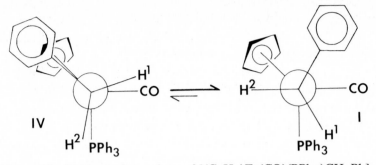

Figure 13. Stable conformations of [(C5H5)Fe(CO)(PPh3)CH2Ph].

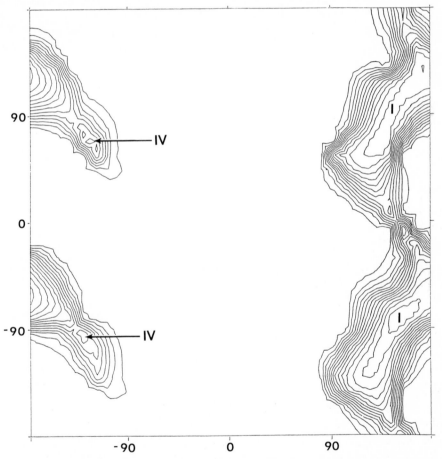

Figure 14. Conformational energy map for [(C5H5)Fe(CO)(PPh3)CH2Ph] allowing for simultaneous rotation about the iron to *alpha* carbon bond and *alpha* carbon to *ipso* carbon bond: Energy (contours separated by 5 Kcal mol^{-1}) *vs* P-Fe-Cα-C$_{ipso}$ (abscissa) and Fe-Cα-C$_{ipso}$-C$_{ortho}$ (ordinate) torsional angles (^{o}). Energy minima are indicated as conformations I and IV.

3.3.2. [(C$_5$H$_5$)Fe(CO)(PPh$_3$)CH$_2$Ph]:

The diastereotopic methylene protons H^1 and H^2 for [(C$_5$H$_5$)Fe(CO)(PPh$_3$)-CH$_2$Ph] appear at δ 2.29 and 2.79 with $^3J_{PH}$ coupling constants of 10.6 and 4.1Hz respectively.[15] In contrast to the mesityl complex above both these coupling constants vary significantly with temperature. Extrapolation of the variable temperature coupling constant data to infinite temperature gives a limiting value of *ca.* 7Hz for both $^3J_{PH}$ coupling constants. These data are consistent only with [(C$_5$H$_5$)Fe(CO)(PPh$_3$)CH$_2$Ph] existing as an equilibrium mixture of conformations I and IV (Figure 13). The limiting values for the $^3J_{PH}$ coupling constants correspond to equal populations of both conformations and allows the relative populations to be estimated as 75 : 25 (I : IV) at ambient temperature. This corresponds to an energy difference of about 0.65 Kcal mol^{-1}.

The calculated[12] conformational energy map (Figure 14) for [(C$_5$H$_5$)Fe(CO)-(PPh$_3$)CH$_2$Ph] is in good agreement with the above model derived from experiment. The planar nature of the phenyl ring allows it to eclipse, face on, the cyclopentadienyl ring without introducing any severe steric interactions.

3.3.3. [(C$_5$H$_5$)Fe(CO)(PPh$_3$)CH$_2$(vinyl)]:

The experimental results obtained for the complex [(C$_5$H$_5$)Fe(CO)(PPh$_3$)-CH$_2$(vinyl)] parallel those for [(C$_5$H$_5$)Fe(CO)(PPh$_3$)CH$_2$Ph]. The diastereotopic methylene protons H^1 and H^2 appear at δ 1.30 and δ 2.04 with $^3J_{PH}$ coupling constants of 11.2 and 3.2Hz respectively. Both these $^3J_{PH}$ coupling constants vary with temperature tending towards 7Hz at infinite temperature. These data are consistent with only conformations I and IV being populated in a ratio of approximately 75 : 25 (Figure 15).[15]

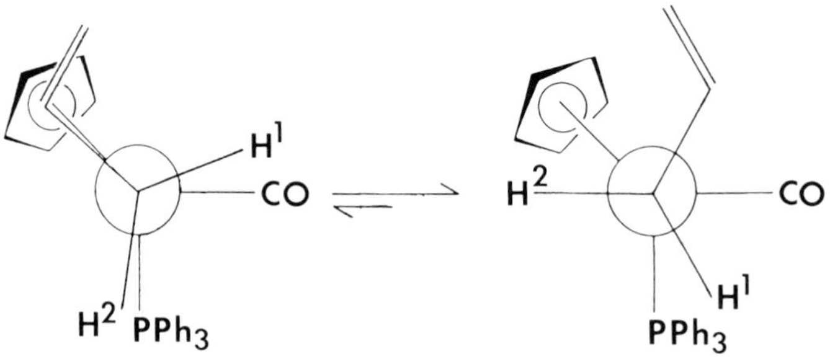

Figure 15: Stable conformations of [(C$_5$H$_5$)Fe(CO)(PPh$_3$)CH$_2$CH=CH$_2$].

3.3.4. [(C5H5)Fe(CO)(PPh3)CH2(1-naphthyl)]:

The diastereotopic methylene protons for [(C5H5)Fe(CO)(PPh3)CH2(1-naphthyl)], H[1] and H[2], appear at δ 3.21 and 2.57 with $^3J_{PH}$ coupling constants of 7.6 and 6.9Hz respectively.[15] These coupling constants varied considerably with temperature having limiting values of 12 and 4Hz at infinite temperature. These data are consistent with the two conformations I and IV being populated at ambient temperatures, but with conformation I becoming increasingly favoured at higher temperatures. Such an unusual situation arises where for two conformations ΔH^O is small but ΔS^O is large so that at higher temperatures $T\Delta S^O$ becomes dominant in determining ΔG^O ($\Delta G^O = \Delta H^O - T\Delta S^O$) and hence the equilibrium position. Molecular models and calculated[12] conformational energies indicate that on moving between conformations I and IV the naphthyl ring turns over (Figure 16) and that rotation about the *alpha* carbon to *ipso* carbon bond is severely restricted in conformation IV but relatively free in conformation I.

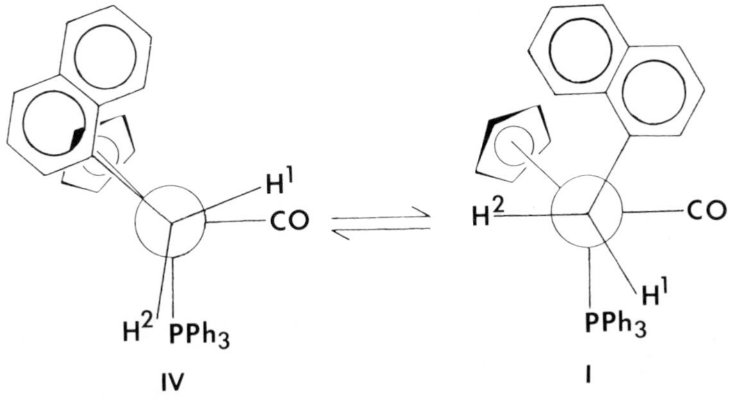

Figure 16: Stable conformations of [(C5H5)Fe(CO)(PPh3)CH2(1-naphthyl)].

4. CONFORMATIONAL ANALYSIS FOR [(C5H5)Fe(CO)(PPh3)CH2XR]

4.1. General Conformational Model:

The preferred conformations for the complexes [(C5H5)Fe(CO)(PPh3)CH2R] where R is an alkyl substituent are determined by steric factors (see Section 3 above). For the complexes [(C5H5)Fe(CO)(PPh3)CH2XR], where X is a heteroatom, then polar effects might contribute to the conformational preferences of these complexes, especially in the cases where XR is small.

The major polar contribution to the overall dipole of the auxiliary [(C5H5)Fe(CO)(PPh3)] will be due to the iron phosphorus bond being substantially polarised [Fe(-)-P(+)], since there is essentially no back-bonding to the phosphorus. Any small polarisation of the cyclopentadienyl to iron bond will align, in part, with the Fe-P dipole, whereas the iron to carbon monoxide bond is not expected to have any significant polarisation. Therefore, for the CH_2XR ligand in the complexes $[(C_5H_5)Fe(CO)(PPh_3)CH_2XR]$ (X = O,S), if polar effects are important then conformations close to V will maximise the overall dipole and those close to II will minimise it (Figure 17). The steric effect of the triphenylphosphine will always be operating to exclude conformations where the XR dips much below the plane C_α-Fe-CO. On the other hand, if steric effects dominate conformation I will be favoured (Figure 17). It should be noted that, in both conformations I and V, XR lies between the cyclopentadienyl and carbon monoxide ligands *i.e.* in zone A (Figure 6).

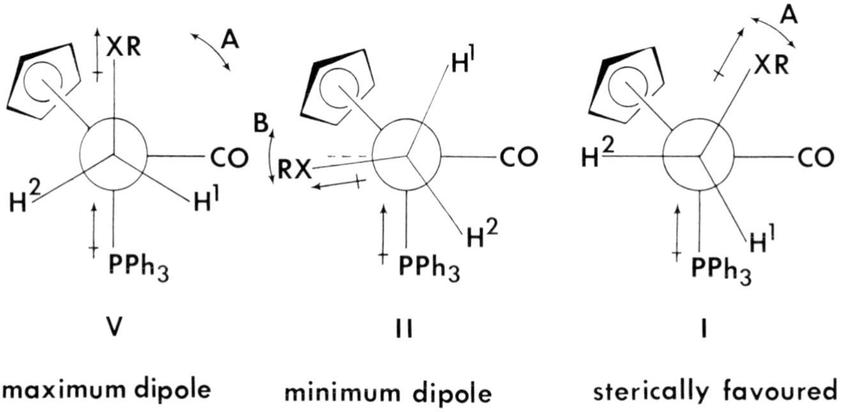

V	II	I
maximum dipole	**minimum dipole**	**sterically favoured**

Figure 17. Favoured conformations of $[(C_5H_5)Fe(CO)(PPh_3)CH_2XR]$.

Thus for very large XR substituents steric effects will dominate and conformation I will be adopted. For small polar XR substituents where steric effects will be less important then conformation V will be favoured in polar solvents, but a conformation close to II will be preferred in non-polar solvents.

4.2. $[(C_5H_5)Fe(CO)(PPh_3)CH_2XR]$ [XR = SiMe$_3$, (PMe$_3$)$^+$, (PPh$_3$)$^+$]:

For the complex $[(C_5H_5)Fe(CO)(PPh_3)CH_2SiMe_3]$ the steric bulk of the trimethylsilyl group restricts it to zone A. The carbon to silicon bond will not be significantly polarised and hence polar effects will not operate in this case. The X-ray crystal structure of $[(C_5H_5)Fe(CO)(PPh_3)CH_2SiMe_3]$ (Figure 18) shows conformation I to be adopted in the solid state.[15] Furthermore, $^3J_{PH}$ coupling constant data (Figure 19) and variable temperature n.m.r. experiments are consistent with only conformation I being populated in solution. This conclusion is supported by the observation of a negative nuclear Overhauser enhancement between the cyclopentadienyl protons and H[1], which requires H[1], H[2] and the circle swept by the

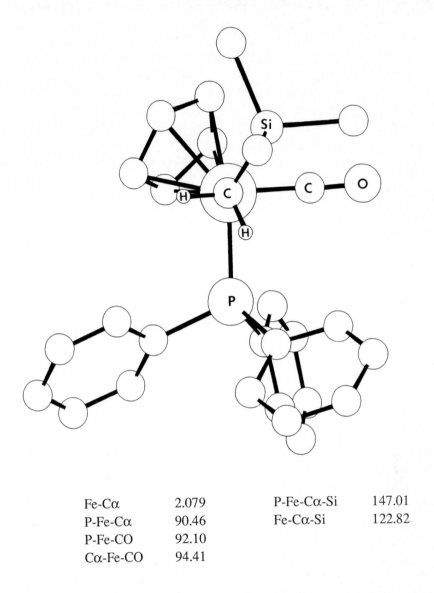

Fe-Cα	2.079	P-Fe-Cα-Si	147.01
P-Fe-Cα	90.46	Fe-Cα-Si	122.82
P-Fe-CO	92.10		
Cα-Fe-CO	94.41		

Figure 18. X-ray crystal structure and selected bond lengths (Å), angles and torsional angles (º) for [(C5H5)Fe(CO)(PPh3)CH2SiMe3]. The diagram shows the Newman projection along the *alpha* carbon to iron bond. Selected protons are removed for clarity.

cyclopentadienyl protons to subtend an obtuse angle, which is only the case in conformations close to I.[15,18,22]

For the complexes [(C5H5)Fe(CO)(PPh3)CH2XR] where XR is the very polar substituent (PMe3)+ or (PPh3)+ any polar effects, due to the positive charge on phosphorus will result in a maximum dipole in conformation II, will be completely overwhelmed by the steric interactions which restrict XR to zone A (Figure 19). Again confirmatory evidence is provided by $^3J_{PH}$ coupling constants, variable temperature n.m.r. experiments and nuclear Overhauser enhancement experiments. In addition the large $^3J_{PP}$ coupling constants of 9.7 [XR = (PMe3)+] and 16.3Hz [XR = (PPh3)+] for these complexes are also consistent with zone A being occupied.[15] The difference in magnitude of the $^3J_{PP}$ coupling constants results from the preferred conformation for XR = (PMe3)+ being close to I, but for the very large group XR = (PPh3)+ this is sterically unattainable and a conformation between I and V is adopted (Figure 19).

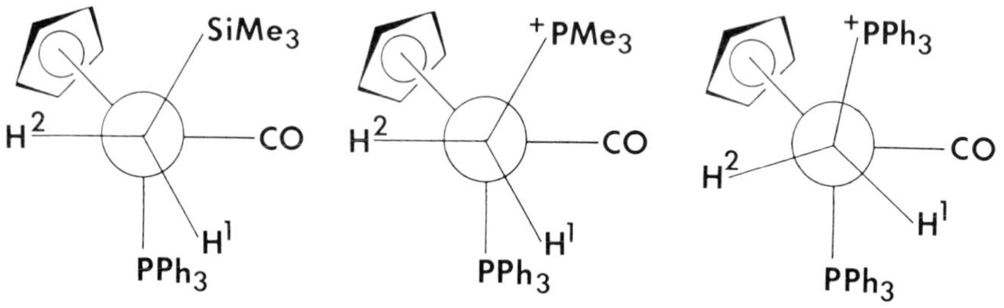

$^3J_{PH}$ H[1] 13.8 H[2] 1.9   H[1] 11.8 H[2] 0   H[1] 9.7 H[2] 1.9

Figure 19. Stable conformations of [(C5H5)Fe(CO)(PPh3)CH2XR] [XR = SiMe3, (PMe3)+, (PPh3)+].

4.3. [(C5H5)Fe(CO)(PPh3)CH2OR] (R = Me, CH2Ph, menthyl):

In the polar solvent dichloromethane the complex [(C5H5)Fe(CO)(PPh3)-CH2OMe] exclusively adopts a conformation close to V with the methoxyl group in zone A.[23] Calculations[12] indicate that, in terms of purely steric effects, conformations placing the methoxyl group in zones A or B are essentially degenerate (Figure 20). A higher energy minimum, which places the methoxyl in zone C but close to eclipsing the carbon monoxide, is also indicated by this calculation. In this latter conformation, relief of interactions between an *alpha* hydrogen in zone B and an *ortho* phenyl hydrogen are balanced by those associated with placing the relatively small methoxyl oxygen in zone C. The conformational preference, therefore, is presumably being determined by the polar solvent stabilising the conformation with the maximum dipole. In the non polar solvent toluene, where a minimisation of the dipole would be expected to be favorable, the complex [$(C_5H_5)Fe(CO)(PPh_3)CH_2OMe$] exists at ambient temperatures predominently in a conformation close to II. The conformation with the methoxyl on the edge of zone C would also minimise the dipole but can be excluded on the basis of nuclear Overhauser enhancement measurements. Conformation I, however, is also populated in toluene indicating a balance between electrostatic and steric effects. Conformation I becomes increasingly favoured at higher temperatures due to the increasingly important contribution made by entropy factors, the OMe group becoming effectively larger at higher temperatures.

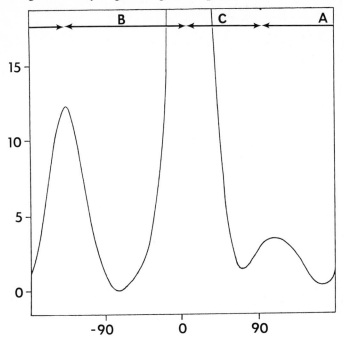

Figure 20. Conformational energy profile for [$(C_5H_5)Fe(CO)(PPh_3)CH_2OMe$]: Energy (Kcal mol^{-1}) *vs* P-Fe-Cα-O torsional angle (^{0}).

The X-ray crystal structure of [$(C_5H_5)Fe(CO)(PPh_3)CH_2OMe$ (Figure 21)[23,24] shows it to adopt a conformation close to II in the solid state consistent with the crystal lattice being a relatively non-polar medium. In contrast, the complex R-(-)-[$(C_5H_5)Fe(CO)(PPh_3)CH_2O$-(+)-menthyl] adopts conformation I (Figure 22) in which the steric interactions associated with the bulky menthyl group are minimised.[7]

The conformational properties of the complex [$(C_5H_5)Fe(CO)(PPh_3)$-CH_2OCH_2Ph] parallel those of [$(C_5H_5)Fe(CO)(PPh_3)CH_2OMe$].[23b]

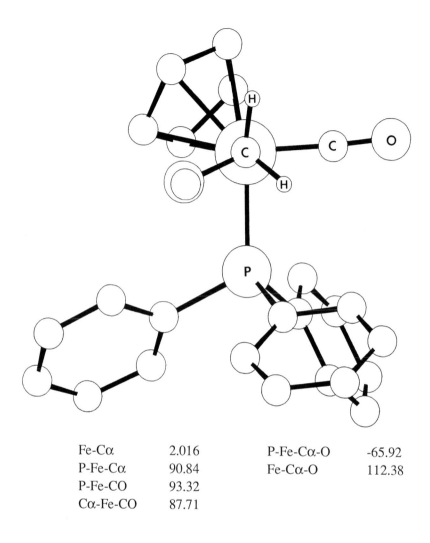

Fe-Cα	2.016	P-Fe-Cα-O	-65.92
P-Fe-Cα	90.84	Fe-Cα-O	112.38
P-Fe-CO	93.32		
Cα-Fe-CO	87.71		

Figure 21. X-ray crystal structure and selected bond lengths (Å), angles and torsional angles (°) for [(C$_5$H$_5$)Fe(CO)(PPh$_3$)CH$_2$OMe]. The diagram shows the Newman projection along the *alpha* carbon to iron bond. Selected protons are removed for clarity.

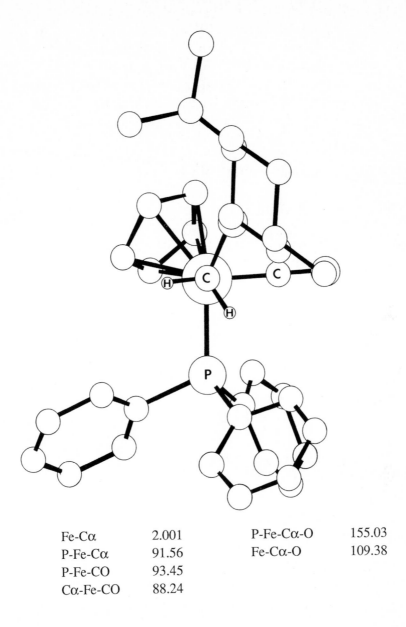

Fe-Cα	2.001	P-Fe-Cα-O	155.03
P-Fe-Cα	91.56	Fe-Cα-O	109.38
P-Fe-CO	93.45		
Cα-Fe-CO	88.24		

Figure 22. Structure and selected bond lengths (Å), angles and torsional angles (°) for R-[(C5H5)Fe(CO)(PPh3)CH2O-(+)-menthyl]. The diagram shows the Newman projection along the *alpha* carbon to iron bond. Selected protons are removed for clarity. For ease in structural comparison, the structure was derived by inverting the X-ray crystal structure of the enantiomer S-[(C5H5)Fe(CO)(PPh3)CH2O-(-)-menthyl] determined by Flood and coworkers.[7]

4.4. [(C5H5)Fe(CO)(PPh3)CH2SR (R = Me, Et, CH2Ph, Ph):

In comparison with the ether complexes above, steric effects are expected to be more important than dipolar effects for the corresponding thioether complexes due to the lower electronegativity and larger size of the sulphur atom. Thus for the complexes [(C5H5)Fe(CO)(PPh3)CH2SR] (R = Me, Et, CH2Ph, Ph) conformation I is observed to be the exclusive conformation in the polar solvent dichloromethane, and favoured over conformation II in the non-polar solvent toluene.[23b]

5. CONFORMATIONAL ANALYSIS FOR [(C5H5)Fe(CO)(PPh3)CHRR']

5.1. General Conformational Model:

For the complexes [(C5H5)Fe(CO)(PPh3)CHRR'] where R and R' are alkyl groups steric effects will determine the preferred conformation. Given that very large groups can only occupy zone A, and that zone C is extremely restricted then the conformational analysis for the complexes [(C5H5)Fe(CO)(PPh3)CHRR'] is as follows. For complexes [(C5H5)Fe(CO)(PPh3)CHRR'] where R = R' the requirement to place the small hydrogen in zone C dominates and hence the two alkyl substituents will occupy zones A and B (Figure 23a). This same requirement

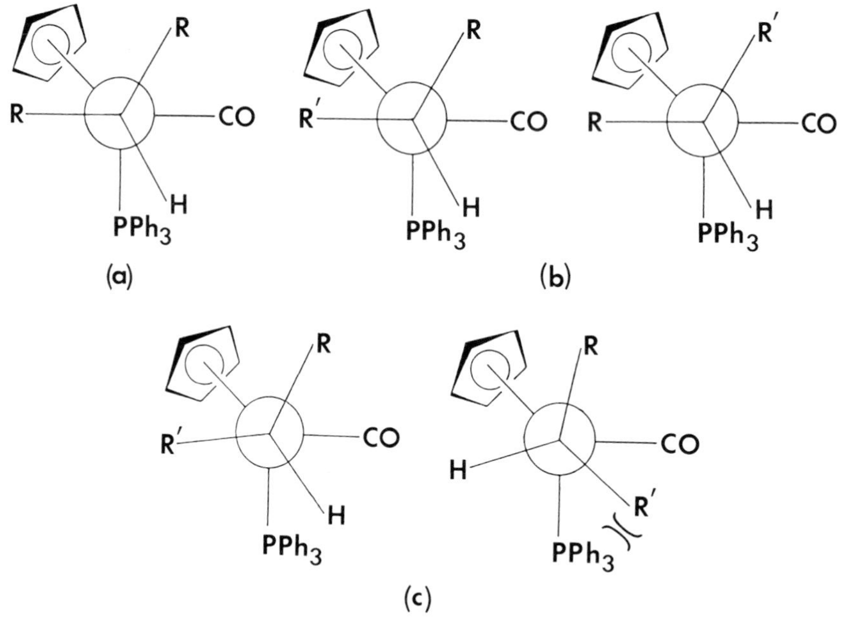

Figure 23: Stable conformations of [(C5H5)Fe(CO)(PPh3)CHRR']: (a) where R = R'; (b) where R and R' have small to medium steric bulk; (c) where R is a very large group and R' is relatively small.

will determine the conformation adopted by the complexes [(C5H5)Fe(CO)(PPh3)-CHRR'] where R and R' are different but have small to medium steric bulk. In this case, for both diastereoisomers of the complexes [(C5H5)Fe(CO)(PPh3)CHRR'], the *alpha* hydrogen will lie in zone C with R and R' in zones A and B, respectively, for one diastereoisomer but in zones B and A, respectively, for the other (Figure 23b). Furthermore, in this case the two diastereoisomers should not differ greatly in energy, the one with the larger group R or R' in zone A being somewhat preferred. However, for the complexes [(C5H5)Fe(CO)(PPh3)CHRR'] where R is a very large group and R' is relatively small then the requirement to place the very bulky group in zone A will determine the conformations. For one of the diastereoisomers this will result in R' occupying zone B, but for the other it will require R' to occupy zone C thus generating unfavourable steric interactions (Figure 23c). In this case, therefore, the former diastereoisomer will be thermodynamically much more stable than the latter. For the complexes [(C5H5)Fe(CO)(PPh3)CHRR'] where R is a small polar substituent such as methoxyl, the steric requirements of R' and the *alpha* hydrogen will overshadow any dipolar effects.

5.2. [(C5H5)Fe(CO)(PPh3)CHMe2]:

Calculations[12] indicate that the complex [(C5H5)Fe(CO)(PPh3)CHMe2] should adopt exclusively the conformation which places the two methyl groups in zones A and B (Figure 24). This prediction is confirmed by experiment: A long range $^4J_{PH}$ coupling of 1.6Hz places one methyl in zone A, while a large $^3J_{PH}$ of 9Hz for the *alpha* hydrogen places it in zone C rather than zone B.[25]

5.3. [(C5H5)Fe(CO)(PPh3)CH(OMe)Et]:

The crystal structure for RR,SS-[(C5H5)Fe(CO)(PPh3)CH(OMe)Et] is shown in Figure 25.[25] The small *alpha* hydrogen is in zone C and this places the ethyl group in zone B and the methoxyl group in zone A. The $^3J_{PH}$ of 4.3Hz for the *alpha* hydrogen is consistent with this being the solution conformation as well, in which the electronegative oxygen atom is *trans* to the phosphorus resulting in a lower $^3J_{PH}$ coupling than for the corresponding alkyl group.[25,26] The conformation adopted by the ethyl and methoxyl fragments (Figure 25) may also be understood in terms of steric effects. The ethyl group in the encumbered zone B is forced to lie with its large methyl *anti* periplanar to the iron. The most favourable orientation of the methoxyl group is with its methyl *anti* periplanar to the methylene carbon thus avoiding interactions between the two methyl groups.

The crystal structure for RS,SR-[(C5H5)Fe(CO)(PPh3)CH(OMe)Et] is shown in Figure 26.[26] Again the small *alpha* hydrogen is in zone C which places the ethyl group in zone A and the methoxyl group in zone B. A large $^3J_{PH}$ coupling constant

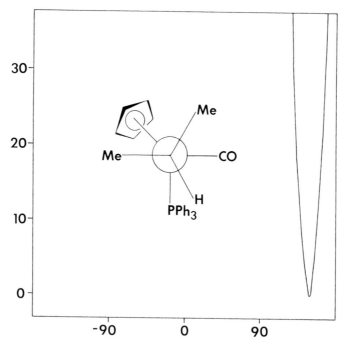

Figure 24. Conformational energy profile for $[(C_5H_5)Fe(CO)(PPh_3)CHMe_2]$: Energy (Kcal mol^{-1}) vs P-Fe-Cα-Cβ torsional angle ($^\circ$).

(10.0Hz) for the *alpha* hydrogen is consistent with this conformation also being adopted in solution. In this case, there is no *trans* electronegative atom effect reducing the coupling constant.[27] The orientations of the ethyl and methoxyl groups can be rationalised as before. Indeed the structures for the two diastereoisomers of $[(C_5H_5)Fe(CO)(PPh_3)CH(OMe)Et]$ are virtually superimposable except for the interchange of the oxygen atom and methylene group.

5.4 $[(C_5H_5)Fe(CO)(PPh_3)CHCMe_2CH_2CH_2O]$:

The crystal structures for the two diastereoisomers of $[(C_5H_5)Fe-(CO)(PPh_3)CHCMe_2CH_2CH_2O]$ are illustrated in Figures 27 and 28.[28] In both cases the conformation is determined by the restriction of the very large quaternary carbon substituent to zone A. This restriction places the oxygen atom and *alpha* hydrogen in zones B and C respectively for the RS,SR-diastereoisomer (Figure 27), but in the reversed positions for the RR,SS-diastereoisomer (Figure 28). Consistent with these conformations the *alpha* hydrogen for the RS,SR-diastereoisomer appears at higher field with a larger $^3J_{PH}$ coupling constant (17.7Hz) than the *alpha*

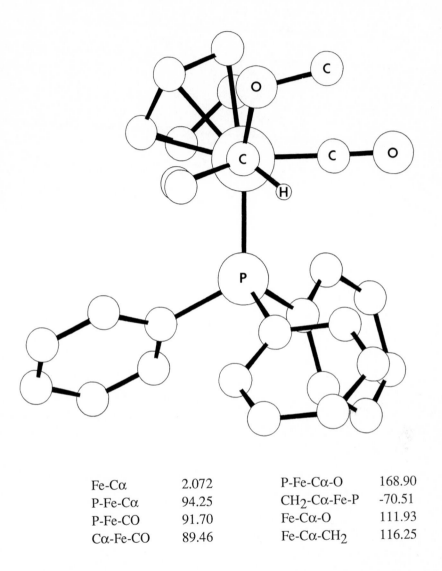

Fe-Cα	2.072	P-Fe-Cα-O	168.90
P-Fe-Cα	94.25	CH₂-Cα-Fe-P	-70.51
P-Fe-CO	91.70	Fe-Cα-O	111.93
Cα-Fe-CO	89.46	Fe-Cα-CH₂	116.25

Figure 25. X-ray crystal structure and selected bond lengths (Å), angles and torsional angles (º) for RR,SS-[(C₅H₅)Fe(CO)(PPh₃)CH(OMe)Et]. The diagram shows the Newman projection along the *alpha* carbon to iron bond. Selected protons are removed for clarity.

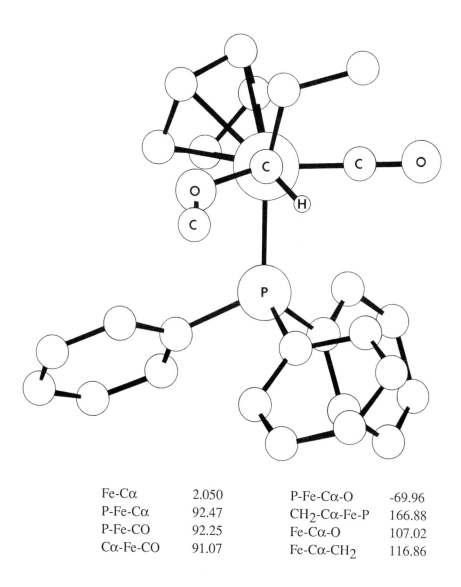

Fe-Cα	2.050	P-Fe-Cα-O	-69.96
P-Fe-Cα	92.47	CH₂-Cα-Fe-P	166.88
P-Fe-CO	92.25	Fe-Cα-O	107.02
Cα-Fe-CO	91.07	Fe-Cα-CH₂	116.86

Figure 26. X-ray crystal structure and selected bond lengths (Å), angles and torsional angles (º) for RS,SR-[(C₅H₅)Fe(CO)(PPh₃)CH(OMe)Et]. The diagram shows the Newman projection along the *alpha* carbon to iron bond. Selected protons are removed for clarity.

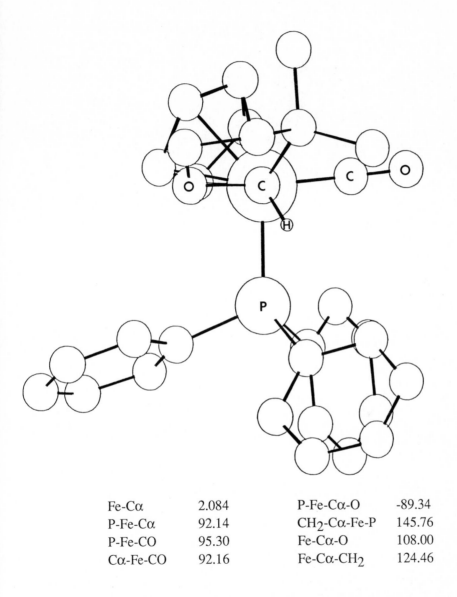

Fe-Cα	2.084	P-Fe-Cα-O	-89.34
P-Fe-Cα	92.14	CH$_2$-Cα-Fe-P	145.76
P-Fe-CO	95.30	Fe-Cα-O	108.00
Cα-Fe-CO	92.16	Fe-Cα-CH$_2$	124.46

Figure 27. X-ray crystal structure and selected bond lengths (Å), angles and torsional angles (º) for RS,SR-[(C$_5$H$_5$)Fe(CO)(PPh$_3$)CHCMe$_2$CH$_2$CH$_2$O]. The diagram shows the Newman projection along the *alpha* carbon to iron bond. Selected protons are removed for clarity.

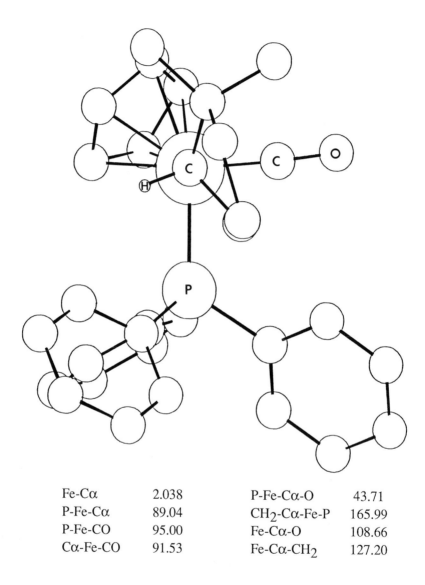

Fe-Cα	2.038	P-Fe-Cα-O	43.71
P-Fe-Cα	89.04	CH₂-Cα-Fe-P	165.99
P-Fe-CO	95.00	Fe-Cα-O	108.66
Cα-Fe-CO	91.53	Fe-Cα-CH₂	127.20

Figure 28. X-ray crystal structure and selected bond lengths (Å), angles and torsional angles (°) for RR,SS-[(C₅H₅)Fe(CO)(PPh₃)$\overline{\text{CHCMe}_2\text{CH}_2\text{CH}_2\text{O}}$]. The diagram shows the Newman projection along the *alpha* carbon to iron bond. Selected protons are removed for clarity.

hydrogen of the RS,SR-diastereoisomer which exhibits only a small $^3J_{PH}$ coupling (5.5Hz).[28]

Figure 29 shows the Newman projections along the iron to phosphorus bond for the two diastereoisomers of [(C$_5$H$_5$)Fe(CO)(PPh$_3$)CHCMe$_2$CH$_2$CH$_2$O]. The structure of the auxiliary [(C$_5$H$_5$)Fe(CO)(PPh$_3$)] in the RS,SR-diastereoisomer is as expected from all other known structures of derived complexes: For the R configuration at iron the triphenylphosphine rotor is clockwise with one phenyl staggered between the carbon monoxide and *alpha* carbon (Figure 29a). For the RR,SS-diastereoisomer, however, the effect of the relatively large oxygen substituent occupying zone C is to drastically change the shape of the auxiliary [(C$_5$H$_5$)Fe(CO)(PPh$_3$)]. For the R configuration at iron the rotor is inverted to anticlockwise, a phenyl is staggered with the carbon monoxide ligand and the cyclopentadienyl is rotated (Figure 29b). These changes have the effect of increasing the size of zone C to accommodate the substituent while decreasing the size of zone B. It is not surprising, therefore, that the RR,SS-diastereoisomer is very much less stable than the RS,SR-diastereoisomer (see Section 9.2).

6. STEREOELECTRONIC EFFECTS.

Calculations of the extended Huckel type[3,29] indicate that for the simplified fragment [(C$_5$H$_5$)Fe(CO)(PH$_3$)] the HOMO lies parallel to the iron phosphorus bond (Figure 30).

If stereoelectronic effects make a significant contribution to the preferred conformation of ligands attached to the auxiliary [(C$_5$H$_5$)Fe(CO)(PPh$_3$)] then it is to be expected that the LUMO on the atom attached to the iron will align preferentially with the HOMO on the iron in order to maximise overlap.[29] If such stereoelectronic effects are important they should manifest themselves in complexes bearing an oxygen substituent on the *alpha* carbon by analogy with the anomeric effect.[30] Stereoelectronic effects would favour, therefore, conformations with the *alpha* oxygen *anti* periplanar to the phosphorus (*syn* periplanar being excluded on steric grounds). Furthermore, in this *anti* periplanar conformation the C-O bond would be longer and the Fe-C bond shorter than the corresponding bonds in the orthogonal conformation where the overlap is minimised. Figure 31 lists all known bond lengths for *alpha* alkoxy alkyl substituents attached to [(C$_5$H$_5$)Fe(CO)(PPh$_3$)]. Examination of these data leads to the conclusion that stereoelectronic effects do not play a significant role in determining conformations. For example, in the two diastereoisomers of [(C$_5$H$_5$)Fe(CO)(PPh$_3$)CH(OMe)Et] the Cα-O bond length (1.437A) is shorter and the Fe-Cα bond length (2.072A) longer in the RS,SR diastereoisomer with the methoxyl *anti* to the phosphorus, than the corresponding bond lengths (1.442 and 2.050A) in the RR,SS diastereoisomer.[25,26] Indeed

(a)

(b)

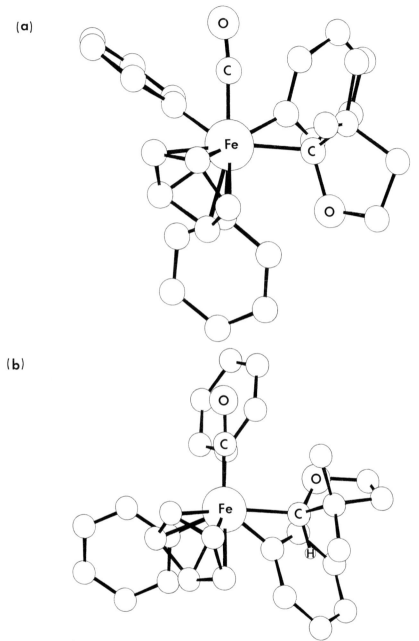

Figure 29. X-ray crystal structures of (a) RS,SR-[(C5H5)Fe(CO)(PPh3) – $\overline{\text{CHCMe2CH2CH2O}}$] and (b) RR,SS-[(C5H5)Fe(CO)(PPh3)$\overline{\text{CHCMe2CH2CH2O}}$]. Each diagram shows the Newman projection along the iron to phosphorus bond. Selected protons are removed for clarity.

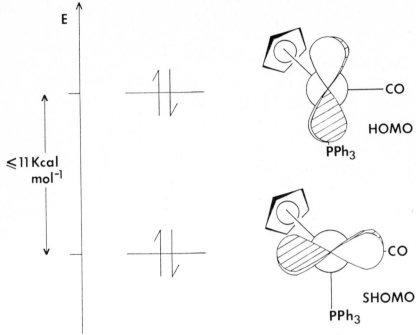

Figure 30: HOMO and SHOMO for [(C$_5$H$_5$)Fe(CO)(PPh$_3$)].

this may indicate that stereoelectronic effects may arise from both the SHOMO and the HOMO on the iron thus resulting in no overall stereoelectronically determined conformational preference. Consistent with this conclusion is the fact that the complex [(C$_5$H$_5$)Fe(CO)(PPh$_3$)CH$_2$OMe] adopts a conformation that places the relatively small methoxyl group *anti* to the carbon monoxide rather than *anti* to the phosphorus (Figure 21).[24]

7. CONFORMATIONAL ANALYSIS FOR [(C$_5$H$_5$)Fe(CO)(PPh$_3$)C(OMe)R]+

7.1. General Conformational Model:

Carbene ligands attached to the auxiliary [(C$_5$H$_5$)Fe(CO)(PPh$_3$)] will be orientated by stereoelectronic effects to lie orthogonal to the iron phosphorus bond to allow maximum overlap between the HOMO on the iron and the empty *p* orbital on the carbene carbon.[29] Such an arrangement, however, eclipses one of the groups on the carbene carbon with the carbon monoxide ligand and steric interactions will work against any stereoelectronic preference. Figure 32 shows the X-ray crystal structure for the symmetrical carbene cationic complex [(C$_5$H$_5$)Fe(CO)(PPh$_3$)CF$_2$]+.[31] For this complex steric interactions between the auxiliary and the fluorine atoms will be small and it adopts a conformation close to orthogonal to the iron-phosphorus bond,

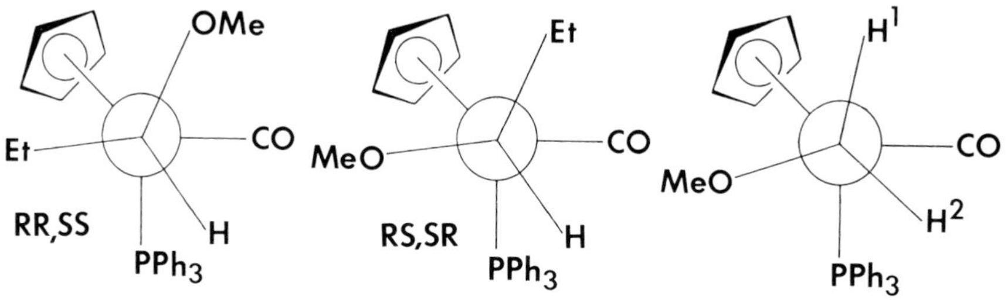

P-Fe-Cα-O | 168.9 | -70.0 | -65.9
Cα-O | 1.437 | 1.442 | 1.434
Fe-Cα | 2.072 | 2.050 | 2.016

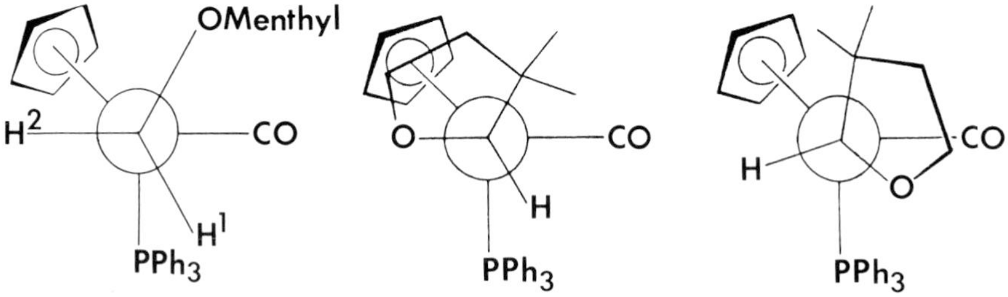

P-Fe-Cα-O | 150.0 | -89.3 | 43.7
Cα-O | 1.355 | 1.461 | 1.449
Fe-Cα | 2.000 | 2.084 | 2.038

Figure 31: Fe-Cα and Cα-O bond lengths (Å) for [(C5H5)Fe(CO)(PPh3)CH(OR)R'].

but rotated slightly to relieve eclipsing interactions. Calculation[12] of the conformational energy profile for the cation [(C5H5)Fe(CO)(PPh3)CF2]+ only taking into account steric effects predicts the solid state conformation to be the sterically most favourable (Figure 33a). There is essentially free rotation[31] about the iron-carbon bond and, therefore, steric and any stereoelectronic effects must be small.

7.2. [(C5H5)Fe(CO)(PPh3)C(OMe)H]+:

The X-ray crystal structure of the complex [(C5H5)Fe(CO)(PPh3)C(OMe)H]+ is shown in Figure 34.[26] The conformation adopted by the methoxycarbene ligand is essentially superimposable on the conformation adopted by the difluorocarbene ligand in [(C5H5)Fe(CO)(PPh3)CF2]+ (Figure 32). Calculation[12] of the conformational energy profile for [(C5H5)Fe(CO)(PPh3)C(OMe)H]+ indicates a small preference on steric grounds for this conformation (Figure 33b). The methoxycarbene complex is orientated with the methoxyl group *anti* to the cyclopentadienyl ligand for steric reasons.

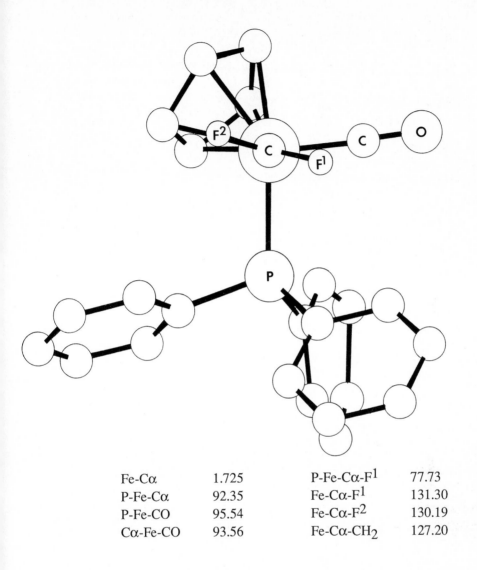

Fe-Cα	1.725	P-Fe-Cα-F^1	77.73
P-Fe-Cα	92.35	Fe-Cα-F^1	131.30
P-Fe-CO	95.54	Fe-Cα-F^2	130.19
Cα-Fe-CO	93.56	Fe-Cα-CH$_2$	127.20

Figure 32. X-ray crystal structure and selected bond lengths (Å), angles and torsional angles (0) for [(C5H5)Fe(CO)(PPh3)CF2]+. The diagram shows the Newman projection along the *alpha* carbon to iron bond. Selected protons are removed for clarity.

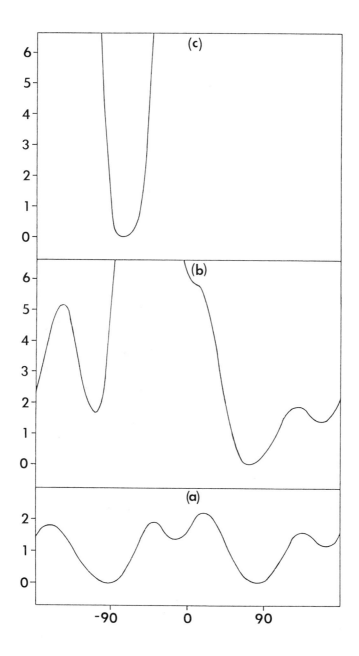

Figure 33. Conformational energy profile for (a) $[(C_5H_5)Fe(CO)(PPh_3)CF_2]^+$:
Energy (Kcal mol^{-1}) *vs* P-Fe-Cα-F^1 torsional angle (^{0}), (b) $[(C_5H_5)Fe(CO)(PPh_3)$
CHOMe]$^+$: Energy (Kcal mol^{-1}) *vs* P-Fe-Cα-O torsional angle (^{0}), (c) $[(C_5H_5)Fe$
$(CO)(PPh_3)C(OMe)Et]^+$: Energy (Kcal mol^{-1}) *vs* P-Fe-Cα-O torsional angle (^{0}).

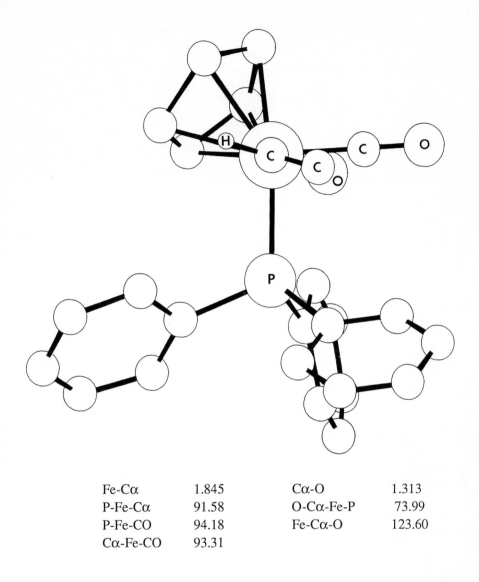

Fe-Cα	1.845	Cα-O	1.313
P-Fe-Cα	91.58	O-Cα-Fe-P	73.99
P-Fe-CO	94.18	Fe-Cα-O	123.60
Cα-Fe-CO	93.31		

Figure 34. X-ray crystal structure and selected bond lengths (Å), angles and torsional angles (o) for [(C5H5)Fe(CO)(PPh3)CHOMe]+. The diagram shows the Newman projection along the *alpha* carbon to iron bond. Selected protons are removed for clarity.

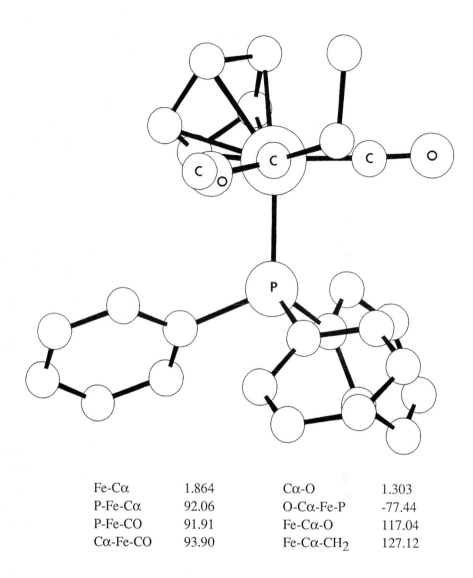

Fe-Cα	1.864	Cα-O	1.303
P-Fe-Cα	92.06	O-Cα-Fe-P	-77.44
P-Fe-CO	91.91	Fe-Cα-O	117.04
Cα-Fe-CO	93.90	Fe-Cα-CH₂	127.12

Figure 35. X-ray crystal structure and selected bond lengths (Å), angles and torsional angles (°) for [(C₅H₅)Fe(CO)(PPh₃)C(OMe)Et]⁺. The diagram shows the Newman projection along the *alpha* carbon to iron bond. Selected protons are removed for clarity.

7.3. [(C5H5)Fe(CO)(PPh3)C(OMe)R]+ (R = Me, Et):

In contrast to the complex [(C5H5)Fe(CO)(PPh3)C(OMe)H]+ described above where steric interactions between the carbene ligand and the iron auxiliary are relatively small, steric effects are dominant in determining the conformation adopted for the complexes [(C5H5)Fe(CO)(PPh3)C(OMe)R]+ (R = alkyl). The calculated[12] conformational energy profile for the complex [(C5H5)Fe(CO)(PPh3)C(OMe)Et]+ is shown in Figure 33c. The minimum energy conformation is the one which places the relatively large alkyl substituent in zone A between the cyclopentadienyl and carbon monoxide ligands. The X-ray crystal structure of the cationic complex [(C5H5)Fe(CO)(PPh3)C(OMe)Et]+ is illustrated in Figure 35.[25,26] In the solid state the complex adopts the predicted minimum energy conformation. That this conformation is also preferred in solution is consistent with its observed stereoselective reduction (see Section 9.2).[32]

A comparison between the conformational analyses for the complex [(C5H5)Fe(CO)(PPh3)C(OMe)H]+ and the corresponding alkyl derivatives [(C5H5)Fe(CO)(PPh3)C(OMe)R]+ reveals that the dominant steric influence is provided by the triphenylphosphine ligand. For carbenes attached to the iron auxiliary the conformation with the carbene orthogonal to the iron phosphorus bond minimises interaction with the triphenylphosphine but maximises eclipsing interactions with the carbon monoxide ligand (Figure 36). Stereoelectronic effects

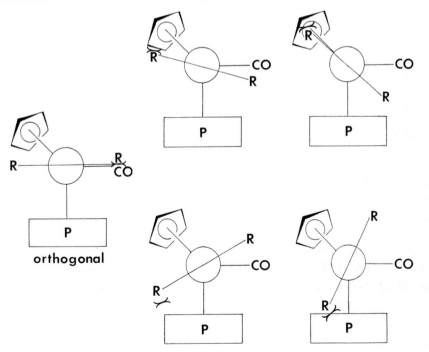

Figure 36: Steric interactions for conformations of [(C5H5)Fe(CO)(PPh3)=CR2]+.

will also favour a conformation close to orthogonal. To diminish these eclipsing interactions the carbene may rotate clockwise to move the eclipsing substituent into zone C or anticlockwise to move it into zone A. A small clockwise rotation is possible since the increasing interactions in zone C are initially compensated for by relief of interactions between the substituent in zone B and an *ortho* hydrogen on the triphenylphosphine. Severe interactions with the triphenylphosphine and the cyclopentadienyl ligand rapidly build up, however, as the angle is increased. For small anticlockwise rotations moderate interactions with an *ortho* hydrogen on the triphenylphosphine in zone B become increasingly important but not severe until relatively large angles, while the eclipsing substituent may freely enter zone A. Hence when both substituents on the carbene are small (e.g. H, F, O, *etc.*) a small clockwise rotation minimises all interactions, but when one substituent is bulky (e.g. alkyl) clockwise rotation cannot relieve the interactions and the preferred conformation places the bulky substituent in zone A.

8. CONFORMATIONAL ANALYSIS FOR [(C5H5)Fe(CO)(PPh3)COR]

8.1. General Conformational Model:

The steric and stereoelectronic effects controlling the preferred conformation of an acyl ligand attached to the iron auxiliary in the complexes [(C5H5)Fe(CO)-(PPh3)COR] will be essentially the same as those operating in the corresponding alkoxy carbene complexes described above. Any dipolar contributions will, however, be opposite since in the acyl complexes the acyl oxygen will bear a partial negative charge, while in the alkoxy carbenes the oxygen bears a partial positive charge (Figure 37). Consistent with this polarisation, the infra-red stretching frequencies of iron acyl carbonyls are close to 1600cm-1.[33] For the formyl complex [(C5H5)Fe(CO)(PPh3)CHO], whose solid state and solution structures are presently unknown, the preferred conformation is expected to resemble that of the carbene complex [(C5H5)Fe(CO)(PPh3)C(OMe)H]+, that is, close to orthogonal with respect to the iron phosphorus bond in the solid state and in nonpolar solvents. The orthogonal conformation minimises the dipolar effect and maximises the stereo-electronic contribution. In polar solvents, however, the conformation which places the formyl oxygen *anti* to the phosphorus will maximise the dipole and thus become more important.

Figure 37: Resonance structures for iron acyl and carbene ligands.

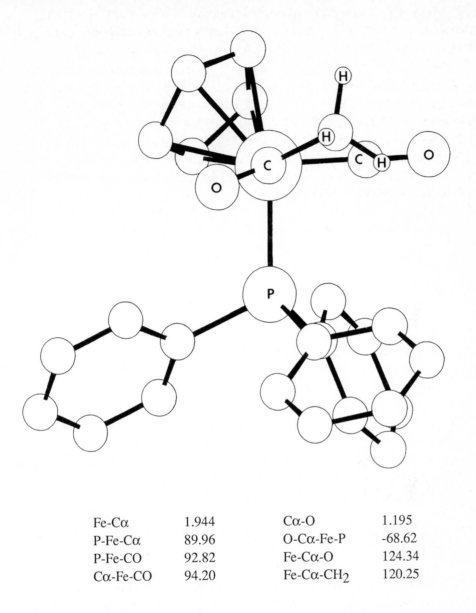

Fe-Cα	1.944	Cα-O	1.195
P-Fe-Cα	89.96	O-Cα-Fe-P	-68.62
P-Fe-CO	92.82	Fe-Cα-O	124.34
Cα-Fe-CO	94.20	Fe-Cα-CH$_2$	120.25

Figure 38. X-ray crystal structure and selected bond lengths (Å), angles and torsional angles (°) for [(C$_5$H$_5$)Fe(CO)(PPh$_3$)COMe]. The diagram shows the Newman projection along the *alpha* carbon to iron bond. Selected protons are removed for clarity.

8.2. [(C5H5)Fe(CO)(PPh3)COMe]:

The X-ray crystal structure for the acetyl complex [(C5H5)Fe(CO)(PPh3)COMe] is shown in Figure 38.[34] The preferred conformation adopted by the acetyl ligand is close to that preferred by the carbene ligand in [(C5H5)Fe(CO)(PPh3)C(OMe)Et]+ (Figure 35), that is, close to orthogonal with the relatively large methyl group in zone A. Calculations on the acetyl complex [(C5H5)Fe(CO)(PPh3)COMe)] itself,[12] and on models (extended Hückel,[2] and *ab initio* SCF MO[35]) indicate that steric interactions with one phenyl ring of the triphenylphosphine are mainly responsible for the orientation approximately orthogonal to the iron phosphorus bond, while interactions with the *ortho* hydrogen of a second proximate phenyl make the conformation with the acetyl oxygen *anti* to the carbon monoxide significantly more stable than the corresponding *syn* complex. The calculated[12] conformational energy profile for the acetyl complex [(C5H5)Fe(CO)(PPh3)COMe] is shown in Figure 39.

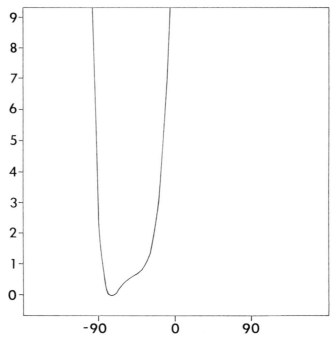

Figure 39: Conformational energy profile for [(C5H5)Fe(CO)(PPh3)COMe]: Energy (Kcal mol^{-1}) vs P-Fe-Cα-O torsional angle (^{o}).

8.3. [(C5H5)Fe(CO)(PPh3)COR (R = alkyl):

X-Ray crystal structures for many acyl complexes of the type [(C5H5)Fe(CO)(PPh3)COR] where R is an alkyl group are known.[6,34,36-42] The preferred conformation always closely resembles that adopted by the acetyl complex [(C5H5)Fe(CO)(PPh3)COMe] with the alkyl group of the acyl occupying zone A but as close to orthogonal as sterically feasible to minimise the acyl oxygen *ortho* hydrogen interactions and maximise the stereoelectronic effects. For example, Figure 40 shows the solid state conformations for the complexes [(C$_5$H$_5$)Fe(CO)-(PPh$_3$)COR] (R = CH$_2$CH$_2$C(OH)(Me)Et,[36] CH$_2$CH$_2$CH(OH)Me,[38] CH$_2$CH-(OH)Et,[39] CH(Me)CH(OCOPh)CH$_2$CH$_2$CH=CH$_2$[42]).

8.4. [(C$_5$H$_5$)Fe(CO)(PPh$_3$)COR] (R = vinyl):

For the iron acetyl complex [(C$_5$H$_5$)Fe(CO)(PPh$_3$)COMe] steric and stereoelectronic factors are responsible for the acyl ligand preferring to lie roughly orthogonal to the iron phosphorus bond. The interactions, in zone B, of the triphenylphosphine with the hydrogens on the *sp^3* methyl carbon that would exist for the *syn* (acyl oxygen to carbon monoxide) conformer determine the orientation as *anti*. Hence in the acryloyl complex [(C$_5$H$_5$)Fe(CO)(PPh$_3$)COCH=CH$_2$], where all carbons in the ligand are *sp^2* hybridised, it may be expected that the *syn* and *anti* conformations, still both favoured on steric and stereoelectronic grounds, will be of similar energy. The calculated[12] conformational energy map, allowing only for van der Waals interactions, for bond rotations about the iron to *alpha* carbon and *alpha* carbon to vinyl group is shown in Figure 41. The *syn* and *anti* acyl oxygen to carbon monoxide conformations are indeed close in energy with the *anti* being slightly favoured. Important features to note about the conformational energy map are that in the *anti* conformation any orientation of the vinyl group within 130° of S-*cis*, in either direction, is sterically feasible whereas in the *syn* conformation the vinyl group, although still S-*cis*, cannot lie coplanar with the acyl carbonyl. In the S-*cis* coplanar arrangement for the *syn* conformation a vinyl hydrogen is interacting with the cyclopentadienyl ligand. Stereoelectronic effects reflecting the stabilisation obtained from orbital overlap in arrangements close to coplanar and entropy factors will, therefore, favour the *anti* conformation over the *syn*.

The similarly calculated[12] conformational energy maps for the corresponding *trans* and *cis* crotonyl complexes display the same features as that for the acryloyl complex above. The *anti* conformation is slightly favoured over the *syn*. In the *anti* conformation, the *trans* methyl vinyl group is sterically free to move 130° away in either direction from S-*cis* coplanar whereas the *cis* methyl vinyl group can only move about 90° in either direction from this arrangement before large steric interactions are encountered. For both these complexes stereoelectronic and entropy factors will favour the *anti* relative to the *syn* conformation.

From the above analysis it can be concluded that for α,β-unsaturated acyl ligands attached to the auxiliary [(C$_5$H$_5$)Fe(CO)(PPh$_3$)] the preferred conformation will be S-*cis* with the acyl oxygen *anti* to the carbon monoxide ligand. Furthermore, in order to avoid eclipsing interactions between the vinyl group and the carbon monoxide ligand, a conformation close to *anti* periplanar but tilted to place the vinyl in zone A

(a)

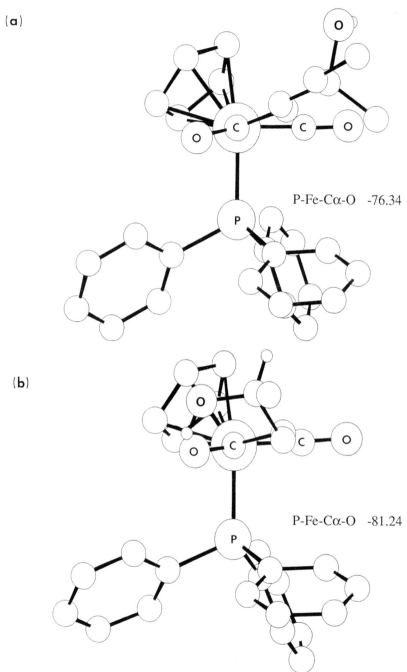

P-Fe-Cα-O   -76.34

(b)

P-Fe-Cα-O   -81.24

Figure 40. X-ray crystal structure and P-Fe-Cα-O torsional angle (^{0}) for (a) RS,SR-[(C$_5$H$_5$)Fe(CO)(PPh$_3$)CH$_2$CH$_2$C(OH)(Me)(Et)], (b) RS,SR-[(C$_5$H$_5$)Fe(CO)(PPh$_3$)-COCH$_2$CH$_2$CH(OH)Me], (c) RR,SS-[(C$_5$H$_5$)Fe(CO)(PPh$_3$)CH$_2$CH(OH)Et],

(c)

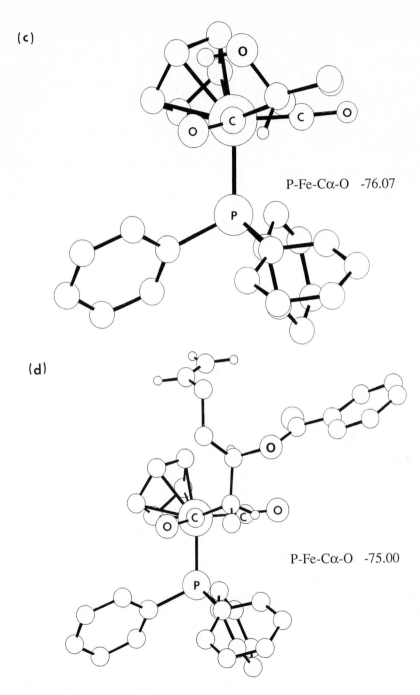

P-Fe-Cα-O   -76.07

(d)

P-Fe-Cα-O   -75.00

(d) RR,SS-[(C5H5)Fe(CO)(PPh3)CH(Me)CH(OCOPh)CH2CH2CH=CH2]. The diagrams show the Newman projection along the *alpha* carbon to iron bond.

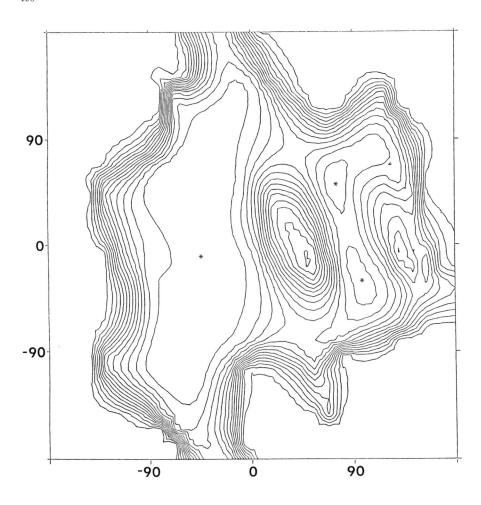

Figure 41: Conformational energy map for $[(C_5H_5)Fe(CO)(PPh_3)COCH=CH_2]$: Energy (contours separated by 5 Kcal mol^{-1}, with * indicating minima) *vs* P-Fe-Cα-O (abscissa) and Fe-Cα-Cβ-Cγ (ordinate) torsional angles (^{0}).

will be adopted. The above predictions are consistent with the observed X-ray crystal structure conformations for the *trans*- (Figure 42)[40] and *cis*- (Figure 43)[41] crotonyl complexes $[(C_5H_5)Fe(CO)(PPh_3)COCH=CHMe]$. That these conformations are also important in solution is evidenced by their stereoselective reactions (see Section 9.5).

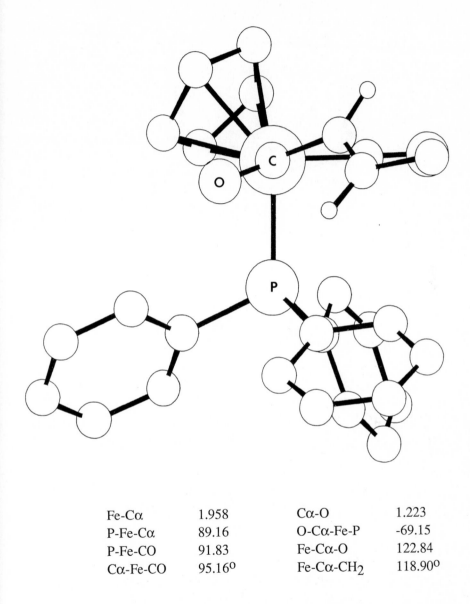

Fe-Cα	1.958	Cα-O	1.223
P-Fe-Cα	89.16	O-Cα-Fe-P	-69.15
P-Fe-CO	91.83	Fe-Cα-O	122.84
Cα-Fe-CO	95.16⁰	Fe-Cα-CH₂	118.90⁰

Figure 42. X-ray crystal structure and selected bond lengths (Å), angles and torsional angles (⁰) for *trans*-[(C₅H₅)Fe(CO)(PPh₃)COCH=CHMe]. The diagram shows the Newman projection along the *alpha* carbon to iron bond. Selected protons are removed for clarity.

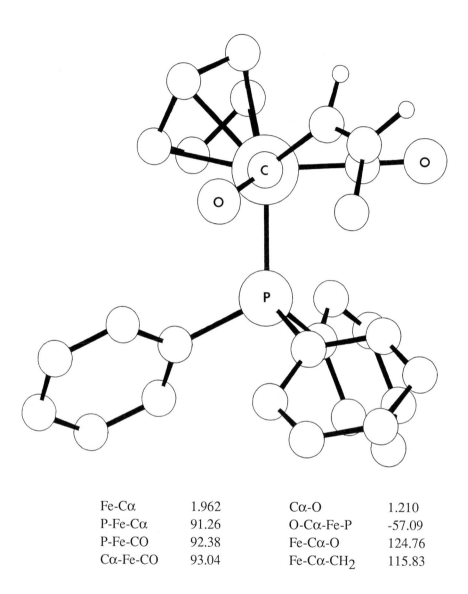

Fe-Cα	1.962	Cα-O	1.210
P-Fe-Cα	91.26	O-Cα-Fe-P	-57.09
P-Fe-CO	92.38	Fe-Cα-O	124.76
Cα-Fe-CO	93.04	Fe-Cα-CH₂	115.83

Figure 43. X-ray crystal structure and selected bond lengths (Å), angles and torsional angles (º) for *cis*-[(C₅H₅)Fe(CO)(PPh₃)COCH=CHMe]. The diagram shows the Newman projection along the *alpha* carbon to iron bond. Selected protons are removed for clarity.

9. STEREOSELECTIVE REACTIONS OF LIGANDS ATTACHED TO THE CHIRAL AUXILIARY [(C5H5)Fe(CO)(PPh3)]

9.1. General Considerations:

The triphenylphosphine ligand is the dominant factor in determining the preferred conformations of ligands attached to the chiral auxiliary. Its massive steric bulk effectively shields the entire space below the plane containing the iron, carbon monoxide and *alpha* carbon. The space above this plane, although it contains the cyclopentadienyl ligand, is relatively unencumbered. Reagents may, therefore, only approach ligands attached to the chiral auxilairy [(C5H5)Fe(CO)(PPh3)] from the direction *anti* to the triphenylphosphine. The triphenylphosphine effectively makes the space between its phenyl rings and the plane containing the iron, carbon monoxide and the *alpha* carbon an excluded volume (Figure 44).[2,34]

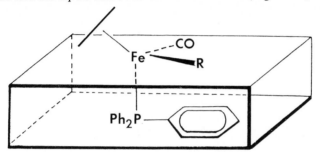

Figure 44: Excluded volume in [(C5H5)Fe(CO)(PPh3)R] structures.

9.2. Nucleophilic Additions to Alkoxy Carbene Ligands:

The cation R,S-[(C5H5)Fe(CO)(PPh3)CCMe2CH2CH2O]+ undergoes a completely stereoselective reduction with sodium borohydride to generate diastereoisomerically pure RR,SS-[(C5H5)Fe(CO)(PPh3)CHCMe2CH2CH2O], which on exposure to a proton source undergoes complete epimerisation to the corresponding RS,SR diastereoisomer (Figure 45).[28] The observed stereoselectivities may be understood in terms of the above conformational analysis. The initial alkoxy carbene ligand is restricted to the conformation with the large quaternary carbon in zone A. Addition of hydride to the *alpha* carbon exclusively from the face away from the triphenylphosphine ligand leads to the RR,SS diastereoisomer stereoselectively. In protic media epimerisation is possible via an elimination/addition mechanism, and occurs completely to generate the RS,SR diastereoisomer (Figure 45). The conformation of both diastereoisomers is determined by the need to place the large *gem* dimethyl group in zone A. In the RR,SS diastereoisomer, formed under the kinetically controlled hydride reduction conditions the oxygen is, therefore, forced to occupy the encumbered zone C. After epimerisation the oxygen substituent is placed in zone B and the *alpha* hydrogen in

194

zone C. A large difference in stability between the two diastereoisomers is evident since at equilibrium the RR,SS diastereoisomer cannot be detected.[28]

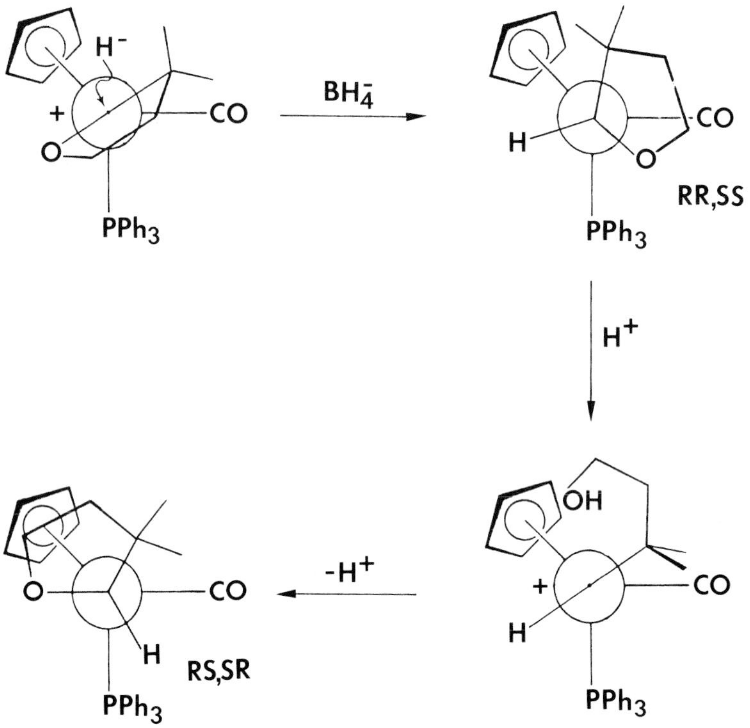

Figure 45: Stereoselective formation of RR,SS-[(C5H5)Fe(CO)(PPh3)-CHCMe2CH2CH2O] and its epimerisation to RS,SR-[(C5H5)Fe(CO)(PPh3)-CHCMe2CH2CH2O].

The methoxy ethyl carbene cation [(C5H5)Fe(CO)(PPh3)C(OMe)Et]+ (Figure 35) undergoes stereoselective addition of hydride (Figure 46) to generate RR,SS-[(C5H5)Fe(CO)(PPh3)CH(OMe)Et] (Figure 25).[32] The same diastereoisomer is also generated stereoselectively by addition of ethyllithium to the parent methoxy carbene complex [(C5H5)Fe(CO)(PPh3)CH(OMe)]+ in tetrahydrofuran as solvent (Figure 46).[24,25,32] In contrast, the epimer, RS,SR-[(C5H5)Fe(CO)(PPh3)CH(OMe)Et] (Figure 26) is generated stereoselectively during the addition of ethyllithium (Figure 46) to the parent methoxy carbene complex [(C5H5)Fe(CO)(PPh3)CH(OMe)]+ (Figure 34) in dichloromethane. The formation of the RR,SS diastereoisomer from both carbenes can be understood in each case in terms of kinetic trapping from the face away from the triphenylphosphine of the most stable conformer, that having the methoxyl group *anti* to carbon monoxide for the methoxy ethyl cation (Section 7.3) but *syn* for the parent cation(Section 7.2). In dichloromethane, however, the ethyllithium would be aggregated into polymers and thus would be less reactive than in the donor solvent tetrahydrofuran.[43] Therefore, in dichloromethane the transition state should be late and product like.[44] Addition of ethyllithium to the methoxy carbene in the *anti* (methoxyl to carbon monoxide) conformation would then be favoured, since the product is generated in its most stable conformation. In contrast, addition to the *syn* conformation initially generates the diastereoisomeric product in a less stable conformation with the methoxyl lying in zone C: subsequent rotation about the iron to *alpha* carbon bond leads to the stable conformation with the methoxyl group in zone A.

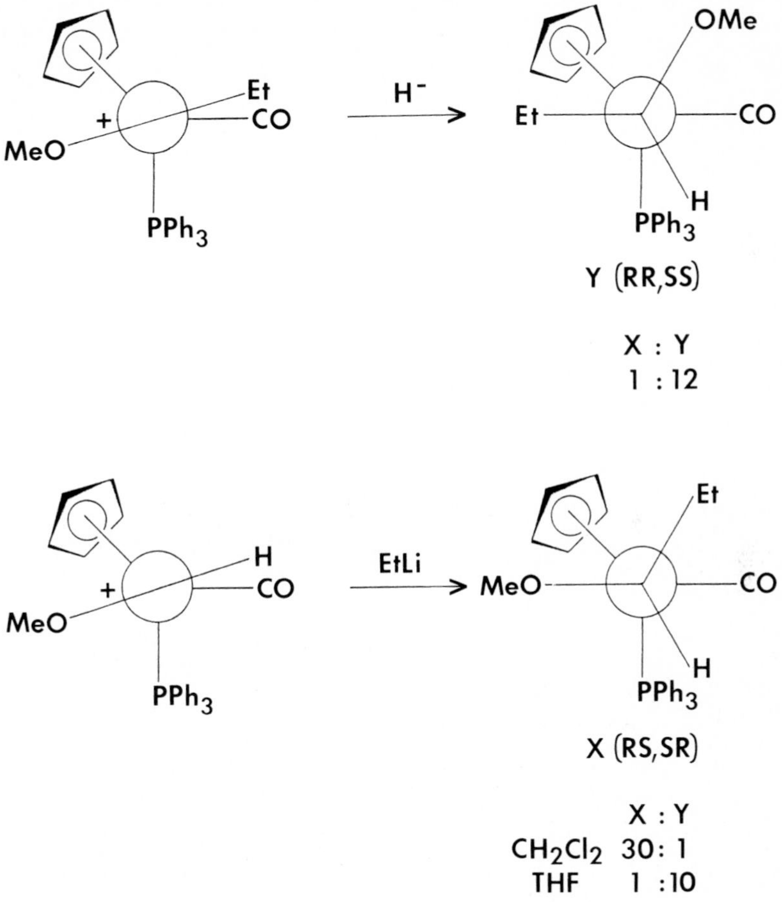

Figure 46: Stereoselective formations of RR,SS- and RS,SR-[(C5H5)Fe(CO)(PPh3)-CH(OMe)Et] (X and Y respectively).

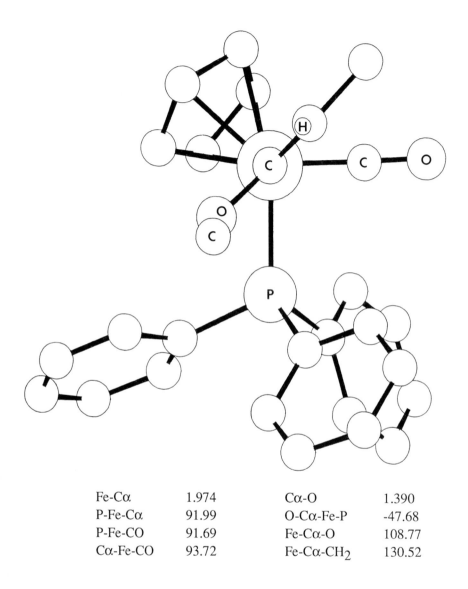

Fe-Cα	1.974	Cα-O	1.390
P-Fe-Cα	91.99	O-Cα-Fe-P	-47.68
P-Fe-CO	91.69	Fe-Cα-O	108.77
Cα-Fe-CO	93.72	Fe-Cα-CH$_2$	130.52

Figure 47. X-ray crystal structure and selected bond lengths (Å), angles and torsional angles (^{o}) for Z-[(C$_5$H$_5$)Fe(CO)(PPh$_3$)C(OMe)=CHMe]. The diagram shows the Newman projection along the *alpha* carbon to iron bond. Selected protons are removed for clarity.

9.3. Alkoxy Vinyl Ligands:

9.3.1. Formation:

The addition of base to the ethyl methoxy carbene complex [(C$_5$H$_5$)Fe(CO)-(PPh$_3$)C(OMe)Et]$^+$ generates the corresponding Z-methoxyvinyl complex (Figure 47) stereoselectively (>100:1).[45,46] For deprotonation, the carbon to hydrogen bond must align closely parallel with the p orbital on the carbene carbon. The required alignment can be readily achieved for formation of the Z-methoxy vinyl but not for formation of the E-methoxyvinyl, due, in the latter case, to interactions between the two methyl groups (Figure 48).

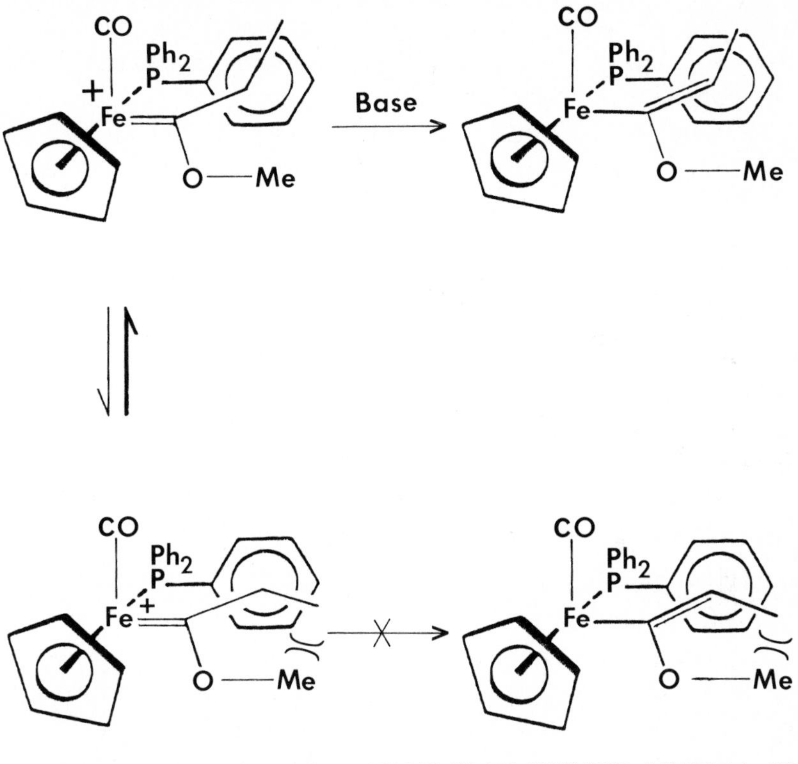

Figure 48. Stereoselective formation of Z-[(C$_5$H$_5$)Fe(CO)(PPh$_3$)C(OMe)=CHMe].

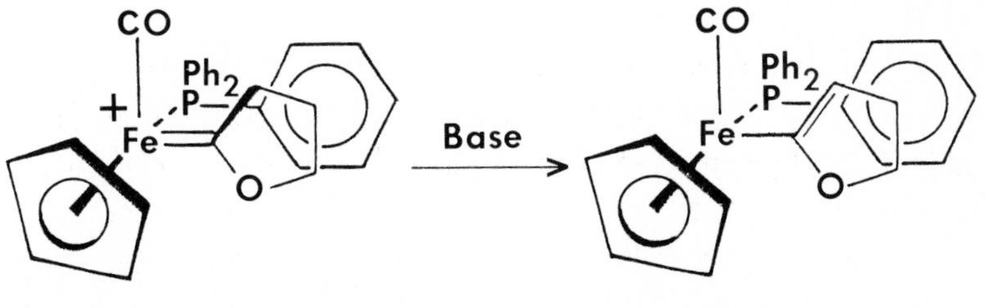

Figure 49. Formation of [(C$_5$H$_5$)Fe(CO)(PPh$_3$)C=CHCH$_2$CH$_2$O].

The cyclic nature of the carbene ligand in the cation [(C5H5)Fe(CO)(PPh3)-=CCH2CH2CH2O]+ means that on deprotonation only the corresponding E-alkoxyvinyl complex is formed (Figure 49).[45,46]

9.3.2. Electrophilic Additions:

Addition of ethyl iodide to Z-[(C5H5)Fe(CO)(PPh3)C(OMe)=CHMe] generates the acyl complex RS,SR-[(C5H5)Fe(CO)(PPh3)COCH(Me)Et] stereoselectively (Figure 50).[45] The stereochemistry of the product was determined by X-ray crystal structure analysis (Figure 51).[45] The observed stereoselectivity is consistent with ethylation occurring onto the face away from the triphenylphosphine of the methoxylvinyl ligand in the *anti* conformation (methoxyl to carbon monoxide).[45-47] The *anti* conformation is favoured since the vinyl methyl group is located in zone A; the *syn* conformation which would place the vinyl methyl in zone B is excluded. Finally the iodide liberated in the nuleophilic addition reaction demethylates the methoxy carbene product to generate the acyl complex.

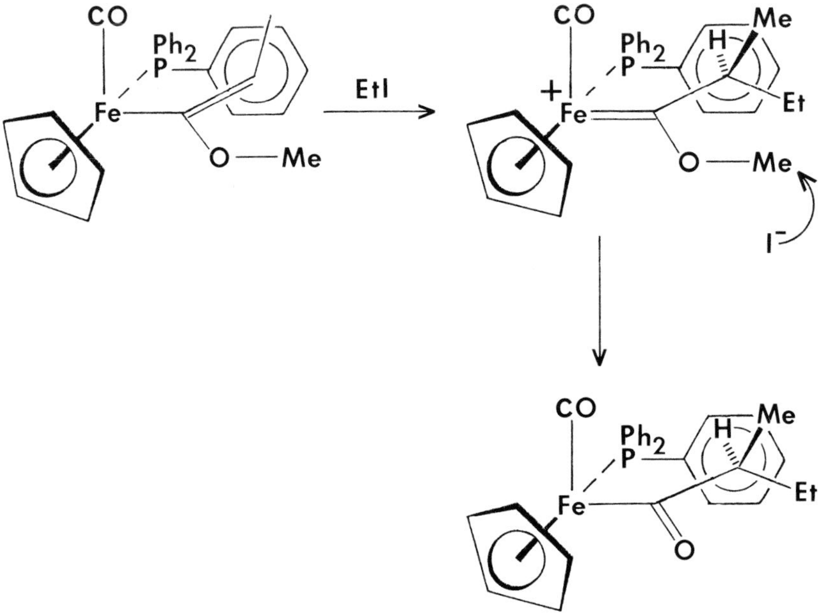

Figure 50: Stereoselective formation of RS,SR-[(C5H5)Fe(CO)(PPh3)COCH(Me)Et] by ethylation of Z-[(C5H5)Fe(CO)(PPh3)C(OMe)=CHMe].

Addition of methyl iodide to E-[(C5H5)Fe(CO)(PPh3)C=CHCH2CH2O] generates RS,SR-[(C5H5)Fe(CO)(PPh3)CCH(Me)CH2CH2O]+ stereoselectively.[46] Again the formation of the RS,SR diastereoisomer is consistent with addition to the

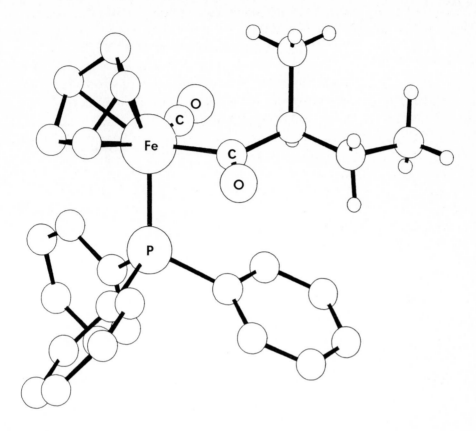

Fe-Cα	1.964	Cα-O	1.207
P-Fe-Cα	89.33	O-Cα-Fe-P	-74.14
P-Fe-CO	92.12	Fe-Cα-O	122.86
Cα-Fe-CO	94.76	Fe-Cα-CH₂	120.05

Figure 51. X-ray crystal structure and selected bond lengths (Å), angles and torsional angles (°) for RS,SR-[(C5H5)Fe(CO)(PPh3)COCH(Me)Et]. The diagram shows the Newman projection along the *alpha* carbon to iron bond. Selected protons are removed for clarity.

anti conformation away from the triphenylphosphine ligand (Figure 52). Although for this alkoxyvinyl complex the *anti* and *syn* conformations are expected to be of similar energy, alkylation of the *anti* conformer will be favoured for two reasons. Firstly, if the reaction is under product development control then formation of the secondary alkyl centre in zone A will be favoured. Secondly, if the reaction is under reagent approach control then the methyl iodide will prefer to approach the *anti* conformer through zone A rather than eclipse the cyclopentadienyl ligand in approaching the *syn* conformer (Figure 53).[48]

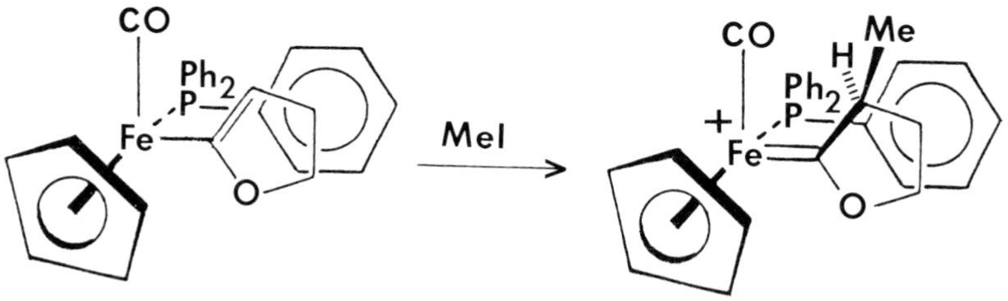

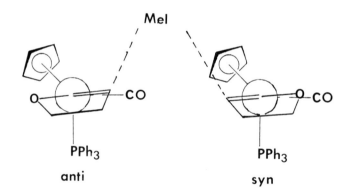

Figure 52: Stereoselective methylation of [(C5H5)Fe(CO)(PPh3)C̅=CHCH2CH2O̅].

Figure 53: Reagent approach to *anti* and *syn* conformers of alkoxy vinyl ligands.

9.4. Alkylation of Acyl Enolates:

In contrast to the methoxy alkyl carbenes described above, the acyl ligands in [(C5H5)Fe(CO)(PPh3)COCH2R] will prefer to undergo deprotonation in the conformation which places the alkyl group R *syn* to the acyl oxygen that is *anti* to the large iron auxiliary;[34,37-41,48,49] the steric interactions of the R group with the oxygen atom being much less than with the iron moiety. Deprotonation thus generates the E-enolates exclusively. Electrophilic addition to these enolates occurs in the *anti* conformation from the face away from the triphenylphosphine ligand (Figure 54). The *syn* and *anti* enolate conformers are expected to have similar energies but the *anti* conformer undergoes preferential alkylation for the reasons described above for the E-alkoxy carbene complex. The X-ray crystal structure of the zirconium enolate {(C5H5)Fe(CO)(PPh3)C[OZr(Cl)(C5H5)2]=CH2} has been determined and in the solid state exists in the *anti* (oxygen to carbon monoxide) conformation.[50]

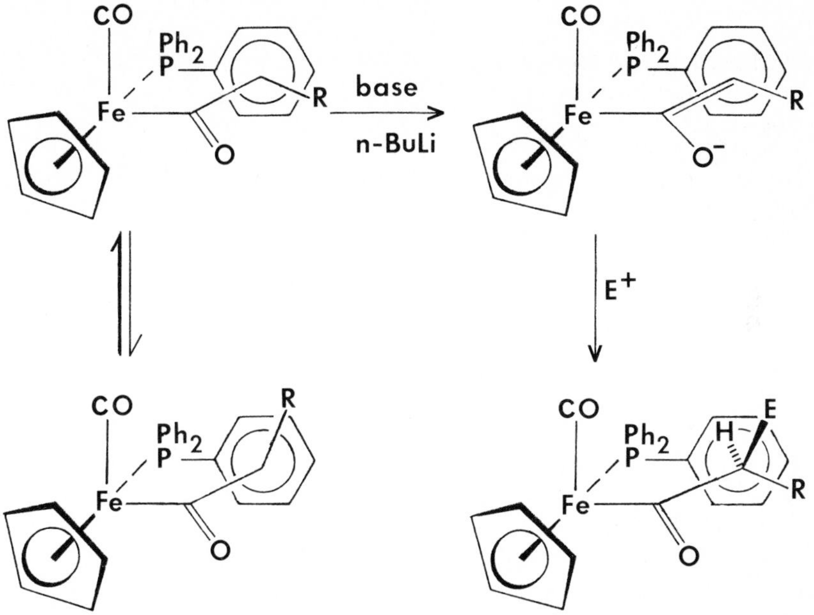

Figure 54: Stereoselective electrophilic additions to enolates derived from the iron acyl complexes [(C5H5)Fe(CO)(PPh3)COCH2R].

For example, sequential deprotonation and methylation of the propanoyl complex [(C5H5)Fe(CO)(PPh3)COCH2CH2CH3] generates stereoselectively RS,SR-[(C5H5)Fe(CO)(PPh3)COCH(Me)Et] (Figures 51 and 55).[48]

Deprotonation of the complex [(C5H5)Fe(CO)(PPh3)COCH2OCH2Ph] generates exclusively the corresponding E-enolate, presumably aided by chelation of the lithium between the two oxygen atoms[51,52] prior to deprotonation.[43] This enolate undergoes completely steroselective methylation to generate RR,SS-[(C5H5)Fe(CO)-(PPh3)COCH(Me)OCH2Ph] (Figure 56).[52,53] Deuteration of the enolate, however, although still preferring to form the RR,SS diastereoisomer of [(C5H5)Fe(CO)-

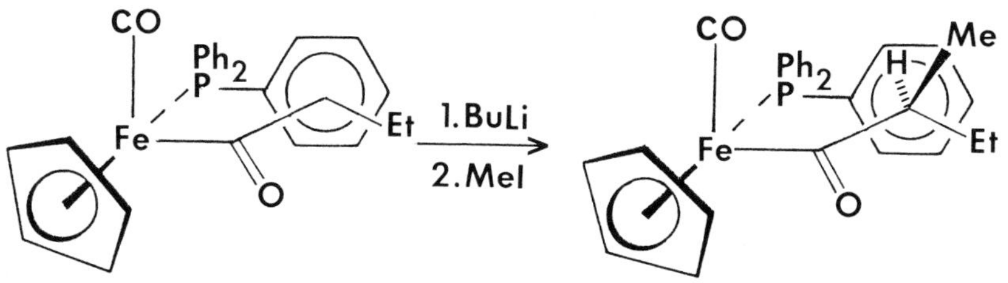

Figure 55: Stereoselective formation of RS,SR-[(C5H5)Fe(CO)(PPh3)COCH(Me)Et] by methylation of the enolate derived from [(C5H5)Fe(CO)(PPh3)COCH2CH2CH3].

(PPh3)COCH(D)OCH2Ph] is less selective (10 : 1). This drop in selectivity is consistent with a deuteron being smaller than methyl iodide and hence less discriminating between the *syn* and *anti* conformers of the enolate. Alternatively, deuteration may occur initially on oxygen with tautomerisation to the acyl product occurring less selectively at the higher temperatures of the work-up.

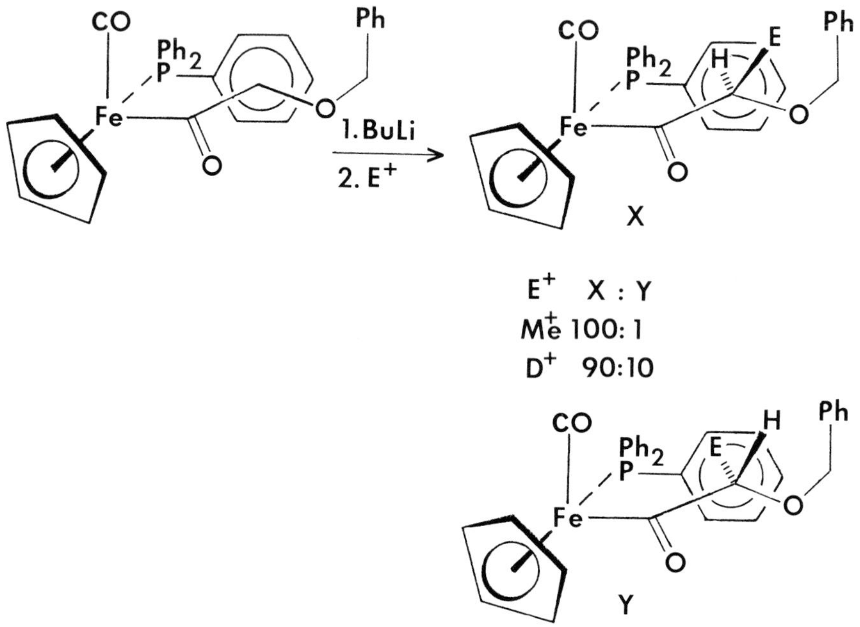

Figure 56: Stereoselective electrophilic additions to the enolate derived from [(C5H5)Fe(CO)(PPh3)COCH2OCH2Ph].

9.5. Tandem Michael Addition and Alkylation Reactions of α,β-Unsaturated Acyls:

The E-α,β-unsaturated acyl complexes E-[(C$_5$H$_5$)Fe(CO)(PPh$_3$)COCH=CHR] undergo Michael addition reactions in the S-*cis* (C=O to C=C) *anti* (oxygen to carbon monoxide) conformation from the face away from the triphenylphosphine ligand.[34,37,40,54] This addition thus generates the corresponding E-enolate, which as before undergoes stereoselective alkylation reactions in the *anti* conformation also from the face away from the triphenylphosphine ligand (Figure 57). The overall consequence of these additions is the stereoselective formation of two carbon carbon bonds onto the same unshielded face of the α,β-unsatutated ligand. The stereo-selectivities observed in both steps are consistent with the present conformational analysis. For example, addition of n-butyllithium followed by methyl iodide to the complex E-[(C$_5$H$_5$)Fe(CO)(PPh$_3$)COCH=CHEt] generates RSS,SRR-[(C$_5$H$_5$)Fe-(CO)(PPh$_3$)COCH(Me)CH(Me)Bu] as a single diastereoisomer.[54b] The relative configurations were assigned on the basis of an X-ray crystal structure analysis (Figure 58).[34]

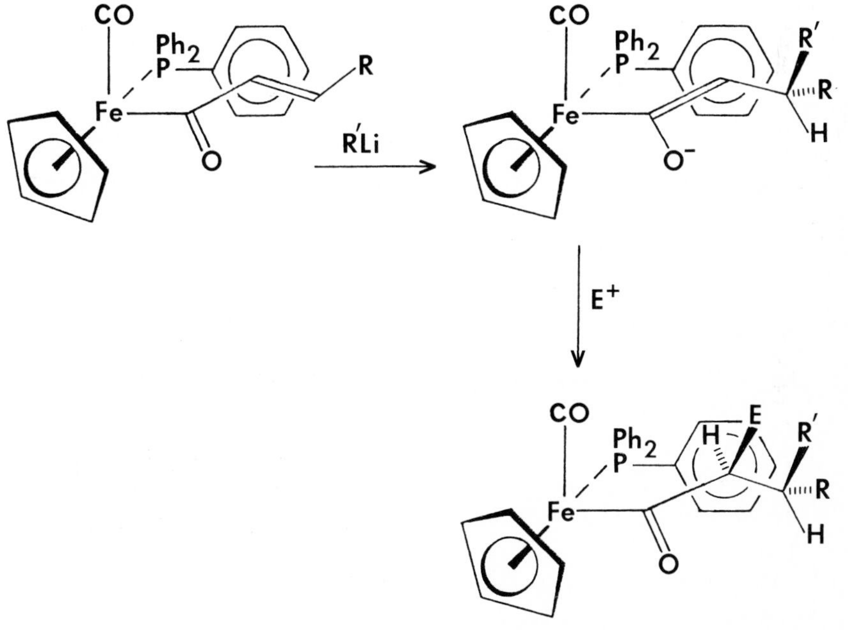

Figure 57: Stereoselective tandem Michael Additions and Alkylations of the E-α,β-unsaturated acyl complexes [(C$_5$H$_5$)Fe(CO)(PPh$_3$)COCH=CHR].

Similarly, sequential addition of methyllithium and methyl iodide to the parent α,β-unsaturated complex [(C$_5$H$_5$)Fe(CO)(PPh$_3$)COCH=CH$_2$] gave RS,SR-[(C$_5$H$_5$)Fe(CO)(PPh$_3$)COCH(Me)Et] exclusively (Figures 51 and 59).[54]

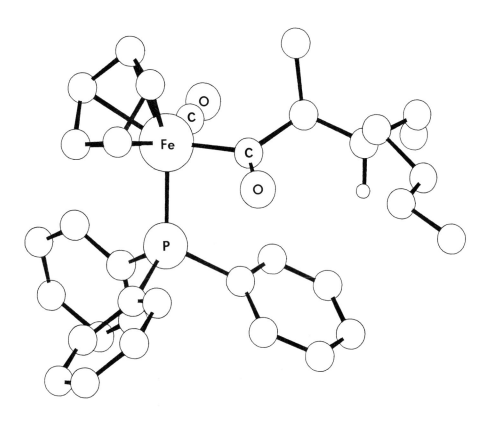

Fe-Cα	1.954	Cα-O	1.230
P-Fe-Cα	89.71	O-Cα-Fe-P	-63.48
P-Fe-CO	92.40	Fe-Cα-O	119.72
Cα-Fe-CO	94.48	Fe-Cα-CH$_2$	122.58

Figure 58. X-ray crystal structure and selected bond lengths (Å), angles and torsional angles (°) for RSS,SRR-[(C$_5$H$_5$)Fe(CO)(PPh$_3$)COCH(Me)CH(Me)Bu]. Selected protons are removed for clarity.

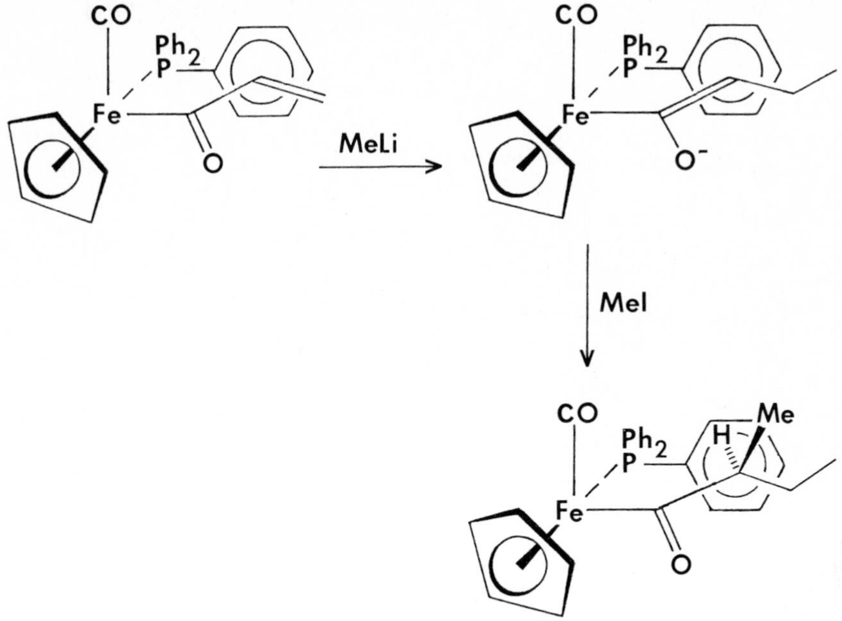

Figure 59: Stereoselective formation of RS,SR-[(C5H5)Fe(CO)(PPh3)COCH(Me)Et] from [(C5H5)Fe(CO)(PPh3)COCH=CH2].

9.6. Stereospecific Rearrangement of α-Alkoxy Acyl Ligands:

In the presence of acid the α-alkoxy acyl complex [(C5H5)Fe(CO)(PPh3)COCH2OCH2Ph] rearranges to the α-metalla benzyl ester [(C5H5)Fe(CO)(PPh3)CH2CO2CH2Ph] (Figure 60).[53] The preferred conformation of the product places the ester group in zone A with proton H^1 being deshielded (δ 0.75) by the proximate phenyl group and exhibiting a large $^3J_{PH}$ coupling constant (10.6Hz). On the other hand H^2 appears at δ 1.30 and shows a small $^3J_{PH}$ coupling constant (2.9Hz).[53]

Acid catalysed rearrangement of a 10 : 1 mixture of the RR,SS and RS,SR diastereoisomers of [(C5H5)Fe(CO)(PPh3)COCH(D)OCH2Ph] gave a corresponding mixture of the diastereoisomers of [(C5H5)Fe(CO)(PPh3)CH(D)CO2CH2Ph].[53] The stereochemistry of the major diastereoisomer of the product followed from the absence of the signal for H^1 in the n.m.r. spectrum while the low field signal for H^2 with the small $^3J_{PH}$ coupling constant remained (Figure 61). This result indicates that the rearrangement is proceeding with complete inversion of configuration at carbon.

Acid catalysed rearrangement of RR,SS-[(C5H5)Fe(CO)(PPh3)-COCH(Me)OCH2Ph] gave RS,SR-[(C5H5)Fe(CO)(PPh3)CH(Me)CO2CH2Ph] as a single diastereoisomer whose stereochemistry followed from conformational analysis

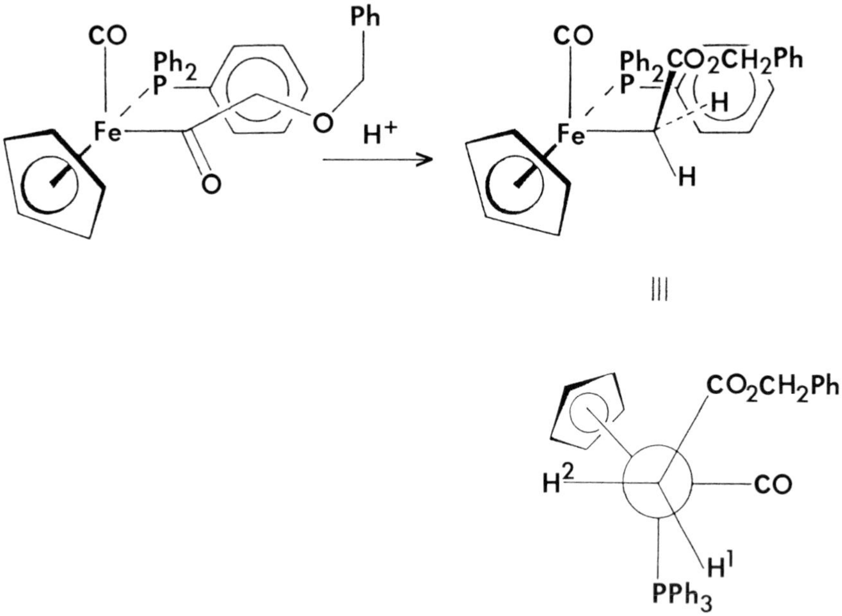

Figure 60. Acid catalysed rearrangement of [(C5H5)Fe(CO)(PPh3)COCH2-OCH2Ph] to the α-metalla benzyl ester [(C5H5)Fe(CO)(PPh3)CH2CO2CH2Ph].

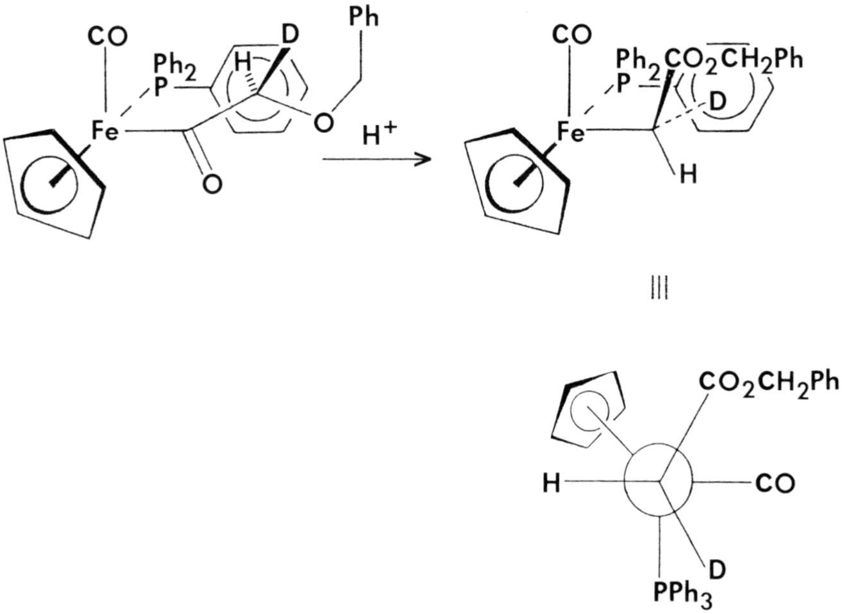

Figure 61. The stereospecific acid catalysed rearrangement of [(C5H5)Fe(CO)(PPh3)COCH(D)OCH2Ph] showing the relative stereochemistry of the product [(C5H5)Fe(CO)(PPh3)CH(D)CO2CH2Ph].

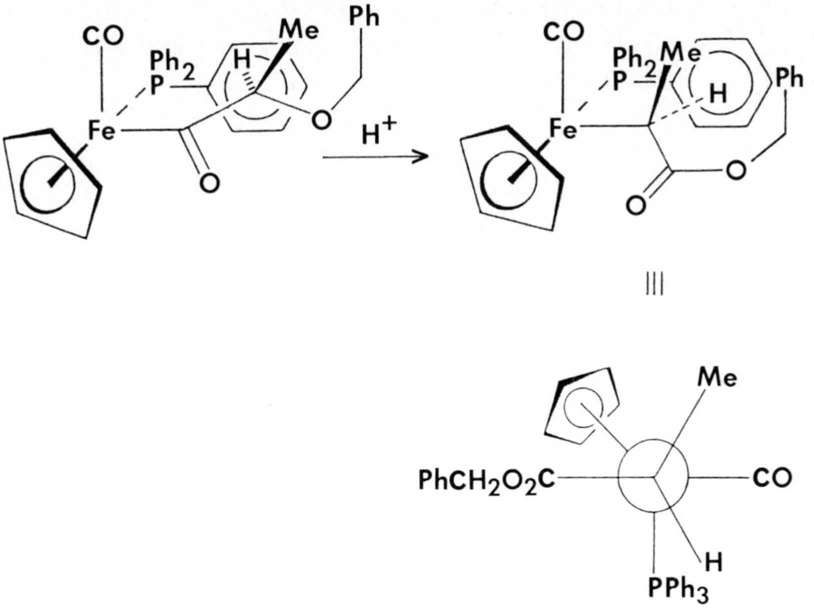

Figure 62. The stereospecific acid catalysed rearrangement of RR,SS-[(C5H5)Fe(CO)(PPh3)COCH(Me)OCH2Ph] showing the relative stereochemistry of the product RS,SR-[(C5H5)Fe(CO)(PPh3)CH(Me)CO2CH2Ph].

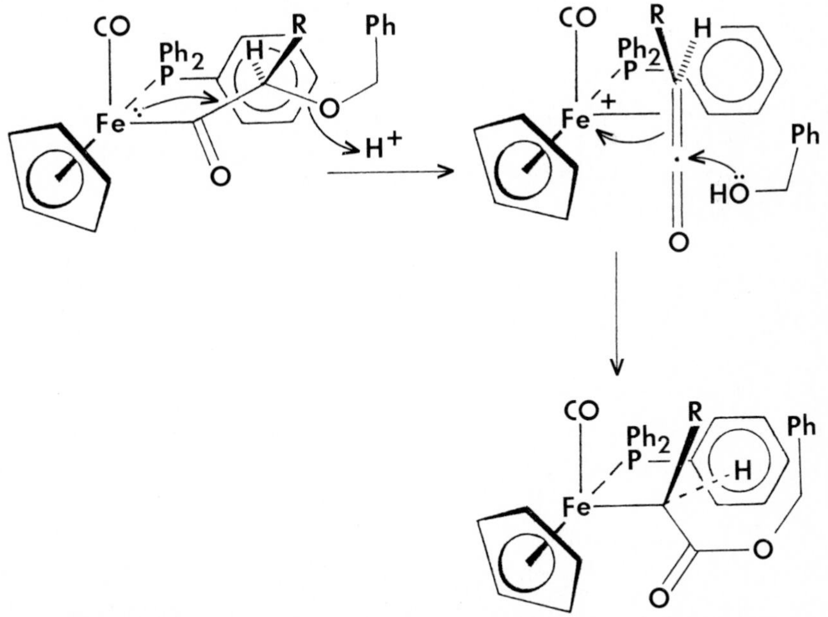

Figure 63. The mechanism of the acid catalysed rearrangement of α-alkoxy iron acyl complexes [(C5H5)Fe(CO)(PPh3)COCH(R)OCH2Ph].

and proton n.m.r. data: The long range $^4J_{PH}$ coupling (1.6Hz) between the methyl and the phosphorus is characteristic of a methyl in zone A, while the chemical shift (δ 1.62) and $^3J_{PH}$ (8.1Hz) of the alpha proton place it in zone C (Figure 62).[53] The observed stereospecificity of this rearrangement is, again, only consistent with complete inversion of configuration at carbon. The observed inversion of configuration during these rearrangements indicates the intermediacy of a ketene carbene complex which is formed stereoselectively and trapped from the uncoordinated face (Figure 63).[53]

10. THE CHIRAL AUXILIARY [(C5H5)Re(NO)(PPh3)]

10.1. Generale Considerations:

The rhenium analogue, [(C5H5)Re(NO)(PPh3)] of the iron chiral auxiliary, [(C5H5)Fe(CO)(PPh3)], has been extensively studied by Gladysz and coworkers with several X-ray crystal structures and stereoselective reactions being reported.[55-66] As with the iron complexes, the rhenium analogues [(C5H5)Re(NO)-(PPh3)R] are close to octahedral in structure with three sites occupied by the triphenylphosphine, nitrosyl and the ligand R, while the remaining three sites are shared by the cyclopentadienyl ligand. Since the nitrosyl ligand in [(C5H5)Re(NO)(PPh3)] complexes is close to linear it will mimic the effect of the carbon monoxide ligand in the iron complexes. The general conformational analysis developed above for the iron chiral auxiliary should, therefore, be applicable to this rhenium analogue. Minor differences between the two systems will arise due to several factors, which are outlined below.

For the alkyl complexes [(C5H5)Re(NO)(PPh3)R] (R = alkyl) longer rhenium to ligand bond lengths,[2,] for example rhenium to triphenylphosphine $ca.$ 2.35A[56] compared to iron to triphenylphosphine $ca.$ 2.20A[2], will result in steric interactions being somewhat less demanding for the rhenium case.

For the carbene complexes [(C5H5)Fe(CO)(PPh3)CH2]+ and [(C5H5)Re-(NO)(PPh3)CH2]+, where the *alpha* carbon is sp^2 hybridised stereoelectronic effects are expected to be much more important for rhenium than for the iron complexes. For both the iron and rhenium auxiliaries the HOMO lies in the plane containing the phosphorus, iron and *alpha* carbon atoms while the SHOMO lies orthogonal to this plane.[29,56-59] However, for the iron auxiliary the energy difference between the HOMO and SHOMO is calculated[29,67] to be $ca.$ 11 Kcal mol[-1] whereas for the rhenium this difference is $ca.$ 48 Kcal mol[-1].[68,69] Predictions derived from PMO theory[70] suggest that for the iron carbene complex [(C5H5)Fe(CO)(PPh3)CH2]+ the conformation with the carbene perpendicular to the iron phosphorus bond will be favoured since this allows maximum overlap between the carbene LUMO and the

HOMO on the iron: However, the orthogonal conformation will be stabilised to a somewhat lesser extent by overlap between the carbene LUMO and the SHOMO on the iron (Figure 64). For the rhenium carbene complex $[(C_5H_5)Re(NO)(PPh_3)CH_2]^+$ maximum stabilisation will also occur when the carbene is perpendicular to the rhenium phosphorus bond, but, in contrast to the iron, the stabilisation due to orbital overlap is very much less in the orthogonal arrangement. This difference manifests itself in a stronger preference for the carbene to lie perpendicular to the metal phosphorus bond (see Section 10.3) and a higher barrier to rotation about the metal to *alpha* carbon bond in rhenium than in iron carbene complexes. For example, essentially free rotation about the iron to *alpha* carbon bond is observed for carbene ligands attached to the iron chiral auxiliary,[31,71] while *syn* and *anti* rotomers can be isolated for carbene ligands attached to the rhenium chiral auxiliary.[57-59]

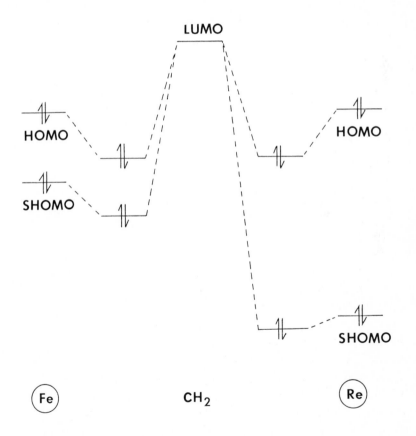

Figure 64: Comparison of molecular orbital interactions for $[(C_5H_5)Fe(CO)(PPh_3)]$ and $[(C_5H_5)Re(NO)(PPh_3)]$ based on PMO theory.

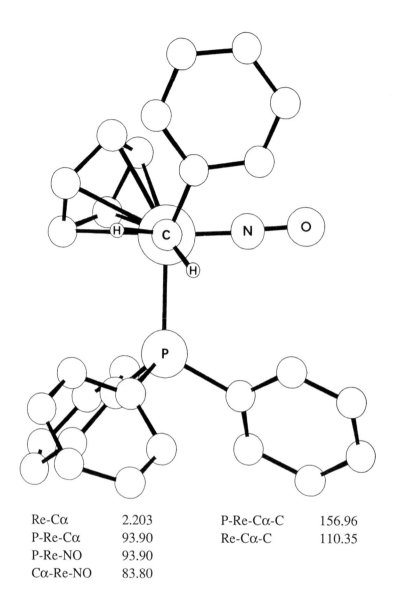

Re-Cα	2.203	P-Re-Cα-C	156.96
P-Re-Cα	93.90	Re-Cα-C	110.35
P-Re-NO	93.90		
Cα-Re-NO	83.80		

Figure 65. X-ray crystal structure and selected bond lengths (Å), angles and torsional angles (⁰) for [(C₅H₅)Re(NO)(PPh₃)CH₂Ph]. The diagram shows the Newman projection along the *alpha* carbon to rhenium bond. Selected protons are removed for clarity.

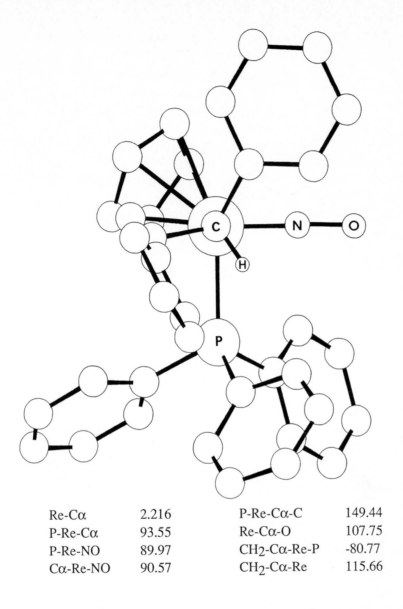

Re-Cα	2.216	P-Re-Cα-C	149.44
P-Re-Cα	93.55	Re-Cα-O	107.75
P-Re-NO	89.97	CH₂-Cα-Re-P	-80.77
Cα-Re-NO	90.57	CH₂-Cα-Re	115.66

Figure 66. X-ray crystal structure and selected bond lengths (Å), angles and torsional angles (°) for RR,SS-[(C₅H₅)Re(NO)(PPh₃)CH(Ph)CH₂Ph]. The diagram shows the Newman projection along the *alpha* carbon to rhenium bond. Selected protons are removed for clarity.

10.2. [(C5H5)Re(NO)(PPh3)R] (R = alkyl):

For the simple alkyl complexes [(C5H5)Re(NO)(PPh3)R] (R = alkyl) stereoelectronic effects will not be important and the preferred conformations will be determined by steric effects. Hence for the complexes [(C5H5)Re(NO)(PPh3)CH2R] the conformation which places the group R in zone A will be preferred. Figure 65 shows the X-ray crystal structure for [(C5H5)Re(NO)(PPh3)CH2Ph] in which the phenyl group does, as expected, occupy zone A.[60] Similarly in complexes of the type [(C5H5)Re(NO)(PPh3)CH2(SR2+)] the (SR2+) group lies in zone A.[61]

The conformational analysis for [(C5H5)Re(NO)(PPh3)CHR2] predicts the conformation with the two identical alkyl groups lying in zones A and B to be preferred. The X-ray crystal structures of the complexes RR,SS-[(C5H5)Re-(NO)(PPh3)CHPhCH2Ph] (Figure 66)[62] and [(C5H5)Re(NO)(PPh3)PHPh2][63] are consistent with this analysis.

The conformations adopted by the two diasteroisomers of [(C5H5)Re(NO)(PPh3)CH(CH3)P(p-tol)3]+ will be determined by the need to restrict the very large phosphine group to zone A (Figure 67). For the RS,SR diastereoisomer this will force the methyl group to occupy the encumbered zone C and epimerisation will lead to the more stable RR,SS diastereoisomer with the methyl group occupying zone B. Consistent with this prediction is the observation of a large $^3J_{PP}$ coupling constant and the lack of any long range $^4J_{PH}$ coupling to the methyl protons, characteristic of a methyl group in Zone A, for both diastereoisomers.[64]

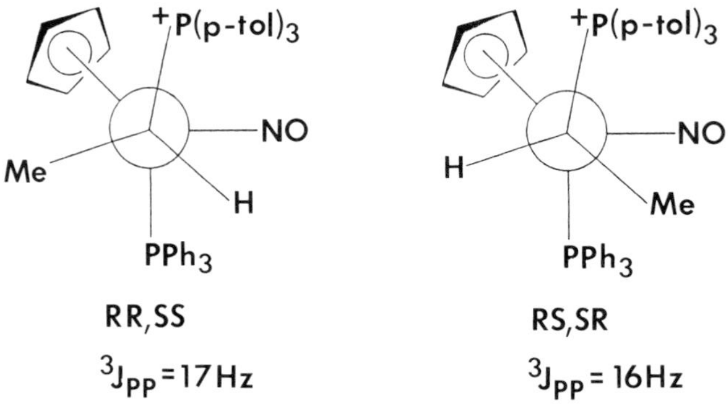

Figure 67: Stable conformations of [(C5H5)Re(NO)(PPh3)CH(CH3)P(p-tol)3]+.

Unfortunately no X-ray crystal structure data is available on *alpha* alkoxy alkyl complexes so that assessment of any stereoelectronic effects similar to the anomeric effect is not possible at this time.

10.3. [(C5H5)Re(NO)(PPh3)COR] (R = H, alkyl):

The X-ray crystal structure for the formyl complex [(C5H5)Re(NO)(PPh3)CHO] is illustrated in Figure 68.[56,65] Both groups on the *alpha* carbon are small and hence the preferred conformation is largely governed by stereoelectronic effects which place the formyl ligand perpendicular to the rhenium phosphorus bond. The oxygen is orientated *anti* to the nitrosyl ligand for steric and dipolar reasons: Unlike a metal carbon monoxide bond the metal nitrosyl bond is expected to be substantially polarised [M(+) - NO(-)] and the *anti* conformation will have a lower dipole than the *syn*.

The greater influence of stereoelectronic effects for rhenium than for iron is apparent from comparison of the X-ray crystal structure of the rhenium acyl [(C5H5)Re(NO)(PPh3)COCH(Me)CH2Ph] (Figure 69)[56,62] with that for the iron acyl [(C5H5)Fe(CO)(PPh3)COCH(Me)Et (Figure 51).[45] In both complexes the *anti* (acyl oxygen to carbon monoxide or nitrosyl) conformation is preferred with the acyl alkyl group occupying zone A. For the rhenium acyl, however, the preferred conformation is much closer to that preferred on stereoelectronic grounds, *i.e.* perpendicular to the metal phosphorus bond, than for the iron acyl.

Consistent with stereoelectronic effects being more important for rhenium than iron are the observed acetyl carbonyl infra-red stretching frequencies, 1545 and 1603 cm^{-1}, for the complexes [(C5H5)Re(NO)(PPh3)COMe] and [(C5H5)Fe(CO)(PPh3)COMe] respectively.[34,56] The weaker carbon oxygen bond in the former case indicates that the resonance structure M(+)=CO(-)Me is a more important contributor for rhenium than for iron. This effect would also account for the greater stability of the rhenium formyl complex [(C5H5)Re(NO)(PPh3)CHO] over the corresponding iron formyl complex [(C5H5)Fe(CO)(PPh3)CHO].[2,56,66]

10.4. [(C5H5)Re(NO)(PPh3)CR2]+:

The X-ray crystal structure of the cationic carbene complex [(C5H5)Re(NO)(PPh3)CHPh]+ is illustrated in Figure 70.[57] As with the rhenium formyl and acyl complexes discussed above, stereoelectronic effects are important in determining the preferred conformation, which is with the carbene close to perpendicular to the rhenium phosphorus bond. The phenyl is orientated *syn* to the nitrosyl presumably to avoid the steric interactions with the cyclopentadienyl ligand, which would be present in the *anti* orientation.

10.5. Stereoselective Reactions:

The stereoselectivities observed for reactions of ligands attached to the rhenium chiral auxiliary parrallel those observed for the corresponding iron complexes. Illustrative examples are given in Figure 71. Nucleophilic additions to cationic

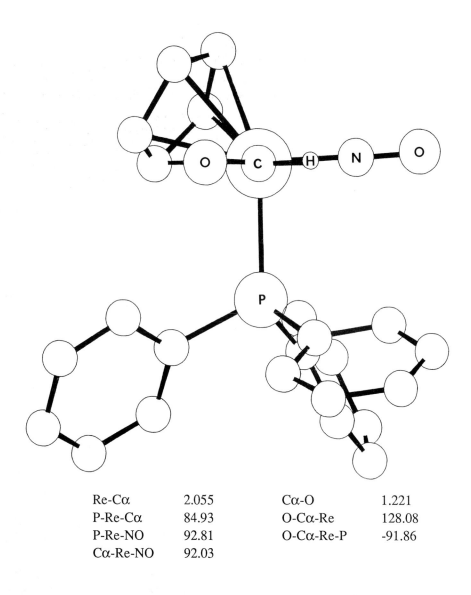

Re-Cα	2.055	Cα-O	1.221
P-Re-Cα	84.93	O-Cα-Re	128.08
P-Re-NO	92.81	O-Cα-Re-P	-91.86
Cα-Re-NO	92.03		

Figure 68. X-ray crystal structure and selected bond lengths (Å), angles and torsional angles (°) for [(C5H5)Re(NO)(PPh3)CHO]. The diagram shows the Newman projection along the *alpha* carbon to rhenium bond. Selected protons are removed for clarity.

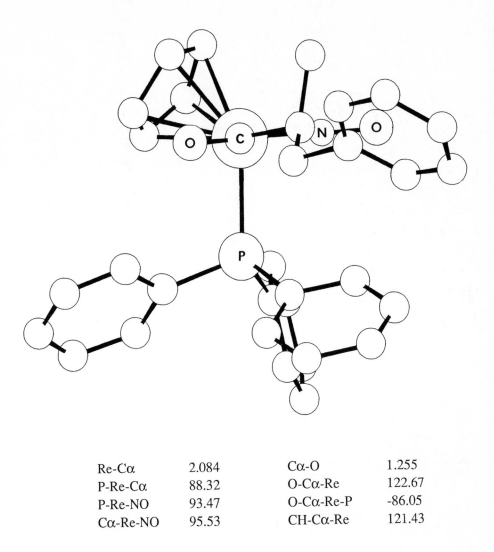

Re-Cα	2.084	Cα-O	1.255
P-Re-Cα	88.32	O-Cα-Re	122.67
P-Re-NO	93.47	O-Cα-Re-P	-86.05
Cα-Re-NO	95.53	CH-Cα-Re	121.43

Figure 69. X-ray crystal structure and selected bond lengths (Å), angles and torsional angles (º) for RS,SR-[(C5H5)Re(NO)(PPh3)COCH(Me)CH2Ph]. The diagram shows the Newman projection along the *alpha* carbon to rhenium bond. Selected protons are removed for clarity.

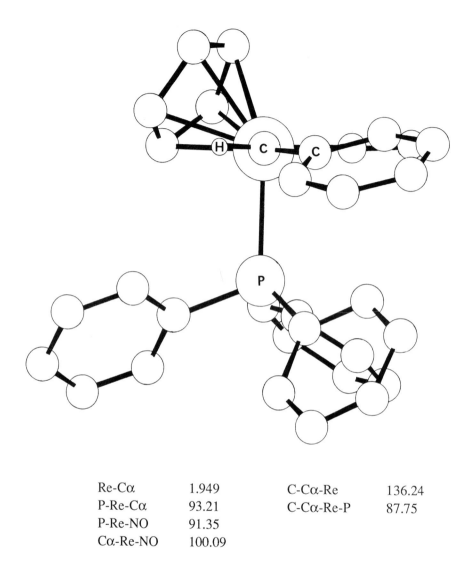

Re-Cα	1.949	C-Cα-Re	136.24
P-Re-Cα	93.21	C-Cα-Re-P	87.75
P-Re-NO	91.35		
Cα-Re-NO	100.09		

Figure 70. X-ray crystal structure and selected bond lengths (Å), angles and torsional angles (º) for [(C5H5)Re(NO)(PPh3)CHPh]+. The diagram shows the Newman projection along the *alpha* carbon to rhenium bond. Selected protons are removed for clarity.

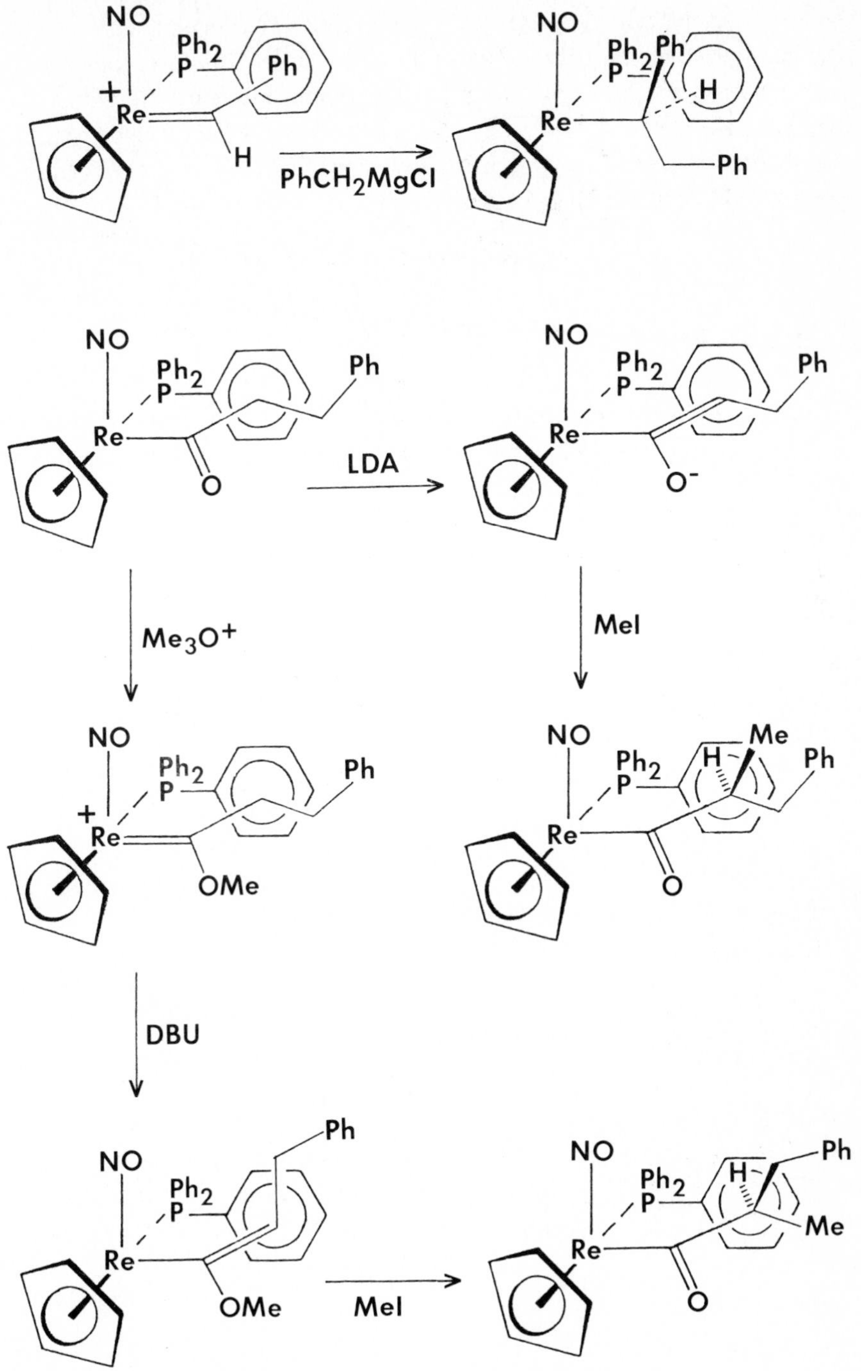

Figure 71. Stereoselective reactions of the ligands attached to the rhenium chiral auxiliary [(C5H5)Re(NO)(PPh3)].

carbene complexes occur onto the face away from the triphenylphosphine ligand.[57,58] Rhenium acyl complexes are deprotonated to generate the corresponding E-enolates, which undergo stereoselective alkylation reactions in the *anti* (enolate oxygen to nitrosyl) conformation from the face away from the triphenylphosphine.[55] Methoxy alkyl carbene cations are deprotonated to the corresponding Z-methoxy vinyl derivatives, which similarly undergo alkylation from the unhindered face in the *anti* (methoxyl to nitrosyl) conformation.[62]

11. CONCLUSIONS

Preferred conformations of ligands attached to the pseudo-octahedral chiral auxiliary $[(C_5H_5)Fe(CO)(PPh_3)]$ are generally determined by steric factors. The relative effective sizes of the ligands attached to the iron chiral auxiliary $[(C_5H_5)Fe(CO)(PPh_3)]$ are triphenylphosphine > cyclopentadienyl > carbon monoxide. The order of steric accessibility of the zones available to ligands attached to the iron chiral auxiliary $[(C_5H_5)Fe(CO)(PPh_3)]$ is, therefore, zone A (between the cyclopentadienyl and carbon monoxide) > zone B (between the cyclopentadienyl and triphenylphosphine) > zone C (between the triphenylphosphine and carbon monoxide). Polar effects only contribute significantly to the conformational preferences in complexes where there are only very small groups (*i.e.* H, O, S, etc., but not alkyl) attached to the *alpha* carbon. Stereoelectronic effects are not important in determining conformations, because the HOMO and the SHOMO on the iron are relatively close in energy and lie orthogonal to each other. Consequently, the preferred conformations of ligands attached to the iron chiral auxiliary can be successfully and easily modelled taking into account only van der Waals interactions. The high stereoselectivities observed during reactions on attached ligands are explained by approach of reagents from the least hindered face away from the triphenylphosphine ligand.

The approach outlined here of using detailed conformational analysis to explain structure - reactivity relationships for complexes derived from $[(C_5H_5)Fe(CO)(PPh_3)]$ should be generally applicable to other systems.

12. ACKNOWLEDGEMENTS: B.P. International Limited are gratefully acknowledged for a Venture Research Award. I (S.G.D.) am indebted to the students and post-doctoral assistants who have developed the work on the iron chiral auxiliary; their names are listed in the references. We thank Drs. Mike Hann, Mike Mingos and Jeff Seeman for helpful discussions. We also thank Professor Robert Bau for providing the X-ray crystal data of S-[(C5H5)Fe(CO)(PPh3)CH2O-(-)-menthyl].

13. REFERENCES

1. a. Green, M.L.H.; Hurley, C.R. *XIX I.U.P.A.C. Congress*, London, **1963**, Abstracts A, p. 195, b. *idem*, *J. Organometallic Chem.* **1967**, *10*, 188, c. Bibler, J.P.; Wojcicki, A. *Inorg. Chem.* **1966**, *5*, 889.

2. a. Seeman, J.I.; Davies, S.G. *J. Am. Chem. Soc.* **1985**, *107*, 6522, b. *idem*, *J. Chem. Soc., Chem. Commun.* **1984**, 1019, c. *idem*, *Tetrahedron Lett.* **1984**, 1845, d. Seeman, J.I.; Davies, S.G. *J. Chem. Soc., Dalton Trans.* **1985**, 2691.

3. Hoffmann, R.; Schilling, B.E.R.; Lichtenberger, D.L. *J. Am. Chem. Soc.* **1979**, *101*, 585.

4. a. Brunner, H.; Schmidt, E. *J. Organometallic Chem.* **1972**, *36*, C18, b. Brunner, H.; Strutz, J. *Z. Naturforch, Teil B*, **1974**, *29* 446, c. Reisner, M.G.; Bernal, I.; Brunner, H.; Muschiol, M. *Angew. Chem., Int. Ed. Eng.* **1976**, *15*, 776.

5. a. Flood, T.C.; Miles, D.L. *J. Am. Chem. Soc*, **1973**, *95*, 6460, b. Flood, T.C.; Di Santi, F.J.; Miles, D.L. *J. Chem. Soc., Chem. Commun.* **1975**, 336, c. Flood, T.C.; Di Santi, F.J.; Miles, D.L. *Inorg. Chem.* **1976**, *15*, 1910.

6. Davies, S.G.; Dordor-Hedgecock, I.M.; Sutton, K.H.; Walker, J.C.; Bourne, C.; Jones, R.H.; Prout, K. *J. Chem. Soc., Chem. Commun.* **1986**, 607.

7. Chou, C.-K.; Miles, D.L.; Bau, R.; Flood, T.C. *J. Am. Chem. Soc.* **1978**, *100*, 7271.

8. The iron acyl complex (C5H5)Fe(CO)(PPh3)COCH3 is available either as a racemate or in enantiomerically pure S-(+) and R-(-) forms from B.P. Chemicals Ltd., New Specialties Business, Belgrave House, 76 Buckingham Palace Road, London, SW1W OSU, U.K.

9. Cahn, R.S.; Ingold, C.; Prelog, V. *Angew. Chem. Int. Ed. Eng.* **1966**, *5*, 385.

10. a. Baird, M.C.; Stanley, K. *J. Am. Chem. Soc.* **1975**, *97*, 6598, b. Sloan, T.E. *Top. Stereochem.* **1981**, *12*, 1.

11. a. Mislow, K. *Acc. Chem. Res.* **1976**, *9*, 26, b. Mislow, K.; Gust, D. *J. Am. Chem. Soc.* **1973**, *95*, 1535, c. Brown. J.M.; Mertis, K. *J. Organometallic Chem.* **1973**, *47*, C5.

12. Chem-X, developed and designed by Chemical Design, Ltd., Oxford, England. These calculations were performed on the basis of X-ray crystal structure data using the default parameters which only take into account van der Waals interactions. Structures for which X-ray crystal data were not available were generated from the X-ray crystal structure of [(C5H5)Fe (CO)(PPh3)Et]. For the conformational energy profiles a negative angle represents a rotation in a clockwise fashion. The decriptors α, β, and γ refer to positions relative to the metal. These calculations are not intended to predict the absolute barrier to rotation, but to provide insight into the inherent steric forces present in these complexes.

13. Eliel, E.L.; Allinger, N.L.; Angyal, S.J.; Morrison, G.A. "Conformational Analysis"; Wiley-Interscience: New York, 1965.

14. This value is quoted in reference 3 and is attributed to J.W. Norton.

15. a. Davies, S.G.; Dordor-Hedgecock, I.M.; Sutton, K.H.; Whittaker, M. *J. Am. Chem. Soc.* **1987**, *109*, 5711, b. *idem*, *J. Organometallic Chem.* **1987**, *320*, C19, c. *idem*, *Chem. Ind. (London)*, **1987**, 338.

16. Waugh, J.S.; Fessenden, R.W. *J. Am. Chem. Soc.* **1957**, *79*, 846.

17. a. Karplus, M. *J. Am. Chem. Soc.* **1963**, *85*, 2870, b. Gorenstein, D.G. "Phosphorus-31 NMR, Principles and Applications"; Academic Press: Orlando, 1984.

18. a. Noggle, J.H.; Schirmer, R.E "The Nuclear Overhauser Effect"; Academic Press: New York, 1971, b. Derome, A.E. "Modern NMR Techniques for Chemistry Research"; Pergamon Press, Oxford, 1987.

19. a. Meinwald, J.; Meinwald, Y.C. *J. Am. Chem. Soc.* **1963**, *85*, 2514, b. Robinson, S.D.; Sahajpal, A. *Inorg. Chem.* **1977**, *16*, 2718, 2722, c. Wakatsuki, Y.; Aoki, K.; Yamazaki, H. *J. Am. Chem. Soc.* **1979**, *101*, 1123, d. Davies, S.G.; Moon, S.D.; Simpson, S.J.; Thomas, S.E. *J. Chem. Soc., Dalton Trans.* **1983**, 1805.

20. The calculation was performed using the Chem-X molecular modelling program.[12] The bond angles, lengths and torsional angles were taken directly from the X-ray crystal structure of [(C5H5)Fe(CO)(PPh3)Et][15] modifying the triphenylphosphine ligand to phenylphosphine: C_{ipso}-P-Fe-$C\alpha$ = 0°, C_{ortho}-C_{ipso}-P-Fe = 90°, C_{ipso}-P-Fe = 109°.

21. Morrison, J.D.; Mosher, H.S. "Asymmetric Organic Reactions"; American Chemical Society: Washington D.C., 1971, p 185.

22. Baird, M.C.; Hunter, B.K. *Organometallics*, **1985**, *4*, 1481.

23. a. Blackburn, B.K.; Davies, S.G.; Whittaker, M. *J. Chem. Soc., Chem. Commun.* **1987**, 1344, b. *idem*, in preparation.

24. Davies, S.G.; Maberly, T.R.; Jones, R.H. unpublished results.

25. a. Davies, S.G.; Maberly, T.R. unpublished results, b. T. R. Maberly, D. Phil. Thesis, Oxford University, 1986.

26.   Davies, S.G.; Maberly, T.R.; Jones, R.H.; Polywka, M.E.C.; Baird, G. in preparation b. G. Baird, D. Phil Thesis, Oxford University, 1985.

27.   Haasnoot, C.A.G.; de Leeuw, F.A.A.M.; Altona, C. *Tetrahedron*, **1980**, *36*, 2783.

28.   a. Ayscough, A.P.; Davies, S.G. *J. Chem. Soc., Chem. Commun.* **1986**, 1648, b. Ayscough, A.P.; Davies, S.G.; Sutton, K.H. unpublished results.

29.   Schilling, B.E.R.; Hoffmann, R.; Faller, J.W. *J. Am. Chem. Soc.* **1979**, *101*, 592 and see references therein.

30.   Kirby, A.J. "Anomeric Effect and Related Stereoelectronic Effects at Oxygen"; Springer-Verlag: Berlin, 1983.

31.   Crespi, A.M.; Shriver, D.F. *Organometallics*, **1985**, *4*, 1830.

32.   Davies, S.G.; Baird, G.J.; Maberly, T.R. *Organometallics*, **1984**, *3*, 1764.

33.   Aktogu, N.; Felkin, H.; Baird, G.J.; Davies, S.G.; Watts, O. *J. Organometallic Chem.* **1984**, *262*, 49.

34.   Davies, S.G.; Dordor-Hedgecock, I.M.; Easton, R.J.C.; Preston, S.C.; Sutton, K.H.; Walker, J.C. *Bull. Chim. Soc. Fr.* **1987**, 608 (see also Bernal, I.; Brunner, H.; Muschiol, M. *Inorg. Chim. Acta*, **1988**, *142*, 235).

35.   Davies, S.G.; Seeman, J.I.; Williams, I.H. *Tetrahedron Lett.* **1986**, *27*, 619.

36.   a. Davies, S.G.; Shipton, M.R. unpublished results, b. M.R. Shipton, Part II Thesis, Oxford University, 1987.

37.   Liebeskind, L.S.; Welker, M.E.; Fengl, R.W. *J. Am. Chem. Soc.* **1986**, *108*, 6328 and see references cited therein.

38.   Brown, S.L.; Davies, S.G.; Warner, P.; Jones, R.H.; Prout, K. *J. Chem. Soc., Chem. Commun.* **1985**, 1446.

39.   Davies, S.G.; Dordor-Hedgecock, I.M.; Warner, P.; Jones, R.H.; Prout, K. *J. Organometallic Chem.* **1985**, *285*, 213.

40.   Davies, S.G.; Dordor-Hedgecock, I.M.; Sutton, K.H.; Walker, J.C.; Jones, R.H.; Prout, K. *Tetrahedron*, **1986**, *42*, 5123.

41.   Davies, S.G.; Easton, R.J.C.; Sutton, K.H.; Walker, J.C.; Jones, R.H. *J. Chem. Soc., Perkin I*, **1987**, 489.

42.   Davies, S.G.; Gravatt, G.L.; Sutton, K.H.; Whittaker, M. unpublished results.

43.   Wakefield, B.J. in "Comprehensive Organic Chemistry" ed D. Barton and W.D. Ollis; Pergamon Press: Oxford, 1979, vol 3, p 943.

44.   Hammond, G.S. *J. Am. Chem. Soc.* **1955**, *77*, 334.

45.   Baird, G.J.; Davies, S.G.; Jones, R.H.; Prout, K.; Warner, P. *J. Chem. Soc., Chem. Commun.* **1984**, 745.

46.   Curtis, P.J.; Davies, S.G. *J. Chem. Soc., Chem. Commun.* **1984**, 747.

47.   Ayscough, A.P.; Davies, S.G. unpublished results.

48.   Brown, S.L.; Davies, S.G.; Foster, D.F.; Seeman, J.I.; Warner, P. *Tetrahedron Lett.* **1986**, *27*, 623.

49.   a. Davies, S.G.; Aktogu, N.; Felkin, H. *J. Chem. Soc., Chem. Commun.* **1982**,

222

1303, b. Davies, S.G.; Watts, O.; Aktogu, N.; Felkin, H. *J. Organometallic Chem.* **1983**, *243*, C51, c. Baird, G.J.; Davies, S.G. *J. Organometallic Chem.* **1983**, *248*, C1, d. Baird, G.J.; Bandy, J.A.; Davies, S.G.; Prout, K. *J. Chem. Soc., Chem Commun.* **1983**, 1202, e. Davies, S.G.; Dordor, I.M.; Walker, J.C.; Warner, P. *Tetrahedron Lett.* **1984**, 2709, f. Davies, S.G.; Dordor, I.M.; Warner, P. *J. Chem. Soc., Chem. Commun.* **1984**, 956, g. Davies, S.G.; Dordor-Hedgecock, I.M.; Warner, P. *Tetrahedron Lett.* **1985**, 2125, h. Ambler, P.W.; Davies, S.G. *Tetrahedron Lett.* **1985**, 2129, i. Davies, S.G.; Easton, R.J.C.; Walker, J.C.; Warner, P. *J. Organometallic Chem.* **1985**, *296*, C40, j. Davies, S.G.; Easton, R.J.C.; Walker, J.C.; Warner, P. *Tetrahedron*, **1986**, *42*, 175.

50. Weinstock, I.; Floriani, I.; Chiesi-Villa, A.; Guastini, C. *J. Am. Chem. Soc.* **1986**, *108*, 8298.

51. For example: a. Gould, T.J.; Balestra, M.; Wittman, M.D.; Gary, J.A.; Rossano, L.T.; Kallmerten, J. *J. Org. Chem.* **1987**, *52*, 3889, b. Heathcock, C.H.; Hagen, J.P.; Jarvi, E.T.; Pirrung, M.C.; Young, S.D. *J. Am. Chem. Soc.* **1981**, *103*, 4972.

52. Davies, S.G.; Wills, M. *J. Organometallic Chem.* **1987**, *328*, C29.

53. Davies, S.G.; Wills, M. *J. Chem. Soc., Chem. Commun.* **1987**, 1647.

54. a. Davies, S.G.; Walker, J.C. *J. Chem. Soc., Chem. Commun.* **1986**, 495, b. Davies, S.G.; Easton, R.J.C.; Gonzalez, A.; Preston, S.C.; Sutton, K.H.; Walker, J.C. *Tetrahedron*, **1986**, *42*, 3987, c. Davies, S.G.; Dordor-Hedgecock, I.M.; Sutton, K.H.; Walker, J.C. *Tetrahedron Lett.* **1986**, *27*, 3787.

55. a. Heah, P.C.; Patton, A.T.; Gladysz, J.A. *J. Am. Chem. Soc.* **1986**, *108*, 1185, b. Heah, P.C.; Gladysz, J.A. *J. Am. Chem. Soc.* **1984**, *106*, 7636, c. Buhro, W.E.; Wong, A.; Merrifield, J.H.; Lin, G.-Y.; Constable, A.C.; Gladysz, J.A. *Organometallics*, **1983**, *2*, 1852.

56. Bodner, G.S.; Patton, A.J.; Smith, D.E.; Georgiou, S.; Tam, W.; Wong, W.-K.; Strouse, C.E.; Gladysz, J.A. *Organometallics*, **1987**, *6*, 1954, and see references cited therein.

57. Kiel, W.A.; Lin, G.-Y.; Constable, A.G.; McCormick, F.B.; Strouse, C.E.; Eisenstein, O.; Gladysz, J.A. *J. Am. Chem. Soc.* **1982**, *104*, 4865.

58. a. Buhro, W.E.; Georgiou, S.; Fernandez, J.M.; Patton, A.T.; Strouse, C.E.; Gladysz, J.A. *Organometallics*, **1986**, *5*, 956, b. Kiel, W.A.; Buhro, W.E.; Gladysz, J.A. *Organometallics*, **1984**, *3*, 879, c. Hatton, W.G.; Gladysz, J.A. *J. Am. Chem. Soc.* **1983**, *105*, 6157.

59. Georgiou, S.; Gladysz, J.A. *Tetrahedron*, **1986**, *42*, 1109.

60. Merrifield, J.H.; Strouse, C.E.; Gladysz, J.A. *Organometallics*, **1982**, *1*, 1204.

61. McCormick, F.B.; Gleason, W.B.; Zhao, X.; Heah, P.C.; Gladysz, J.A.

Organometallics, **1986**, *5*, 1778.

62. Smith, D.E.; Gladysz, J.A. *Organometallics*, **1985**, *4*, 1480.

63. Buhro, W.E.; Georgiou, S.; Hutchinson, J.P.; Gladysz, J.A. *J. Am. Chem. Soc.* **1985**, *107*, 3346.

64. Crocco, G.L.; Gladysz, J.A. *J. Chem. Soc., Chem. Commun.* **1986**, 1154.

65. Wong, W.-K.; Tam, W.; Strouse, C.E.; Gladysz, J.A. *J. Chem. Soc., Chem. Commun.* **1979**, 530.

66. Tam, W.; Lin, G.-Y.; Wong, W.-K.; Kiel, W.A.; Wong, V.K.; Gladysz, J.A. *J. Am. Chem. Soc.* **1982**, *104*, 141.

67. This value is taken from the EHMO calculations performed by Hoffmann and coworkers for the unsubstituted analogue $[(C_5H_5)Mo(CO)(PH_3)]^+$ and is the energy difference between the a'' (HOMO) and 2a' (SHOMO) molecular orbitals for $[(C_5H_5)Fe(CO)(PPh_3)]$.[29]

68. Fenske, R.F.; Milletti, M.C.; Arndt, M. *Organometallics*, **1986**, *5*, 2316.

69. This value is the calculated energy difference between the HOMO and SHOMO on $[(C_5H_5)Re(NO)(PH_3)CH_3]$[68] where there is expected to be no substantial orbital interactions from the metal to alkyl ligand and should, therefore, act as a satisfactory model for the fragment $[(C_5H_5)Re(NO)(PH_3)]$

70. Fukui, K. *Angew. Chem. Int. Ed. Eng.* **1982**, *21*, 801.

71. Manganiello, F.J.; Radcliffe, M.D.; Jones, W.M. *J. Organometallic Chem.* **1982**, *228*, 273, b. Brookhart, M.; Tucker, J.A.; Flood, T.C.; Jensen, J. *J. Am. Chem. Soc.* **1980**, *102*, 1203, c. Davison, A.; Reger, D.L. *J. Am. Chem. Soc.* **1972**, *94*, 9237, d. Brookhart, M.; Tucker, J.R.; Husk, G.R. *J. Am. Chem. Soc.* **1983**, *105*, 258, e. Boland-Lussier, B.E.; Churchill, M.R.; Hughes, R.P.; Rheingold, A.L. *Organometallics*, **1982**, *1*, 628.

STERIC AND ELECTRONIC EFFECTS ON THE PHOTOCHEMICAL REACTIONS OF METAL-METAL BONDED CARBONYLS

D.J. Stufkens

STERIC AND ELECTRONIC EFFECTS ON THE PHOTOCHEMICAL REACTIONS OF METAL-METAL BONDED CARBONYLS

D.J. Stufkens

1. LIST OF ABBREVIATIONS

bpy	2,2'-bipyridyl
bquin	2,2'-biquinoline
Cp	η^5-cyclopentadienyl
Cp*	η^5-pentamethylcyclopentadienyl
Cy	cyclohexyl
depe	1,2-bis(diethylphosphino)ethane
dien	diethylenetetramine
dmpe	1,2-bis(dimethylphosphino)ethane
dmpm	bis(dimethylphosphino)methane
dppe	1,2-bis(diphenylphosphino)ethane
dppm	bis(diphenylphosphino)methane
DTBQ	3,5-di-tert-butyl-o-quinone
EPA	mixture of ethanol, iso-pentane and ether (2:5:5)
esr	electron spin resonance
Et	ethyl
ETC	Electron Transfer Catalysis
HPLC	High Performance Liquid Chromatography
i-Pr	iso-Propyl
ir	infrared
LF	Ligand Field
MCH	Methylcyclohexane

Me	Methyl
2-MeTHF	2-Methyltetrahydrofuran
MLCT	Metal to Ligand Charge Transfer
n-Bu	normal-Butyl
OEt	Ethoxy
OMe	Methoxy
O-i-Pr	iso-Propoxy
OPh	Phenoxy
PE	Photoelectron
Ph	Phenyl
phen	1,10-phenanthroline
PVC	Polyvinylchloride
py	pyridine
R-DAB	1,4-diaza-1,3-butadiene
R-PyCa	pyridine-2-carbaldehyde-imine
t-Bu	tertiary-Butyl
TCNE	Tetracyanoethylene
tetraphos	1,1,4,7,10,10-hexaphenyl-1,4,7,10-tetra-phosphadecane
THF	Tetrahydrofuran
triphos	1,1,1-tris(di-phenylphosphinomethyl)ethane
uv	ultraviolet
vis	visible

2   INTRODUCTION

The photophysics and photochemistry of transition metal coordination compounds have been the subject of many studies during and after the development of Ligand Field theory. Already in 1967 (ref. 1) Adamson presented his semiempirical rules for the photoreactions of Cr(III) complexes which stimulated further research in this field. On the other hand, systematic photochemical studies on organometallic complexes have been scarce before the seventies. Moreover, these studies were often performed for synthetic purposes and hardly gave mechanistic information. More fundamental and systematic studies in this field were started about fifteen years ago. At that time new spectroscopic techniques, and at a later stage, flash photolysis came in use for photochemical investigations. The development of time-resolved spectroscopy made it possible to follow photochemical reactions by studying the uv/vis (ref. 2), ir (ref. 3) or resonance Raman (ref. 4) spectral changes. Besides, various solid media came in use for the identification of primary photoproducts, such as 2-MeTHF glasses, inert gas matrices and PVC films. More recently liquid Xenon has been introduced as a superinert and transparent solvent for the stabilization and characterization of reactive intermediates in photochemical reactions (ref. 5). In the field of Organometallic Photochemistry these new techniques have mainly been applied for mechanistic studies on transition metal carbonyls for which the ground- and excited state properties are best known. The same holds for the metal-metal bonded systems, for which systematic mechanistic studies have mainly been confined to the carbonyl dimers. In this review only these latter complexes will be discussed. Moreover, most attention will be paid to mechanistic aspects without giving a general survey of all reactions studied.

The photochemical reactions of these metal-metal bonded dimers take place from an excited state, which is strongly metal-metal or metal-ligand antibonding. The primary photoprocesses are therefore dissociative in nature and steric factors are not very important. They play, however, an important role in the secondary thermal reactions of the primary photoproducts.

The photochemistry of metal-metal bonded complexes was reviewed in 1979 by Geoffroy and Wrighton (ref. 6) and a few years ago by Meyer and Caspar (ref. 7). More specific reactions of these systems have been described by Stiegman and Tyler (ref. 8,9) and by Kobayashi et al. (ref. 2) and reactions with special ligands by Kreiter (ref. 10).

The results of recent studies show that a close agreement exists between the fundamental photoprocesses of the various metal-metal bonded dimers. It is therefore not useful to discuss here the mechanistic details of all these complexes. This will only be done for one representative group of complexes, $M_2(CO)_{10}$ (M=Mn,Re) and their derivatives. For the other dimers only those reactions which are specific for that group of complexes, will be discussed.

3      $M_2(CO)_{10}$ (M=Mn,Re)

3.1   Structural and Spectroscopic Properties

From these complexes, $Mn_2(CO)_{10}$ is the first binuclear carbonyl complex ever synthesized (ref. 11). According to their crystal structures, all species

$M_2(CO)_{10}$ (M=Mn,Tc,Re) and also $ReMn(CO)_{10}$ are isomorphous (ref. 12-16). The molecules have approximate D_{4d} symmetry and a schematic representation of their structure is shown in Fig. 1.

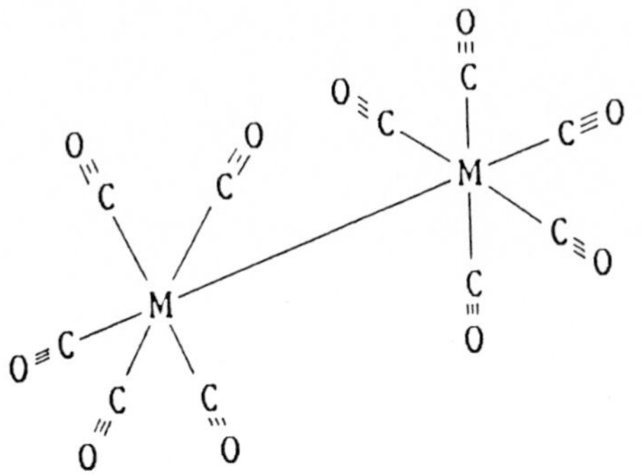

Fig. 1.   Structure of $M_2(CO)_{10}$ (M=Mn,Re).

    The crystal structure of $ReMn(CO)_{10}$ has recently been redetermined (ref. 16). It shows a short Re-Mn distance of 2.909(1)Å compared to the Mn-Mn bond length in $Mn_2(CO)_{10}$ (2.9038(6)Å) and the Re-Re bond length in $Re_2(CO)_{10}$ (3.0413(11Å)). This short Re-Mn distance is consistent with the M-M bond dissociation energies of these complexes, viz. D(Mn-Mn) = 94(13) kJ mol^{-1}, D(Re-Re)=187(5) kJ mol^{-1} and D(Re-Mn)=210(10) kJ mol^{-1} (ref. 17) and with the net overlap population of the metal fragments, derived theoretically (ref. 18). It also agrees with the results from the uv Photoelectron (PE) spectra of these complexes which show the highest Ionization Potential for the metal-metal bonding orbital σ_b(M-M') in the case of $ReMn(CO)_{10}$ (ref. 19-22). All m.o. calculations (ref. 18,21,23-25) and PE data are consistent with the relative ordering of orbitals shown in Fig. 2. The PE spectra give Ionization Energies for σ_b (M-M') which are 0.3-0.5 eV lower than those of the metal dπ-orbitals. These latter orbitals are very close in energy both in the m.o. calculations and PE spectra, which means that the interaction between these orbitals across the metal-metal bond is only a weak one.

    The electronic absorption spectra of these complexes (ref. 23,26,27) are characterized by an intense band (II) between 300 and 350 nm (ε=13,000 – 22,000lmol^{-1}cm^{-1}). In the case of $Mn_2(CO)_{10}$ a second, weaker band (I) is observed at lower energy (~390 nm, see Fig. 3), while the spectrum of

$Re_2(CO)_{10}$ shows an extra intense band (III) at 275 nm.

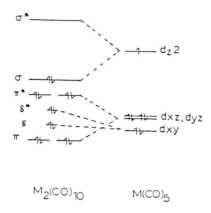

Fig. 2.   Qualitative m.o. diagram of $M_2(CO)_{10}$ (M=Mn,Re).

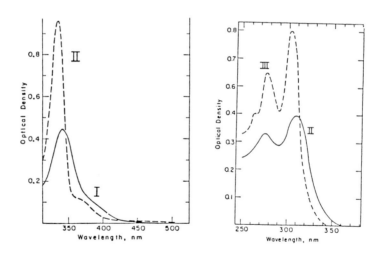

Fig. 3.   Absorption spectra of $Mn_2(CO)_{10}$ (left) and $Re_2(CO)_{10}$ (right) in EPA at 298 K (—) and 77 K (---) (reproduced with permission from ref. 27).

Band II has been assigned to the strongly allowed $\sigma_b \rightarrow \sigma^*$(M-M') transition (ref. 23,26). Such a transition is characteristic for all two-electron metal-metal bonded homo- and heterodinuclear species. Support for this assignment comes from polarization measurements according to which this band appears to be polarized along the metal-metal bond (ref. 26). Band I has been assigned to the x,y polarized metal $d\pi \rightarrow \sigma^*(M-M')$ transition. The fact that this transition falls lower than $\sigma_b \rightarrow \sigma^*(M-M')$ in $Mn_2(CO)_{10}$ has been attributed by Levenson and Gray (ref. 23) to differences in interelectronic repulsion effects in the two excited states. Recently, Meyer and Caspar have proposed an alternative assignment of this band to a d-d transition in the equatorial plane (ref. 7). This assignment was based on the observation of release of an equatorial CO-ligand upon irradiation into this band and on the presence of a d-d transition at about the same wavelength in the mononuclear $Mn(CO)_5$ I complex. The band at 275 nm in the spectrum of $Re_2(CO)_{10}$ has been assigned to $\sigma_b,d\pi \rightarrow \pi^*(CO)$ transitions. Substitution of one or more CO ligands by a nucleophile such as PPh_3 or t-BuNC causes a shift of the $\sigma_b \rightarrow \sigma^*$ transition to lower energy (ref. 27,28). This effect is consistent with the observed weakening of the metal-metal bond in the crystal structures (ref. 29).

A special group of substitution products are those in which two carbonyls of one metal fragment are replaced by a chelating α-diimine ligand such as 2,2'-bipyridine(bpy), 1,10-phenanthroline(phen), pyridine-2-carbaldehyde-imine (R-PyCa) or 1,4-diaza-1,3-butadiene (R-DAB) (see Fig. 17). Since these ligands possess a low-lying π* orbital, absorption spectra of their $(CO)_5MM'(CO)_3$(α-diimine) (M,M'=Mn,Re) complexes are characterized by intense metal to α-diimine charge transfer (MLCT) transitions in the visible region (ref. 30-32). Irradiation of these complexes into this MLCT band causes a homolytic splitting of the metal-metal bond (ref. 30,33-36) and, based on this observation, Morse and Wrighton (ref. 30) assigned this band to a $\sigma_b \rightarrow \pi^*$(α-diimine) transition. The metal-metal bond will then be weakened in the excited state which explains the photochemistry. There are, however, strong arguments against this assignment. First of all there is a striking similarity between the energies and intensities of these bands and those of the mononuclear d^6-[$M(CO)_4L$ (M=Cr,Mo,W)] (ref. 37-43) and d^8-[$M'(CO)_3L$ (M'=Fe,Ru)] (ref. 44-45) analogues. Secondly, the metal-metal bond is hardly affected by this electronic transition since the metal-metal vibration does not show a resonance Raman effect upon excitation into this band (ref. 32). These low-energy structured bands have therefore been assigned to transitions from different metal $d\pi$ orbitals, not involved in the metal-metal bond, to the lowest π* orbital of the α-diimine ligand (ref. 32). It will be shown in a later section how the photochemistry of these complexes can still be explained in the light of this reassignment.

3.2 Primary Photoprocesses

What sort of photochemistry might be expected for these $M_2(CO)_{10}$ complexes given that the low-lying electronic transitions are of the $d\pi$-σ^* and
$\pi_b \rightarrow \sigma^*$ type? During both transitions, the metal-metal anti-bonding orbital is
populated which causes a weakening or breaking of the metal-metal bond.
Especially the $\sigma_b \rightarrow \sigma^*$ transition yields an excited state in which the metal-metal
bond is completely broken. Photolysis of these dimers will therefore lead to
rupture of the metal-metal bond and to formation of two 17-electron metal
radicals. This reaction does indeed occur and homolytic splitting of the
metal-metal bond (eq. 1) has long been assumed to be the only primary
photoprocess (ref. 6,27,46-50).

$$M_2(CO)_{10} \quad \overset{h\nu}{\underset{\leftarrow}{\rightarrow}} \quad 2\ M(CO)_5 \tag{1}$$

Wrighton and co-workers were the first to perform a quantitative study of
the photochemistry of these complexes and some of their derivatives in
CCl_4 (ref. 27,46). The disappearance quantum yields of the reactants in eq. 2,
which was assumed to be a radical trap reaction, appeared to be high (0.4-0.7).
The observation of a clean cross-coupling

$$M_2(CO)_{10} \quad \overset{h\nu}{\underset{CCl_4}{\overset{\rightarrow}{\leftarrow}}} \quad 2\ M(CO)_5Cl \tag{2}$$

reaction between $Mn_2(CO)_{10}$ and $Re_2(CO)_{10}$ to give $ReMn(CO)_{10}$ (ref. 27) also
led to the conclusion that metal-metal bond homolysis was the only primary
photoprocess. Moreover, for the reaction of $Re_2(CO)_{10}$ with I_2, the
disappearance quantum yield of $Re_2(CO)_{10}$ appeared to decrease from 0.64 in
isopentane to 0.30 in Nujol (ref. 27). This viscosity dependence confirmed the
formation of free $M(CO)_5$ radicals upon photolysis. The $Mn(CO)_5$ radicals could
in fact be detected with esr by irradiating a solution of $Mn_2(CO)_{10}$ in the
cavity of an esr spectrometer in the presence of nitrosodurene as a spin trap
(ref. 51). Photolysis of $Mn_2(CO)_{10}$ in the presence of 0.1 M PPh_3 gave
$Mn_2(CO)_8(PPh_3)_2$ as the major photoproduct and only a small amount of
$Mn_2(CO)_9(PPh_3)$ (ref. 27). $Mn_2(CO)_8(PPh_3)_2$ was proposed to be formed out of
two $Mn(CO)_4(PPh_3)$ radicals after homolytic splitting of the metal-metal bond
and substitution of a CO ligand of the $Mn(CO)_5$ radical by PPh_3. $Mn_2(CO)_9$

(PPh$_3$) could have been formed both by reaction of the Mn(CO)$_5$ and Mn(CO)$_4$(PPh$_3$) radicals and by photosubstitution of CO in Mn$_2$(CO)$_{10}$ by PPh$_3$. If the latter reaction were of importance here, Mn$_2$(CO)$_9$(PPh$_3$) would have been a major reaction product at short irradiation times. A few years later, however, it was established that the relative amounts of Mn$_2$(CO)$_8$(PPh$_3$)$_2$ and Mn$_2$(CO)$_9$(PPh$_3$) were strongly dependent on the phosphine concentration used (ref. 7,52) (vide infra).

Meyer and co-workers (ref. 53) were the first who proposed the formation of two primary photoproducts with very different lifetimes. They studied the photodecomposition of Mn$_2$(CO)$_{10}$ in cyclohexane and THF with conventional flash photolysis and observed regeneration of Mn$_2$(CO)$_{10}$ by both a rapid and slow process. The short-lived intermediate was proposed to be the Mn(CO)$_5$ radical. Although the second intermediate had an absorption band at 500 nm, consistent with that of Mn$_2$(CO)$_9$ reported later (ref. 55,56), its lifetime ($\tau_{1/2}$=26 sec.) agreed with that of a product observed when the solvent was insufficiently deoxygenated before use (ref. 55).

A few years later Fox and Poë (ref. 54) reported the concentration and light intensity dependence of the quantum yield of the photochemical reaction of Mn$_2$(CO)$_{10}$ with CCl$_4$. The results of this study could only be explained by assuming the formation of a second primary photoproduct, presumably an isomer of Mn$_2$(CO)$_{10}$, out of the excited species Mn$_2$(CO)$^*_{10}$.

Definite proof for the occurrence of two primary photoprocesses was given by two independent laser flash photolysis studies (ref. 55,56). The transient absorption spectrum, recorded by Rothberg et al. (ref. 56) 25 ps after photolysis of Mn$_2$(CO)$_{10}$ in ethanol, showed two maxima with λ_{max}780 and 480 nm, respectively. The band at 780 nm was assigned to the Mn(CO)$_5$ radical by comparison with pulse radiolysis (ref. 57) and matrix isolation (ref. 58) data. In the CO-matrix study (ref. 58) the Mn(CO)$_5$ radical had been synthesized by uv photolysis of HMn(CO)$_5$ and fully characterized by ir and uv/vis spectroscopy. The band at 480 nm in the transient spectrum (ref. 56) is not present in the spectrum of Mn(CO)$_5$. It was assigned to Mn$_2$(CO)$_9$(EtOH), formed by loss of CO from Mn$_2$(CO)$_{10}$ and coordination of a solvent molecule (eq. 6). Its absorption band is close to that of Mn$_2$(CO)$_9$(pyr) (pyr=pyridine) (ref. 59). Since the transients were already observed 25 ps after the flash, they were both assigned to primary photoproducts (eq. 3-4).

$$M_2(CO)_{10} \quad \xrightarrow{h\nu} \quad \begin{array}{l} \longrightarrow \quad 2\ M(CO)_5 \qquad\qquad (3) \\ \\ \longrightarrow \quad M_2(CO)_9 + CO \qquad (4) \end{array}$$

$$2\ M(CO)_5 \quad \xrightarrow{k_1} \quad M_2(CO)_{10} \qquad\qquad (5)$$

$$M_2(CO)_9 + L \quad \xrightarrow{k_2} \quad M_2(CO)_9 L \qquad\qquad (6)$$

Yesaka et al. (ref. 55) obtained similar results with nanosecond flash photolysis of $Mn_2(CO)_{10}$ in cyclohexane. By adding L=EtCN or MeCN to the solution they observed the formation of the stable $Mn_2(CO)_9 L$ product out of the $Mn_2(CO)_9$ intermediate (eq. 6). With the use of the transient absorbances and molar absorptivity of $Mn_2(CO)_9(MeCN)$, the concentrations of the primary photoproducts could be determined. From these data quantum yields of around 0.3 were derived for both reactions (3) and (4), which means that Mn-CO bond scission is roughly as important as Mn-Mn bond breaking. The lifetimes of the intermediates are very different. The $Mn(CO)_5$ radical is short-lived and dimerizes to the parent compound, according to reaction (5), with a second-order rate constant k_1 of about $109\ mol^{-1}\ s^{-1}$. This value is somewhat solvent dependent. The $Mn_2(CO)_9$ intermediate reacts back to $Mn_2(CO)_{10}$ (eq 6 for L=CO) with a rate constant, k_2, of about $10^5 mol^{-1}\ s^{-1}$. For L=RCN or PR_3 this rate constant k_2 is much larger (10^7-$10^9\ mol^{-1}\ s^{-1}$). This result is consistent with previous observations (ref. 60,61) on the reaction of $Cr(CO)_5$ with MeCN in cyclohexane for which the rate constant is much larger ($1.6 \times 10^8\ mol^{-1}\ s^{-1}$) than for the backreaction with CO ($3 \times 10^6\ mol^{-1}\ s^{-1}$). This difference in lifetime between the two primary photoproducts largely determines the observed concentration dependence of the photosubstitution reactions of $Mn_2(CO)_{10}$ by phosphines and nitriles (ref. 7,52,55) (vide infra).

Photolysis of $Mn_2(CO)_{10}$ and $Re_2(CO)_{10}$ in the gas-phase (ref. 62) gave mass spectrometric evidence that, in addition to $M(CO)_5$ fragments, previously reported by Freedman and Bersohn (ref. 49), partially decarbonylated dimetal photofragments are also produced. Thus, also in the gas-phase, loss of CO and M-M bond scission are primary photoprocesses of these complexes. A similar result has been reported recently by Seder et al. (ref. 63) for the photodis-sociation reaction of $Mn_2(CO)_{10}$ in the gas-phase. These authors reinvestigated this reaction with time-resolved ir spectroscopy for a better identification of the reaction products. Time-resolved ir spectroscopy is a new and very valuable technique for the study of photochemical reactions of metal carbonyls. Irradiation of the complexes normally takes place with a nanosecond laser and ir spectra are measured point by point directly after the flash with a CO or diode laser as ir light source (ref. 3). With the use of this time-resolved technique the reaction products in the gas-phase were identified as $Mn(CO)_5$ and $Mn_2(CO)_9$ (ref. 63).

The same technique was used by Church et al. (ref. 64) to study the formation and decay kinetics of $Mn(CO)_5$ and $Mn_2(CO)_9$ in solution. The relative yields of these species and the reformation rates of $Mn_2(CO)_{10}$ indicated that the initial concentrations of $Mn(CO)_5$ and $Mn_2(CO)_9$ were approximately the same, in agreement with the findings of Yesaka et al. (ref. 55).

The $Mn_2(CO)_9$ species can also be stabilized and characterized well with conventional ir spectroscopy by photolysis of $Mn_2(CO)_{10}$ in low-temperature alkane (ref. 52) and Ar (ref. 65) matrices. Irradiation of $Mn_2(CO)_{10}$ in a 3-methylpentane glass at 77 K resulted in the formation of $Mn_2(CO)_9$ as the only photoproduct (ref. 52). The $Mn(CO)_5$ radicals cannot be observed here since they recombine instantaneously to the parent compound in the cage of the matrix. The ir spectrum of $Mn_2(CO)_9$ shows a band at 1760 cm^{-1} belonging to a (semi) bridging CO group. Warming up the matrix to 298 K led to regeneration of $Mn_2(CO)_{10}$. Irradiation of $Mn_2(CO)_{10}$ in the alkane matrix in the presence of PPh_3 also yielded $Mn_2(CO)_9$, which product reacted with the PPh_3 to $Mn_2(CO)_9(PPh_3)$ upon raising the temperature. The 1760 cm^{-1} band was not observed upon photolysis in a 2-MeTHF matrix due to the formation of $Mn_2(CO)_9(2\text{-MeTHF})$. Similarly, photolysis of $Mn_2(CO)_{10}$ in an Argon matrix at 12 K yielded $Mn_2(CO)_9$ as the only photoproduct. When this reaction was performed with planepolarized light, no dichroism was observed for the bridging CO vibration of $Mn_2(CO)_9$ at 1764 cm^{-1} (ref. 65).

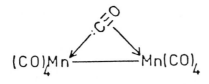

Fig. 4.   Proposed structure of $Mn_2(CO)_9$ (reproduced with permission from ref. 65).

It was therefore concluded that the band has an associated transition dipole oriented approximately 45° from the photoactive transition dipole of $Mn_2(CO)_{10}$. According to the authors, a structure of $Mn_2(CO)_9$ with a semi-bridging carbonyl group (Fig. 4) is most plausible. This structure is fully consistent with that of $(CO)_4Mn(\sigma,\sigma,\eta^2\text{-R-PyCa})Mn(CO)_3$ formed by photolysis of $(CO)_5MnMn(CO)_3(\sigma,\sigma\text{-R-PyCa})$ (see Fig. 20).

According to Yesaka et al. (ref. 55) and Church et al. (ref. 64) the quantum yields of $Mn(CO)_5$ and $Mn_2(CO)_9$ formation are nearly the same upon 350 nm excitation. This result, however, does not agree with the findings of Hepp and Wrighton (ref. 52) who estimated from their study of the photochemical reactions of $Mn_2(CO)_{10}$ with PPh_3 and CH_3CN, in the absence and presence of CCl_4, a quantum yields ratio of about 2:1 for $Mn(CO)_5$ and $Mn_2(CO)_9$ formation. This discrepancy might be caused by the fact that Hepp and Wrighton, contrary to the other two groups, used polychromatic light for their photolysis experiments. Quite recently, Kobayashi et al. reported the excitation wavelength dependence of the quantum yields of $Mn(CO)_5$ and $Mn_2(CO)_9$ formation (ref. 66). At 266 nm these quantum yields are 0.16 and 0.84, respectively, at 337 nm, 0.30 and 0.70; and at 355 nm, 0.49 and 0.44. They were calculated with the assumption that their sum is equal to unity.

Recently it has been established, with nanosecond flash photolysis (ref. 67), that also in the case of $Re_2(CO)_{10}$ two primary photoprocesses occur. Contrary to $Mn_2(CO)_{10}$, however, the quantum yield ratio between the homolysis and CO loss reaction is now 0.13 upon 355 nm excitation (~1.0 for $Mn_2(CO)_{10}$). The absorption band of $Re(CO)_5$ is observed at 550 nm, which is a large shift with respect to the corresponding band of $Mn(CO)_5$. The absorption band of $Re_2(CO)_9$ has its maximum at 420 nm. The second-order rate constants for reactions (5) and (6) are of the same order of magnitude as for $Mn_2(CO)_{10}$.

How can we explain the occurrence of these two primary photoprocesses in the light of the electronic transitions of these complexes discussed in the preceeding section? Stiegman and Tyler (ref. 9) and Meyer and Caspar (ref. 7) have pointed out that the bonding of the $M_2(CO)_{10}$ complexes in their σ_b^2 ground- and $\sigma_b\sigma^*$ singlet and triplet excited states is analogous to that of H_2. Coulson and Fischer (ref. 68) have calculated the relative energies of the σ_b^2 and $\sigma_b\sigma^*$ states of H_2 as a function of the H-H normal coordinate. A qualitative energy diagram, based on these results and applied for the $M_2(CO)_{10}$ complexes, is shown in Fig. 5.

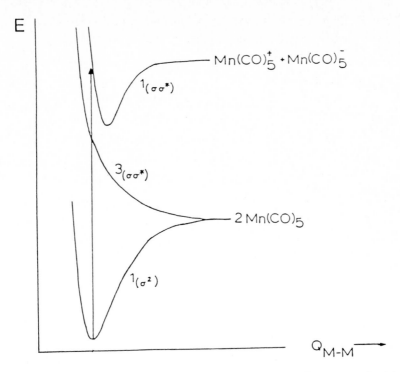

Fig. 5. Schematic energy level diagram of $Mn_2(CO)_{10}$ (reproduced with permission from ref. 7).

The $^1\sigma_b\sigma^*$ state correlates with the $Mn(CO)_5^+$ and $Mn(CO)_5^-$ ions. Irradiation into the $\sigma_b \rightarrow \sigma^*$ transition might therefore lead to a heterolytic splitting of the metal-metal bond, provided that the $^1\sigma_b\sigma^*$ state is reached above its dissociation limit. Only in one case has such a heterolytic splitting been proposed, so far (ref. 69). If such a disproportionation reaction does not take place, fast intersystem crossing to the repulsive $^3\sigma_b\sigma^*$ state will lead to homolytic splitting of the metal-metal bond.

Irradiation into the low-energy $d\pi \rightarrow \sigma^*$ transition causes a deficiency in the $d\pi$ orbitals and therefore a weakening of the metal-CO bonds. At the same time the occupation of σ^* leads to a weakening of the metal-metal bond. Occupation of $^3d\pi\sigma^*$, after intersystem crossing from $^1d\pi\sigma^*$, may therefore be responsible for the formation of both $M(CO)_5$ and $M_2(CO)_9$ from that state. Alternatively, the scission of the metal-metal bond upon low-energy excitation can take place from the $^3\sigma_b\sigma^*$ state after a surface crossing from $^3d\pi\sigma^*$ to this state. As has been mentioned in the preceeding section, Meyer and Caspar (ref. 7) assigned the low-energy band of $Mn_2(CO)_{10}$ to a $d \rightarrow d$ transition in the xy plane of the complex in analogy with similar transitions in mononuclear Mn-complexes. Occupation of such a dd state leads to loss of an equatorial CO group. Although there is no compelling evidence for loss of an equatorial instead of an axial CO group as the primary process, all photoproducts formed by loss CO are equatorially substituted. More important, only eq. $Re_2(CO)_9(N_2)$ is formed upon irradiation of $Re_2(CO)_{10}$ in a N_2-matrix (ref. 70).

According to Kobayashi et al. (ref. 66) (vide supra) the quantum yield of metal-CO scission increases with the energy of excitation with respect to that of metal-metal bond breaking. This effect is most probably due to the presence of $d\pi \rightarrow \pi^*(CO)$ transitions at higher energy.

3.3 Secondary Thermal Reactions

As noted in the preceeding section, the $Mn(CO)_5$ radical has fully been characterized by ir and uv/vis spectroscopy after photolysis of $HMn(CO)_5$ in a CO-matrix (ref. 58). The corresponding $Re(CO)_5$ radical has been generated and characterized by depositing Re atoms in a CO/Ar matrix (ref. 71). In solution, at room temperature, these $M(CO)_5$ species cannot be observed with conventional spectroscopic techniques since their recombination is nearly diffusion-controlled (ref. 72). They have, however, been studied with esr using a spin trap (ref. 51). Substituted radicals $Mn(CO)_4(PR_3)$ were identified in a similar way (ref. 73,74). These substituted radicals $Mn(CO)_{5-n}L_n$ normally have much longer lifetimes than the $Mn(CO)_5$ radical especially in the case of bulky ligands L. Walker et al. (ref. 72) subjected a series of $Mn_2(CO)_8(PR_3)_2$ complexes to flash photolysis and measured the second order recombination rate constants k_r of the radicals $Mn(CO)_4(PR_3)$ as a function of R. All k_r's appeared to be smaller than the value 9×10^8 $Mol^{-1}s^{-1}$ determined for $Mn_2(CO)_{10}$ by factors ranging from 9 to more than 200. For the trialkylphosphines, k_r decreased steadily with increasing cone angle. On the other hand, electronic effects hardly influenced k_r since closely similar values were obtained for the radicals with $P(n-Bu)_3$ and $P(OPh)_3$. These ligands have approximately the same cone angle but they differ significantly in electronic character. Apparently, only steric effects influence the recombination rate constant. A similar influence of steric effects on the rate constant has been observed for the intramolecular recombination reaction of the $(CO)_4Re\overparen{LL}Re(CO)_4$ biradicals (ref. 75).

Brown and co-workers (ref. 48) have studied in detail the esr spectra of the persistent radicals $Mn(CO)_3$ $(P(n-Bu)_3)_2$ and $Mn(CO)_3(P(OEt)_3)_2$ which were produced by photolysis of $Mn_2(CO)_8(PR_3)_2$ in the presence of PR_3 (reactions 7-10).

$$Mn_2(CO)_8L_2 \underset{\leftarrow}{\overset{h\nu}{\rightarrow}} 2Mn(CO)_4L \qquad (7)$$

$$Mn(CO)_4L + L \rightarrow Mn(CO)_3L_2 + CO \qquad (8)$$

$$Mn(CO)_3L_2 + Mn(CO)_4L \underset{h\nu}{\overset{\rightarrow}{\leftarrow}} Mn_2(CO)_7L_3 \qquad (9)$$

$$2Mn(CO)_3L_2 \underset{h\nu}{\overset{\rightarrow}{\leftarrow}} Mn_2(CO)_6L_4 \qquad (10)$$

As already mentioned, the second primary photoproduct $Mn_2(CO)_9$ has been fully characterized by flash photolysis with ir (ref. 63,64) and uv/vis (ref. 55,56) detection and by conventional ir spectroscopy after photolysis of $Mn_2(CO)_{10}$ in low-temperature alkane (ref. 52) and Ar (ref. 65) matrices. More recently, the corresponding $Re_2(CO)_9$ species has been characterized (ref. 67,70). These intermediates have a much longer lifetime in solution than the $M(CO)_5$ species, and in the case of $Mn_2(CO)_9$, the rate constant for recombination to give $Mn_2(CO)_{10}$ is a factor 10^4 smaller than that of $Mn(CO)_5$ (ref. 76). This rate constant becomes even smaller for the substituted intermediates $Mn_2(CO)_7(PR_3)_2$ obtained by photolysis of $Mn_2(CO)_8(PR_3)_2$ (ref. 76). Both steric and electronic properties of the PR_3 ligand control the magnitude of the rate constant, which varies from 2.9×10^5 $Mol^{-1}s^{-1}$ in the case of $Mn_2(CO)_9$ to 1.4×10^2 $Mol^{-1}s^{-1}$ for $Mn_2(CO)_7(P(i-Pr)_3)_2$.

This difference in lifetime between the two primary photoproducts strongly influences the secondary thermal reactions taking place. Both species react with nucleophiles such as phosphines but their products differ. $Mn_2(CO)_9$ gives rise only to the formation of $Mn_2(CO)_9(PR_3)$ whereas $Mn(CO)_5$ yields both $Mn_2(CO)_9(PR_3)$ and $Mn_2(CO)_8(PR_3)_2$ (reactions 11-14).

$$Mn_2(CO)_9 + PR_3 \rightarrow Mn_2(CO)_9(PR_3) \qquad (11)$$

$$Mn(CO)_5 + PR_3 \rightarrow Mn(CO)_4(PR_3) + CO \qquad (12)$$

$$2Mn(CO)_4(PR_3) \rightarrow Mn_2(CO)_8(PR_3)_2 \qquad (13)$$

$$Mn(CO)_4(PR_3) + Mn(CO)_5 \rightarrow Mn_2(CO)_9(PR_3) \qquad (14)$$

Due to its fast recombination reaction, $Mn(CO)_5$ does not react with PR_3 at low concentrations of PR_3. Thus, the rate of recombination of $Mn(CO)_5$ is unaffected when $Mn_2(CO)_{10}$ is irradiated in the presence of 10^{-3} Mol dm^{-3} PR_3 (ref. 7,55) and only $Mn_2(CO)_9(PR_3)$ is formed out of $Mn_2(CO)_9$. At higher concentrations ($> 10^{-2}$ Mol dm^{-3}) both $Mn_2(CO)_9(PR_3)$ and $Mn_2(CO)_8(PR_3)_2$ are formed (ref. 7,52,55). So, at low concentrations of the substituting ligand only the long-lived intermediate $Mn_2(CO)_9$ reacts with PR_3. A remarkable observation is that irradiation of $Mn_2(CO)_{10}$ in MeCN solvent (ref. 52) and in cyclohexane in the presence of 0.14 Mol dm^{-3} EtCN (ref. 55) yields $Mn_2(CO)_9(RCN)$ only. During this reaction, the recombination rate constant of $Mn(CO)_5$ to give $Mn_2(CO)_{10}$ is not affected (ref. 55) and addition of CCl_4, as radical scavenger for $Mn(CO)_5$, does not suppress the quantum yield of $Mn_2(CO)_9(RCN)$ formation (ref. 52). So, RCN does not substitute CO in $Mn(CO)_5$. This result agrees with the observation (ref. 52) that oxidation of $Mn(CO)_5^-$ in CH_3CN yields $Mn_2(CO)_{10}$ as the only product. On the other hand, the $Mn(CO)_5$ radicals are more readily oxidized in CH_3CN than in a nondonor solvent (ref. 77,78) which points to the formation of a 19-electron ($19e^-$) species $Mn(CO)_5$ (CH_3CN). These $19e^-$ intermediates will be shown to play a dominant role in the disproportionation of $Mn_2(CO)_{10}$ and other metal-metal bonded complexes in the presence of nitrogen or oxygen donor ligands (vide infra).

The question remains whether the substitution of CO in $M(CO)_5$ is dissociative or associative. It has long been assumed that this reaction proceeds by facile dissociative loss of CO and rapid uptake of L (ref. 47,50,79). Poë and co-workers, however, pointed out that CO dissociation is relatively slow and that rapid substitution can only proceed via an associative mechanism (ref. 80). Direct evidence for such a mechanism was first obtained by these authors from their study of the competition between chlorine atom abstraction from CCl_4 and ligand substution in the case of $Re(CO)_5$ (ref. 80). From their results it appeared that the rate constant for CO dissociation is negligible and that substitution of CO by PR_3 is exclusively first order in $[PR_3]$ and associative in nature. A similar study on $Mn(CO)_5$ confirmed this result and showed that the first-order rate constant for dissociative loss of CO must be less than 90 $Mol^{-1}s^{-1}$ (ref. 81). A recent study of the CO substitution in the analogous $V(CO)_6$ radical showed that in this case the substitution reaction is also associative in nature (ref. 82).

3.4   Atom Transfer

The $M(CO)_5$ radicals readily react with organic halides such as CCl_4 to give $M(CO)_5X$ (X=Cl,Br) (eq. 15):

$$M(CO)_5 + CCl_4 \rightarrow M(CO)_5Cl + CCl_3 \qquad (15)$$

The occurrence of such halogen atom transfer reactions has been used as evidence that the photolytic metal-metal bond cleavage is homolytic (ref. 27,83). $M_2(CO)_9$, on the other hand, hardly reacts with these halides as can be concluded from the photochemical reaction of $Mn_2(CO)_{10}$ in an alkane glass in the presence of CCl_4 (ref. 52). $Mn_2(CO)_9$ is then formed as the only photoproduct and raising the temperature to 298 K yields only negligible amounts of $Mn(CO)_5Cl$. On the other hand, Fox and Poë have stated that $Mn_2(CO)_9$ is not completely inert towards CCl_4 (ref. 54). These authors determined the quantum yield of the photochemical reaction of $Mn_2(CO)_{10}$ with CCl_4 as a function of $[CCl_4]$ and their results could be explained only by assuming a slow reaction from a second intermediate $(Mn_2(CO)_9)$ in addition of $Mn(CO)_5$.

The reaction of the $M(CO)_5$ radicals with the organic halides is fast as can be seen from the flash photolysis reaction of $Mn_2(CO)_{10}$ in neat CCl_4 (ref. 55). The decrease of the transient absorption intensity of $Mn(CO)_5$(827 nm) after the flash then follows first-order kinetics which means that the radicals react with CCl_4 before recombination to $Mn_2(CO)_{10}$ takes place. The rate constant of this chlorine atom transfer reaction (k_T) (ref. 55) was 9×10^5 $M^{-1}s^{-1}$ compared to a value of $9 \times 10^8 M^{-1}s^{-1}$ for the recombination reaction (ref. 72). Nearly the same value of k_T was obtained by Meckstroth et al. (ref. 84) and Herrick et al. (ref. 85).

For the reaction of $Re(CO)_5$ with CCl_4 a still higher rate constant of 5×10^7 $M^{-1}s^{-1}$ was found (ref. 84). Hepp and Wrighton (ref. 86) determined the relative rates of halogen atom transfer and oxidation for the radicals $(\eta^5-C_5H_5)W(CO)_3$, $Mn(CO)_5$ and $Re(CO)_5$. From their study it appeared that the reactions of these radicals with CCl_4 do not proceed by prior reduction of CCl_4 to CCl_4^- but as an atom abstraction, which may be considered an innersphere redox process. Brown and co-workers performed a detailed flash photolysis study of the rate constants of halogen atom transfer to a series of $M(CO)_4L$ (M=Mn, L=CO, PR_3; M=Re, L=CO, PR_3, $AsEt_3$) and $Mn(CO)_3(PR_3)_2$ radicals (ref. 85,87). The results for the Mn-radicals are collected in Table 1.

TABLE 1.

Atom transfer rate constants for various pairs of organic halides and manganese carbonyl radicals (reproduced with permission from ref. 85)

metal carbonyl radical	$\theta,^a$ deg	organic halide	[RX], M	k_T, M^{-1} s^{-1}
Mn(CO)$_5$·	95	CBr$_4$	1.02×10^{-3}	$(1.5 \pm 0.2) \times 10^9$
Mn(CO)$_5$·		CHBr$_3$	0.92	$(1.04 \pm 0.03) \times 10^7$
Mn(CO)$_5$·		C$_6$H$_5$CH$_2$Br	1.34	$(4.8 \pm 0.5) \times 10^5$
Mn(CO)$_5$·		CH$_2$Br$_2$	neat (13.8)	$(7 \pm 1) \times 10^3$
Mn(CO)$_5$·		CCl$_4$	0.174	$(1.4 \pm 0.1) \times 10^6$
Mn(CO)$_4$P(n-Bu)$_3$·	132	CCl$_4$	0.332	$(1.8 \pm 0.2) \times 10^6$
Mn(CO)$_4$P(O-i-Pr)$_3$·	130	CCl$_4$	0.415	$(2.2 \pm 0.1) \times 10^4$
Mn(CO)$_4$P(i-Bu)$_3$	143	CCl$_4$	0.217-0.021	$(8.9 \pm 1.0) \times 10^4$
Mn(CO)$_4$P(i-Pr)$_3$·	160	CCl$_4$	0.260-0.052	$(2.8 \pm 0.2) \times 10^4$
Mn(CO)$_4$P(C$_6$H$_{11}$)$_3$·	170	CCl$_4$	0.104	$(2.0 \pm 0.4) \times 10^4$
Mn(CO)$_4$P(i-Pr)$_3$·		CH$_2$Br$_2$	2.87-0.97	$(8.6 \pm 0.9) \times 10^2$
Mn(CO)$_4$P(i-Pr)$_3$·		CHCl$_3$	neat (12.4)	$(1.6 \pm 0.2) \times 10^1$
Mn(CO)$_4$P(C$_6$H$_{11}$)$_3$·		CHCl$_3$	neat (12.4)	7.3 ± 0.7
Mn(CO)$_3$[P(i-Bu)$_3$]$_2$·		CCl$_4$	0.174	$(3.0 \pm 0.6) \times 10^4$
Mn(CO)$_3$[P(i-Bu)$_3$]$_2$·		C$_6$H$_5$CH$_2$Cl	0.01-0.1	$(8.0 \pm 1.6) \times 10^{-1}$
Mn(CO)$_3$[P(i-Bu)$_3$]$_2$·		CH$_2$Cl$_2$	0.010	$(4.0 \pm 0.8) \times 10^{-2}$
Mn(CO)$_3$[P(i-Pr)$_3$]$_2$·		CCl$_4$	2.5×10^{-3}, 5.0×10^{-3}	$(1.0 \pm 0.2) \times 10^3$
Mn(CO)$_3$[P(n-Bu)$_3$]$_2$·		C$_6$H$_5$CH$_2$Cl	1.5×10^{-3}	$(1.0 \pm 0.2) \times 10^3$

^{a}Phosphine cone angle.

According to this table, k_T decreases with the size of the PR$_3$ ligand (Tolman cone angle) and upon going from Mn(CO)$_5$ to Mn(CO)$_4$PR$_3$ and Mn(CO)$_3$(PR$_3$)$_2$. An increase of electron donation by the ligand causes an increase in k_T. This is e.g. apparent from the higher rate constant for Mn(CO)$_4$(P(n-Bu)$_3$) as compared with Mn(CO)$_4$(P(O-iPr)$_3$). The phosphine ligands in these radicals have approximately the same cone angle but P(n-Bu)$_3$ is a more electron-donating ligand. Large variations in k_T are found for the various organic halides, the reactivity decreasing in the order CX$_4$>CHX$_3$>C$_6$H$_5$CH$_2$X>-CH$_2$X$_2$. This ordering corresponds to a decrease in C-X bond dissociation energy. A similar decrease in C-X bond dissociation energy is responsible for the higher values of k_T for the reactions with CBr$_4$ compared to those with CCl$_4$. In the case of CBr$_4$, the reaction with Mn(CO)$_5$ even approaches the diffusion controlled limit. Similar trends have been observed for the halogen atom transfer reactions of the $Re(CO)_4L$ radicals (ref. 87).

Apart from halogen atom transfer reactions, hydrogen abstraction from $HSn(n-Bu)_3$ has also been observed for the $Re(CO)_4L$ (ref. 87) and $Mn(CO)_3-(PR_3)_2$ (ref. 88) radicals. As in the case of the halogen atom transfer reactions, a similar dependence of k_T on the steric and electronic properties of L was observed in spite of the fact that the dependence on the electronic properties of L was smaller here.

3.5   Electron Transfer

The $M(CO)_5$ radicals are both better oxidants and reductants than their metal-metal bonded precursors since the d_z2-orbital of these radicals is more stable than σ^* and less stable than σ_b of the dimers. They are therefore expected to show more efficient electron transfer reactions with both electron donors and acceptors. To our knowledge only reactions with oxidizing agents have been reported sofar. Both inner- and outersphere electron-transfer from the radicals to a large variety of oxidants has been observed. It has been noted in a preceding section that the substitution reactions of the $M(CO)_5$ radicals are associative in nature. Moreover, $Mn(CO)_5$ is more readily oxidized in CH_3CN than in a non donor solvent (ref. 77,78). This means that $19e^-$ species $M(CO)_5L$ are easily formed in coordinating solvents and in the presence of Lewis bases. These $19e^-$ species are still better reductants than the $M(CO)_5$ radicals and they play a dominant role in electron transfer reactions to the parent dimers giving rise to catalytic disproportionation reactions. These disproportionation reactions have attracted much attention during the last five years and will be discussed in more detail in the next section.

Hepp and Wrighton (ref. 86) have studied the electron transfer reactions of the $M(CO)_5$ radicals (and of $(\eta^5-C_5H_5)W(CO)_3$) with several oxidants in competition with chlorine abstraction from CCl_4. They concluded from their results that oxidation of these radicals is a viable reaction pathway. For the oxidation of $Mn(CO)_5$ by ferricenium ion, a rate constant of $7 \times 10^6 M^{-1} s^{-1}$ was obtained. These rate constants correlated well with the potential of the oxidant redox couple and were dependend on the solvent used. Thus, oxidation of the radicals was more efficient in CH_3CN than in CH_2Cl_2 in agreement with the formation of a more reducing $19e^-$ intermediate adduct $M(CO)_5(CH_3CN)$.

If a Lewis base is sufficiently π electron deficient, adduct formation to $M(CO)_5$ can be followed by innersphere electron transfer. This mechanism is often referred to as spin-trapping. As noted before, spin-trapping agents have been employed to detect and identify the $M(CO)_5$ radicals (ref. 51,73,74). The spin-trap is normally an organic nitroso compound such as nitrosodurene $(2,3,5,6-Me_4C_6H-NO)$ or nitroso-t-butane $(C(CH_3)_3NO)$ wich forms a spin adduct with the $M(CO)_5$ radical having the nitroxide structure R - N - $M(CO)_5$. The single electron of the $M(CO)_5$ radical has been transferred from the energetically unfavourable d_{z^2}-orbital at Mn to the low-lying π^*_{NO} orbital of RNO. These paramagnetic nitroxides are, specially in the case of nitroso-t-butane, stable for hours at room temperature and their esr spectra have been studied in detail. (ref. 51,73,74).

Krusic et al. (ref. 108) have shown that tetracyanoethylene (TCNE) can also act as a spin-trap for $Mn(CO)_5$ radicals. At low concentrations $(10^{-3}M)$ of $Mn_2(CO)_{10}$ and TCNE, in THF, a complex esr spectrum is obtained which is the superposition of the spectra of the TCNE radical anion and of another radical. If $Mn_2(CO)_{10}$ is added to the solution in excess, the TCNE radical anion disappears and only the spectrum of the second species is observed. The spectrum could be assigned to the adduct of the TCNE radical anion to $Mn(CO)_5$ (see Fig. 6). The same radical was observed when $NaMn(CO)_5$ was slowly added to a solution of TCNE.

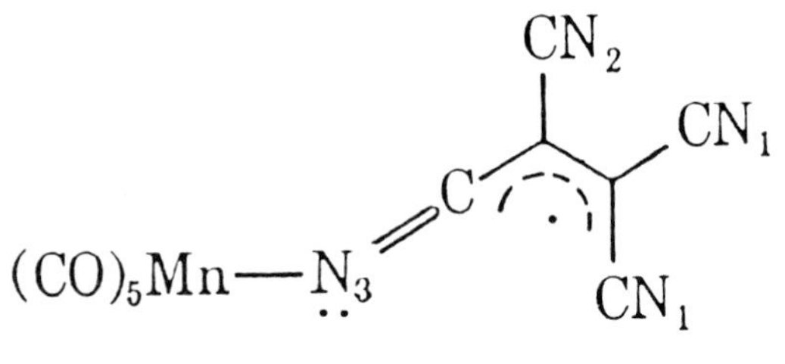

Fig. 6   Proposed structure of $(TCNE)Mn(CO)_5$ (reproduced with permission from ref. 108).

Many other innersphere electron transfer reactions from the $M(CO)_5$ radicals to Lewis bases possessing a low-energy π^* orbital have been observed (ref. 36,89-107). From these reactions, those with 2-dicarbonyls (α-diketones and ortho-quinones) (ref. 90-101) have been studied most extensively.

Alberti and Camagi (ref. 91,92) were the first; they studied in detail the products formed by irradiation of $Mn_2(CO)_{10}$ in the presence of one of several 1,2-dicarbonyls. The paramagnetic complexes formed appeared to be very stable since their esr spectra did not change in the temperature range -60° to +120°C.

The g- factors of the complexes are very close to the free electron value g_e (2.0023) which means that the unpaired electron occupies an orbital which is mainly ligand π^* in character. Metal-centered Mn-radicals such as $Mn(CO)_3(P-(n-Bu)_3)_2$ (ref. 47) have g- values as high as 2.03. Furthermore, the reaction is accompanied with evolvement of CO and the esr-spectra are characteristic of symmetric radicals down to -100°C. From this evidence, the photoproducts were proposed to have the structure of the tetracarbonyl complex shown in Fig 7.

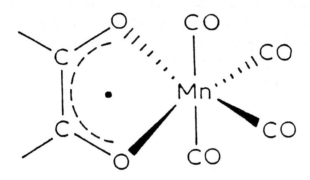

Fig. 7.   Proposed structure of (1,2-dicarbonyl)Mn(CO)$_4$ (reproduced with permission from ref. 91).

A similar symmetrical structure was postulated by Foster et al. (ref. 94) for the photoproduct of 9,10-phenanthrenequinone.

Tumanskij et al. however, who studied the photochemical reaction of $Mn_2(CO)_{10}$ with 3,6-di-tert-butyl-1,2-benzoquinone(3,6-DTBQ) in n-pentane at -80°C, observed in the esr spectrum of the photoproduct inequivalence for the protons at the 4- and 5-positions of the reduced ligand (ref. 209). They therefore proposed the photoproduct to be the radical complex $Mn(CO)_5(3,6-DTBQ)$. Warming up the solution caused the formation of the symmetrical complex $Mn(CO)_4(3,6-DTBQ)$ in hydrocarbons and $Mn(CO)_3(3,6-DTBQ)(S)$ in coordinating solvents (S). A recent esr study by Vlček on the reaction of 3,5-di-tert-butyl-1,2-benzoquinone(DTBQ, ref. 210) confirms this conclusion as do the preliminary ir data on this latter reaction (ref. 109).

Much attention has been paid to the substitution reactions of these radical complexes. Alberti et al. (ref. 91,92) studied the substitution of CO by a series of N-,P-,As- and Sb-donor ligands. All ligands with the exception of PPh_3 and $P(OEt)_3$ gave rise to the formation of a single new species in which one CO ligand was substituted. With PPh_3 and $P(OEt)_3$ two species were obtained: at lower temperatures a product with one PR_3 group, at higher temperatures a complex with two equivalent PR_3 groups. A structure was proposed in which each substituting ligand occupies an axial position. Definite proof for this proposal was obtained by Creber and Wan who were able to isolate the very stable radical complex $(DTB\dot{Q})Re(CO)_4$ and several of its substitution products (ref. 95,100). The separation of the radical complex from the reactants was accomplished by using a general-purpose HPLC-ESR apparatus. After evaporation of the eluent, a paramagnetic solid was obtained which was studied with esr, ir, uv/vis and emission spectroscopy. From the ir spectra, the axial position of the substituting ligands was definitely established.

Creber et al. (ref. 100) also studied the kinetics of the substitution reactions and concluded that these reactions are dissociative in nature, contrary to the substitution of the $M(CO)_5$ radicals. Ho et al. (ref. 97) prepared an optically active radical complex by reacting the $(DT\dot{B}Q)Re(CO)_4$ complex with an optically active phosphine ligand. Thermal substitution of this ligand by another ligand appeared to proceed stereoselectively (ref. 101).

Analogous tetracarbonyl radical complexes were obtained by irradiation of $Mn_2(CO)_{10}$ and $Re_2(CO)_{10}$ in the presence of the dithiokethone bis(ethoxythio-carbonyl) sulfide (ref. 105). In a later esr experiment on the Mn-complex, ^{13}C hyperfine coupling in natural abundance from the four CO ligands was observed (ref. 106).

A completely different product was obtained by reaction of $Mn_2(CO)_{10}$ with 2,6,-di-R-p-quinone (R=t-butyl, methyl)(ref. 94). Because of the para positions of the carbonyl groups, no chelate formation can take place. Instead, a $Mn(CO)_5$ adduct was formed for which the above structure was proposed on the basis of its esr spectrum. These results were not confirmed by Vlček (ref. 210), who assigned the esr signal of the photoproduct of the reaction of $Mn_2(CO)_{10}$ with 2,6-di-tert-butyl-1,4-benzoquinone(p-DTBQ) to the ion-pair $[Mn^I(CO)_{6-n}(S)_n]^+[p\text{-}DT\dot{B}Q]^-$ (S=solvent molecule, n=1-3).

Mn- and Re-complexes containing the radical anion of the α-diimine ligand di-t-butyl-1,4-diaza-1,3-butadiene (see Fig. 8) were prepared in the same way as the radical complexes of the 1,2-dicarbonyls (ref. 103). The Re-complex is much more persistent than the Mn-one and could be detected practically unaltered after several days. A striking difference between these radical complexes and those of the 1,2-dicarbonyls is their inertness with respect to substitution of CO by PR_3.

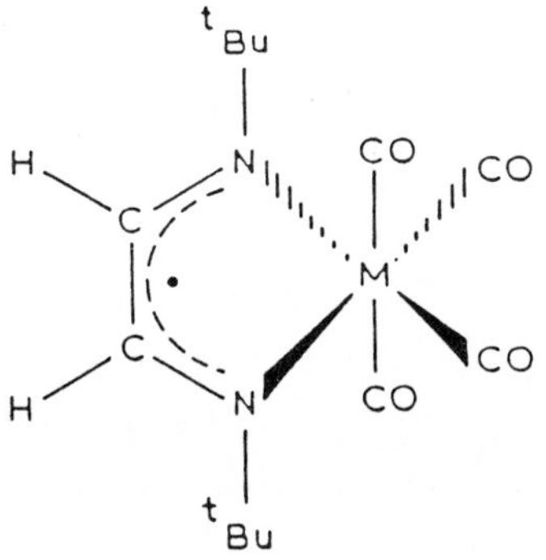

Fig. 8.   Proposed structure of (t-Bu-DȦB)M(CO)$_4$ (M=Mn,Re) (reproduced with permission from ref. 103).

α-Diimine radical complexes have also been prepared by irradiation of the metal-metal bonded complexes having the general formula $XM(CO)_3L$ (M=Mn, Re; X= $Mn(CO)_5$ or $Re(CO)_5$ (ref. 36), $Co(CO)_4$ (ref. 110), $(\eta^5\text{-}C_5H_5)Fe(CO)_2$ (ref. 111), $SnPh_3$ (ref. 104,112); L = α-diimine). The main reaction of these complexes, with the exception of $Ph_3SnMn(CO)_3L$, is a homolytic splitting of the metal-metal bond (vide infra). The solvated radicals, $LM(CO)_3S$, formed have been detected and identified with esr (ref.36,104). These esr spectra show the characteristic features of a radical anion with the unpaired electron localized mainly at the α-diimine ligand. The solvated radical is in equilibrium with its dimer $M_2(CO)_6L_2$ (eq. 16) and the solvent molecule can easily be replaced by a nucleophile (eq. 17).

$$(16)$$

$$(17)$$

The position of the equilibrium depends on the temperature and the stability of the radical with respect to the dimer. Raising the temperature causes an increase of radical concentration. The stability of the radical is mainly determined by the electronic and steric properties of L. Since the electron resides in the lowest π* orbital of L, the energy of this orbital influences the stability of the radical with respect to the dimer. An increase of stability is therefore observed going from L=bpy (2,2'-bipyridine) to L=R-DAB (1,4-diaza-1,3-butadiene) wich is the order of decreasing energy of the lowest π* orbital of L (ref. 104).

Just as in the case of the metal centered radicals $Mn(CO)_3(PR_3)_2$ (ref. 48), formation of the dimer is prevented by bulky substituents at the coordinating nitrogen atoms of L. The radical complex (t-Bu-DAB)Re(CO)$_3$S with bulky t-Bu-groups appeared to be, in fact, very stable, especially at higher temperatures (+70°C). The esr spectrum of this radical, showing extensive hyperfine splittings due to [185,187]Re and to two equivalent ^{14}N and ^{1}H nuclei, is presented in Fig. 9.

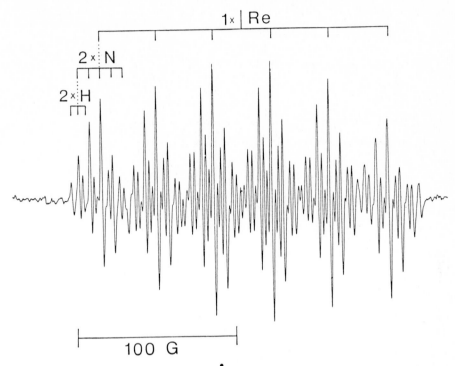

Fig. 9.   ESR spectrum of (t-Bu-DÅB)Re(CO)$_3$ in cyclohexane (343 K).

The corresponding Mn-complexes are much less stable and only (t-Bu-DÅB)Mn(CO)$_3$S could be studied with esr without using a spin-trap (ref. 104). The solvent molecules can easily be replaced by nucleophiles. Thus, adding P(OPh)$_3$ to a solution of (t-Bu-DÅB)Re(CO)$_3$S in benzene leads to the formation of the persistent radical (t-Bu-DÅB)Re(CO)$_3$(P(OPh)$_3$) in which the phosphite ligand is coordinated to Re in an axial position, just as in the o-quinone complexes (vide supra). Recall that these substituted radical complexes cannot be prepared by thermal substitution of a CO-ligand in (t-Bu-DÅB)Re(CO)$_4$ (ref. 103)

3.6   Disproportionation

A type of reaction which is very interesting, mechanistically, is the photochemical disproportionation of the dimers. These reactions, involving a formal change in the metal atom oxidation states of ±1, are formulated as follows (eq 18):

$$L_nM - ML_n \quad \xrightarrow[L']{h\nu} \quad ML_n^- + ML_mL'^+_{(n-m+1)} \quad (18)$$

These disproportionation reactions often occur thermally as well as photochemically. Much has been done to unravel the mechanisms of these reactions and especially the detailed studies of Stiegman and Tyler have to be mentioned here (ref. 8,9).

Hieber and Schropp (ref. 113) were the first who showed that $Mn_2(CO)_{10}$ reacts both thermally and photochemically with nitrogen bases (B) according to eq. 19.

$$Mn_2(CO)_{10} \quad \xrightarrow[B]{\Delta \text{ or } h\nu} \quad Mn(B)_6^{2+} + Mn(CO)_5^- \quad (19)$$

However, for ammonia and n-butylamine the Mn(I) complexes $[Mn(CO)_3(NH_3)_3^+]$ and $[Mn(CO)_5(NH_2(n\text{-butyl})^+]$ were obtained instead. Allen et al. (ref. 114) also observed the formation of $Mn(B)_6^{2+}$ and $Mn(CO)_5^-$ upon photolysis of $Mn_2(CO)_{10}$ in pyridine and other polar solvents. They proposed a heterolytic splitting of the metal-metal bond in $Mn_2(CO)_{10}$ with formation of $[Mn(CO)_5^+]$ and $[Mn(CO)_5^-]$ as the primary photoprocess. Both Hieber and Allen assumed that the disproportionation was due to solvent effects, polar solvent molecules inducing a dipole in the metal-metal bond. Excitation of this dipole then results in heterolysis.

McCullen and Brown (ref. 115) pointed out that only homolysis (and loss of CO) had been observed as a primary photoprocess and that the presence of a polar donor solvent would not alter these reactions. They observed, apart from $Mn_2(CO)_9(py)$, the formation of $[Mn(CO)_3(py)_3^+]$ and $[Mn(CO)_5^-]$ upon irradiation of $Mn_2(CO)_{10}$ in pyridine. By further photolysis, $[Mn(CO)_3(py)_3^+]$ was converted into $[Mn(py)_6^{2+}]$ in agreement with the results of Hieber and Allen. The formation of $[Mn(CO)_3(py)_3^+]$ was inhibited by adding the radical scavenger galvinoxyl and therefore a free-radical pathway was proposed for the reaction (eq 20-27):

$$Mn_2(CO)_{10} \;\rightleftharpoons\; 2MnCO)_5 \tag{20}$$

$$Mn(CO)_5 \;\rightleftharpoons\; Mn(CO)_4 + CO \tag{21}$$

$$Mn(CO)_4 + N \;\rightleftharpoons\; Mn(CO)_4N \tag{22}$$

$$Mn(CO)_4N \;\rightleftharpoons\; Mn(CO)_3N + CO \tag{23}$$

$$Mn(CO)_3N + N \;\rightleftharpoons\; Mn(CO)_3N_2 \tag{24}$$

$$Mn(CO)_3N_2 + Mn_2(CO)_{10} \;\rightarrow\; Mn(CO)_3N_2^+ + Mn_2(CO)_{10}^- \tag{25}$$

$$Mn_2(CO)_{10}^- \;\rightarrow\; Mn(CO)_5 + Mn(CO)_5^- \tag{26}$$

$$Mn(CO)_3N_2^+ + N \;\rightarrow\; Mn(CO)_3N_3^+ \tag{27}$$

N=pyridine, amine

Tyler and co-workers (ref. 116,117) reinvestigated the disproportionation of $Mn_2(CO)_{10}$ with nitrogen donor ligands. They showed that the reaction is not brought about by any special bulk properties of the solvent by irradiating low concentrations ($\sim 10^{-2}$M) of $Mn_2(CO)_{10}$ and pyridine in hexane. The same $Mn(CO)_3(py)_3^+$ and $Mn(CO)_5^-$ ions were formed as when performing the reaction in neat pyridine. This means that pyridine (and other polar solvents) do not change the primary photoprocess but rather act as ligands inducing the disproportionation. A second important observation was that the quantum yield of the disproportionation reaction drastically increases upon going from the monodentate triethylamine (ϕ= 0.10) to the terdentate ligand diethylenetriamine (ϕ = 10-20). The very high quantum yield for the latter ligand points to efficient electron transfer only from a radical with three N-donor atoms bonded to the metal. The authors proposed the 19e$^-$ intermediate $Mn(CO)_3N_3$ as the key feature for the electron transfer reaction in agreement with their previous results for $(\eta^5$-$C_5H_5)Mo_2(CO)_6$ (ref. 118). The increase of quantum yield upon going from a mono- to a terdentate ligand is due to the fact that these latter chelating ligands effectively increase the concentration of the 19e$^-$ intermediate. Two experiments (ref. 117) provided additional evidence for the 19e$^-$ intermediate. First, the reaction of PMe_3 with $Mn(CO)_3(depe)$ (depe = 1,2-bis(diethylphosphino)ethane) in the presence of $Mn_2(CO)_{10}$, in the dark, gave disproportionation products. PMe_3 attacks $Mn(CO)_3(depe)$ giving the 19e$^-$ complex $Mn(CO)_3(depe)$ (PMe_3), which then reduces $Mn_2(CO)_{10}$. Secondly, the cation formed by reaction of tetraphos (1,1,4,7,10,10-hexaphenyl -1,4,7,10-tetraphosphadecane) with $Mn_2(CO)_{10}$ is $Mn(CO)_3(tetraphos-P,P',P'')^+$ and not $Mn(CO)_2(tetraphos- P,P',P'',P''')^+$, which means that the chain reaction involves electron transfer from the species $Mn(CO)_3L_3$ and not from a 17e$^-$ intermediate $Mn(CO)_2L_3$. Based on these observations, Tyler and co-workers proposed the radical chain mechanism shown in eq. 28-33. Eqs. 21 and 22 and also eqs. 23 and 24 have now been

$$Mn_2(CO)_{10} \xrightarrow{h\nu} 2Mn(CO)_5 \tag{28}$$

$$Mn(CO)_5 + N \rightleftarrows Mn(CO)_4N + CO \tag{29}$$

$$Mn(CO)_4N + N \rightleftarrows Mn(CO)_3N_2 + CO \tag{30}$$

$$Mn(CO)_3N_2 + N \rightleftarrows Mn(CO)_3N_3 \tag{31}$$

$$Mn(CO)_3N_3 + Mn_2(CO)_{10} \rightarrow Mn(CO)_3N_3^+ + Mn_2(CO)_{10}^- \tag{32}$$

$$Mn_2(CO)_{10}^- \rightarrow Mn(CO)_5 + Mn(CO)_5^- \tag{33}$$

combined to give eqs. 29 and 30, respectively, since substitution of $Mn(CO)_5$ and its derivatives has been shown to be associative in nature (vide supra) (ref. 80,81).

Meyer and Caspar (ref. 7) have questioned this mechanism since there is clear evidence that acetonitrile and pyridine do not react with $Mn(CO)_5$ to give $Mn(CO)_4N$ (eq. 29). Thus, oxidation of $Mn(CO)_5^-$ in acetonitrile only yields $Mn_2(CO)_{10}$ (ref. 52) and the photochemical reaction of $Mn_2(CO)_{10}$ with MeCN (ref. 55) and EtCN (ref. 52) only gives $Mn_2(CO)_9N$ out of the primary photoproduct $Mn_2(CO)_9$, and no $Mn_2(CO)_8N_2$ out of $Mn(CO)_5$. Caspar and Meyer therefore proposed a reaction mechanism (eq. 34-40) in which the intermediate species $Mn(CO)_4N$ is formed by photolysis of $Mn_2(CO)_9N$ and not by substitution of $Mn(CO)_5$.

$$Mn_2(CO)_{10} \xrightleftharpoons{h\nu} 2Mn(CO)_5 \tag{34}$$

$$Mn_2(CO)_{10} \xrightleftharpoons{h\nu} Mn_2(CO)_9 \xrightarrow{N} Mn_2(CO)_9N \tag{35}$$

$$Mn_2(CO)_9N \xrightarrow{h\nu} Mn(CO)_4N + Mn(CO)_5 \tag{36}$$

$$Mn(CO)_4N + N \rightarrow Mn(CO)_3N_2 + CO \tag{37}$$

$$Mn(CO)_3N_2 + N \rightarrow Mn(CO)_3N_3 \tag{38}$$

$$Mn(CO)_3N_3 + Mn(CO)_5 \rightarrow Mn(CO)_3N_3^+ + Mn(CO)_5^- \tag{39}$$

$$Mn(CO)_3N_3 + Mn_2(CO)_{10} \rightarrow Mn(CO)_3N_3^+ + Mn(CO)_5^- + Mn(CO)_5 \tag{40}$$

In both the reaction schemes of Stiegman and Tyler (ref. 116) and Meyer and Caspar (ref. 7) the 19e$^-$ radical $Mn(CO)_3L_3$ is the key intermediate and any factor that affects the rate of electron transfer from this species, or its concentration, will affect the disproportionation process. Thus, if L is bulky, $Mn(CO)_3L_3$ cannot be formed or its formation is slow and disproportionation inefficient. Similarly, if L is a poor donor as e.g. PR_3, then the electron transfer step is slow. In general, disproportionation is observed with good electron donors such as N- or O-donor ligands. Phosphines only show disproportionation if they are a chelating ligand or in the presence of such a ligand (e.g. PMe_3 with $Mn(CO)_3(depe)$). The latter situation also occurs if the substituted metal carbonyls $(CO)_5MnMn(CO)_3(\alpha\text{-diimine})$ are irradiated in the presence of a N- or O-donor ligand, or in the presence of a basic phosphine (ref. 119). This ligand forms an adduct with the primary photoproduct $Mn(CO)_3(\alpha\text{-diimine})$ which then reduces the parent compound in a radical chain reaction.

19e$^-$ radical complexes not only play a dominant role in the photodisproportionation of $Mn_2(CO)_{10}$ and other metal-metal bonded complexes such as $(\eta^5\text{-}C_5H_5)_2Mo_2(CO)_6$ and $(\eta^5\text{-}C_5H_5)_2Fe_2(CO)_4$ (vide infra) (ref. 8,9), they can also act as strong reductors with respect to other complexes. Thus, irradiation of $Mn_2(CO)_{10}$ in the presence of diethylenetriamine (dien) causes the reduction of $(\eta^5\text{-}C_5H_5)_2Mo_2(CO)_6$ into $(\eta^5\text{-}C_5H_5)Mo(CO)_3^-$ according to eq. 41 (ref. 120). Furthermore, highly reducing 19e$^-$ radicals cannot only be prepared photochemically but also by (electrochemical) reduction (ref. 121).

$$Mn_2(CO)_{10}+dien+(\eta^5\text{-}C_5H_5)_2Mo_2(CO)_6 \xrightarrow{h\nu} 2Mn(CO)_3(dien)^+ +2(\eta^5\text{-}C_5H_5)Mo(CO)_3^- \quad (41)$$

3.7 Substitution

Photosubstitution of one or more CO ligands in $M_2(CO)_{10}$ only occurs under special conditions. Especially for strong O- and N- donor ligands substitution is in competition with the disproportionation discussed in the preceeding section. Thus, irradiation of $Mn_2(CO)_{10}$ in the presence of mono (N)- and bidentate (NN) N-donor ligands only yields $Mn_2(CO)_9(N)$ and $Mn_2(CO)_8(NN)$, respectively, at low ligand concentration (ref. 116). At higher ligand concentration only the ionic disproportionation products are formed and few or none of the substituted dimer products. With tridentate ligands, no substitution products are observed at all, even at very low ligand concentrations. The reason for the concentration dependence is the formation of the reducing 19e⁻ intermediate $Mn(CO)_3N_3$ at high ligand concentration, according to reactions 29-31 or 36-38.

In the $Mn_2(CO)_9(N)$ complexes the nitrogen bases appear to occupy an equatorial site (ref.77,78). An analogous eq-$Re_2(CO)_9(N_2)$ photoproduct was observed upon short-wavelength irradiation of $Re_2(CO)_{10}$ in liquid Xenon doped with N_2 (ref. 70). Irradiation of the same complex in a N_2-matrix yielded $Re_2(CO)_9$ which was then converted to eq- $Re_2(CO)_9(N_2)$ by warming the matrix to 15-20 K.

Eq-$Re_2(CO)_9(OH_2)$ was formed as the initial photoproduct by irradiation of $Re_2(CO)_{10}$ with H_2O in THF (ref. 122) and eq-$Mn_2(CO)_9$(2-MeTHF) by irradiation of $Mn_2(CO)_{10}$ in a 77K/2-MeTHF matrix (ref. 52). If this latter reaction is performed in the presence of PPh_3, the phosphine replaces the loosely bound 2-MeTHF molecule upon warming the matrix.

Stable disubstituted $Mn_2(CO)_8(N)_2$ (N=nitrogen donor ligand) complexes are neither formed by irradiation of $Mn_2(CO)_{10}$ nor by photolysis of $Mn_2(CO)_9(N)$. In the case of Re, however, 1,2-eq,eq-$Re_2(CO)_8(N)_2$ complexes are formed by irradiation of eq-$Re_2(CO)_9(N)$ (ref. 123). In the case of CH_3CN the disubstituted complex is photochemically rather stable, but the corresponding complex 1,2-eq,eq-$Re_2(CO)_8(py)_2$ complex reacts further photochemically to give the interesting hydrogen bridged complex $(\mu-H)Re_2(CO)_7(py)(NC_5H_4)$, shown in Fig. 10.

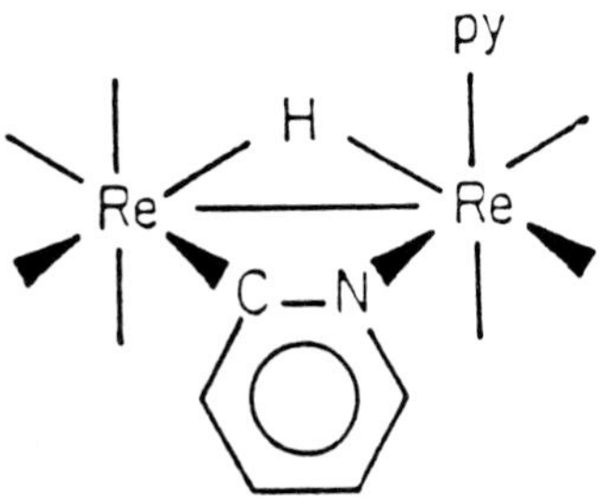

Fig. 10. Structure of 1,1-$(\mu-H)Re_2(CO)_7(py)(NC_5H_4)$ (reproduced with permission from ref. 123).

This photoproduct possesses one normally coordinated pyridine and one pyridine which is orthometallated. Both nitrogen atoms coordinate to the same Re atom. Thermal treatment of 1,2-eq,eq-$Re_2(CO)_8(py)_2$ affords the analogous complex $(\mu-H)Re_2(CO)_8(NC_5H_4)$ for which an X-ray structure has been determined (ref. 124). Also, in the case of eq-$Re_2(CO)_9(RNH_2)$ (R=CH_3,C_2H_5) no 1,2-eq,eq-$Re_2(CO)_8(RNH_2)_2$ is formed but instead 1,1-ax,eq-$Re_2(CO)_8$ $(RNH_2)_2$.

There is a close relationship between the reaction of $Re_2(CO)_{10}$ with pyridine and those with water (ref. 122) and with alkenes (ref. 125-128). The eq-$Re_2(CO)_9(OH_2)$ complex, initially formed decomposes rapidly with 366 nm irradiation to give $HRe(CO)_5$ and $Re_4(CO)_{12}(OH)_4$. This decomposition reaction is thought to involve $Re_2(CO)_8(OH_2)_2$ as an unstable intermediate. Indirect evidence for such an intermediate is given by the thermal reaction of $Re_2(CO)_8(CH_3CN)_2$ with H_2O which produces the same $Re_4(CO)_{12}(OH)_4$. $Re_2(CO)_8(OH_2)_2$ is thermally less stable than $Re_2(CO)_8(py)_2$. By loss of H_2O, and oxidative addition of an OH bond, $(\mu-H)(\mu-OH)Re_2(CO)_8$ is formed (eq. 42), in close agreement with the reaction of pyridine. Although the doubly bridged intermediate could not be detected, a completely analogous complex has been formed as a stable photoproduct by irradiation of $Re_2(CO)_8(L-L)$ (L-L=bridging diphosphine) in the presence of water (vide infra). A stable $(\mu-H)(\mu-OH)$-$(Os_3(CO)_{10}$ complex has also been formed by reaction of $Os_3(CO)_{10}$-$(CH_3CN)_2$ with water (ref. 120,121).

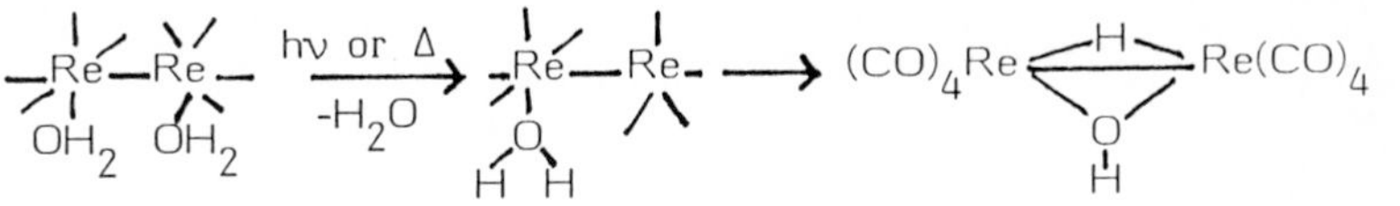

$$\underset{OH_2 \quad OH_2}{-Re-Re-} \xrightarrow[-H_2O]{h\nu \; or \; \Delta} \underset{O \atop H \; H}{-Re-Re-} \longrightarrow (CO)_4Re \underset{O \atop H}{\overset{H}{\rlap{\Large\diagdown\diagup}}} Re(CO)_4 \quad (42)$$

Such reactions as observed for pyridine and water may also be expected for other relatively labile ligands possessing a functionality (C-H,O-H) suitable for oxidative addition. In fact, photolysis of $Re_2(CO)_{10}$ (and derivatives) in the presence of alkenes (ref. 125-128) yields stable $(\mu-H)(\mu-alkenyl)Re_2(CO)_8$ complexes having the structure shown in figure 12 (vide infra).

For phosphines, which are poorer donors, competition between substitution and disproportionation only occurs for chelating tridentate and tetradentate ligands such as triphos and tetraphos (ref. 117). There is, however, still a concentration dependence in the product formation for these ligands. At low ligand concentration ($\sim 10^{-3}$M) only $Mn_2(CO)_9$ (PR_3) is formed out of the primary photoproduct $Mn_2(CO)_9$ (ref. 7,52,55). At these low concentrations the $Mn(CO)_5$ radicals do not react with PR_3 due to their diffusion-controlled recombination reaction. At higher phosphine concentration ($\geqslant 10^{-2}$M) the $Mn(CO)_5$ radicals are substituted also by the phosphine giving rise to the formation of both $Mn_2(CO)_9(PR_3)$ and $1,2\text{-}Mn_2(CO)_8(PR_3)_2$ according to reactions 11-14. Because of its steric bulk PPh_3 occupies an axial position in $ax\text{-}Mn_2(CO)_9(PPh_3)$ (ref. 77,78) instead of the electronically more favourable equatorial site. In $Mn_2(CO)_8(PPh_3)_2$ the phosphines also occupy the axial positions.

The reactions of phosphines with $Re_2(CO)_{10}$ also yield the axially substituted complexes $ax\text{-}Re_2(CO)_9(PR_3)$ and $1,2\text{-}ax,ax\text{-}Re_2(CO)_8(PR_3)_2$. Brown and co-workers (ref. 126) found, however, that stable $1,2\text{-}eq,eq\text{-}Re_2(CO)_8(PR_3)_2$ complexes can be obtained by thermal reaction of $(\mu\text{-}H)(\mu\text{-}alkenyl)\text{-}Re_2(CO)_8$ with small phosphines such as PMe_3, $P(OMe)_3$ and $P(OPh)_3$. This geometry is preferred electronically in the absence of steric effects. With larger ligands, such as PPh_3 or $P(n\text{-}Bu)_3$ the initially obtained major product is $1,2\text{-}ax,eq\text{-}Re_2(CO)_8(PR_3)_2$, which slowly isomerizes to $1,2\text{-}ax,ax\text{-}Re_2(CO)_8(PR_3)_2$. This isomerization appeared to be nondissociative.

Cheng and co-workers have photolyzed $Re_2(CO)_{10}$ under vacuum in the presence of PPh_3 (ref. 131) and $P(OPh)_3$ (ref. 132). In contrast to previous studies they observed, in the case of PPh_3, formation of three sterically crowded triply substituted complexes $Re_2(CO)_7(PPh_3)_3$ and even a four substituted $Re_2(CO)_6(PPh_3)_4$ product. With the smaller $P(OPh)_3$ ligand even the photoproduct $Re_2(CO)_4(P(OPh)_3)_6$ was obtained.

Of special interest are the photochemical reactions of phosphine bridged complexes. Lemke and Kubiak (ref. 193) synthesized the diphosphine-bridged complex $Mn_2(CO)_6(dmpm)_2$ (dmpm=bis(dimethylphosphino)methane) and observed an interesting photoelectron-transfer reaction for this compound. Brown and co-workers studied the preparation and photochemistry of the corresponding mono-phosphine bridged Re-complexes $1,2\text{-}eq,eq\text{-}Re_2(CO)_8(L\text{-}L)$ (Fig. 11a) (L-L=dmpm; dppm=bis(diphenylphosphino)methane; dmpe=bis(dimethyl-phosphino)ethane; dppe=bis(diphenylphosphino)ethane) (ref. 128,133,134).

Just as in the case of $Re_2(CO)_8(PR_3)_2$ the complexes $(\mu\text{-}H)(\mu\text{-}alkenyl)\text{-}Re_2(CO)_8$ (Fig. 12) are excellent precursors for the synthesis of the bridged species (eq. 43,44).

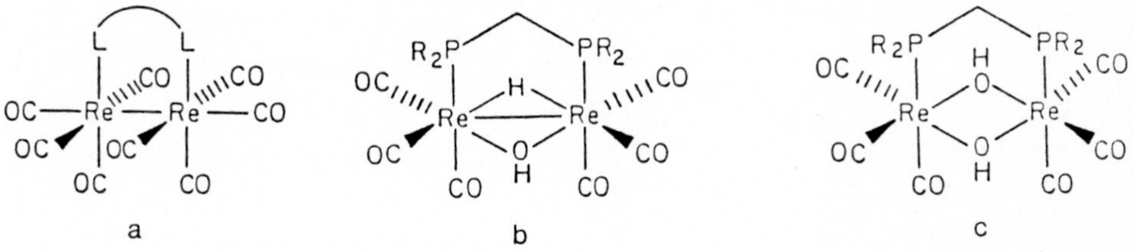

Fig. 11.   Structures of 1,2-eq,eq-$Re_2(CO)_8$(L-L) (a),(μ-H)(μ-OH)$Re_2(CO)_6$(L-L) (b) and (μ-OH)$_2Re_2(CO)_6$(L-L) (c) (from ref. 133).

$$Re_2(CO)_{10} + 1 \text{ hexene} \xrightarrow{h\nu} (\mu\text{-H})(\mu\text{-CH=CHC}_4\text{H}_9)Re_2(CO)_8 \qquad (43)$$

$$(\mu\text{-H})(\mu\text{-CH=CHC}_4\text{H}_9)Re_2(CO)_8 \xrightarrow[25°C]{\text{L-L}} 1,2\text{-}Re_2(CO)_8(\text{L-L}) \qquad (44)$$

Irradiation of $Re_2(CO)_8$(dppm) in wet toluene or benzene yielded (μ-H)(μ-OH)$Re_2(CO)_6$(dppm) and (μ-OH)$_2Re_2(CO)_6$(dppm) having the structures b and c of Fig. 11, respectively (ref. 133). The crystal structure of complex b was determined. For the complexes with a bridging ligand containing a two-carbon chain backbone structure (dppe or dmpe) no such products could be isolated, although they were still present in the reaction mixture. The reactions with methanol were faster than with water, but only in the case of $Re_2(CO)_8$(dppm) as parent compound could (μ-H)(μ-OCH$_3$)$Re_2(CO)_6$(dppm) and (μ-OCH$_3$)$_2Re_2(CO)_6$(dppm) be isolated. In a later article (ref. 75) Lee et al. showed, by laser flash photolysis, that CO loss is an inefficient primary photoprocess in these systems when compared to Re-Re bond homolysis. This observation supports the radical mechanism proposed for the above reactions. The formation of (μ-H)(μ-OR')$Re_2(CO)_6$(L-L), (μ-OR')$_2Re_2(CO)_6$(L-L) and, in one case, of (μ-H)$_2Re_2(CO)_6$(L-L) could only be explained by photochemical formation of a biradical followed by substitution of CO at both radical centers by H$_2$O or methanol. Since this substitution is a rather slow process with respect to recombination of the biradical, the above reactions, as well as the raction of $Re_2(CO)_8$(L-L) with alkenes (ref. 128) and alkynes (ref. 134) are comparatively slow. Only upon short-wavelength irradiation (254 nm) is CO loss expected to play an important role since then $(\mu\text{-H})(\mu\text{-OR})Re_2(CO)_6(L\text{-}L)$ is the main product.

Brown and co-workers studied the photochemical reactions of $M_2(CO)_{10}$ with simple mono-olefins. $Mn_2(CO)_{10}$ did not react; however, $Re_2(CO)_{10}$ gave stable complexes $(\mu\text{-H})(\mu\text{-alkenyl})Re_2(CO)_8$ (see Fig. 12) by reaction with ethylene, terminal olefins or 2-butene (ref. 125-127). The structures of these complexes are similar to that of $(\mu\text{-H})(\mu\text{-}C_5H_4N)Re_2(CO)_8$ (see Fig. 10) and of the intermediate $(\mu\text{-H})(\mu\text{-OH})Re_2(CO)_8$ (eq. 42). The bridging alkenyl group is σ bonded to one Re and π bonded to the other. Just as in the case of the reactions with pyridine and water, the reactions are thought to proceed through a radical pathway, producing a $1,2\text{-}Re_2(CO)_8(\eta^2\text{-olefin})_2$ intermediate that thermally converts into the observed product.

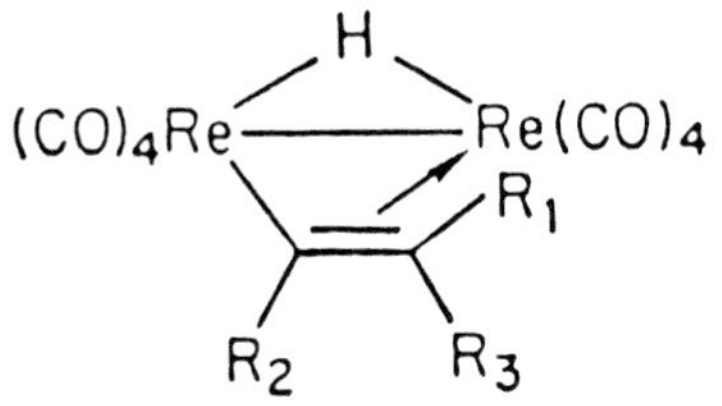

Fig. 12. Proposed structure of $(\mu\text{-hydrido})(\mu\text{-alkenyl})Re_2(CO)_8$ (reproduced with permission from ref. 126).

A characteristic property of these complexes is the ease with in which the olefin can be eliminated thermally. Treatment with pyridine, $P(OMe)_3$ and $P(OPh)_3$ yields $1,2\text{-eq,eq-}Re_2(CO)_8L_2$. With other olefins, hydride- and alkenyl-exchanged hydrido-alkenyl species are formed. Analogous $(\mu\text{-H})$-$(\mu\text{-alkenyl})$ complexes were obtained by photochemical reaction of $Re_2(CO)_9L$, $1,2\text{-}Re_2(CO)_8L_2$ and $Re_2(CO)_8(L\text{-}L)$ (L=phosphine; L-L=bridging diphosphine) with alkenes (ref. 128).

An interesting photochemical reaction of these $(\mu\text{-H})$ $(\mu\text{-alkenyl})Re_2(CO)_8$ complexes takes place in the presence of excess olefin (ref. 127). Thus, uv photolysis of $(\mu\text{-H})(\mu\text{-ethenyl})Re_2(CO)_8$ with ethylene affords $(\mu\text{-H})(\mu\text{-butenyl})$-

$Re_2(CO)_8$. The mechanism proposed for this reaction is loss of CO, followed by coordination of ethylene, insertion of ethylene into the Re-H or Re-ethenyl σ bond, recoordination of CO, C-C or C-H reductive elimination to yield $Re_2(CO)_8$(1-butene) and, finally, oxidative addition of a vinylic C-H bond of the coordinated butene (see Fig. 13). During this reaction 1-butene and trans-3-hexene are produced catalytically.

Kreiter and co-workers have studied in detail the photochemical reactions of $M_2(CO)_{10}$ with conjugated and cumulated dienes. This work has been reviewed quite recently (ref. 10). Contrary to the behaviour of monoolefines, dienes react photochemically with $Mn_2(CO)_{10}$.

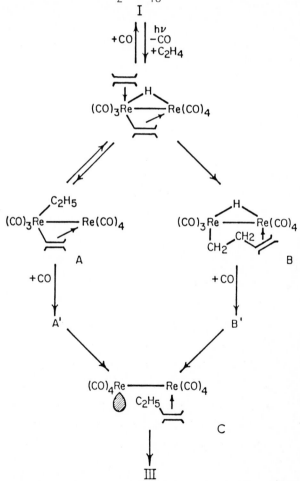

Fig. 13. Proposed mechanism for the photochemical transformation of $(\mu\text{-}H)(\mu\text{-}CH=CH_2)Re_2(CO)_8$ (I) into $(\mu\text{-}H)(\mu\text{-}CH=CH\text{-}C_2H_5)Re_2(CO)_8$ (III) (reproduced with permission from ref. 127).

Thus, irradiation of $Mn_2(CO)_{10}$ at 253 K in the presence of 1,3-butadiene affords a mixture of four compounds from which, the one shown in Fig. 14a, is the major reaction product at the onset of the reaction (ref. 136,137).

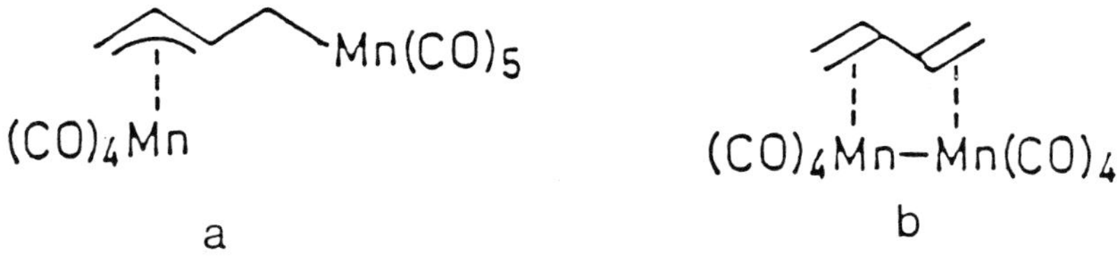

Fig. 14. Proposed structures of $Mn_2(CO)_9(C_4H_6)$ (a) and $Mn_2(CO)_8(C_4H_6)$ (b) (reproduced from ref. 136).

The formation of this complex has been rationalized by attack of the $Mn(CO)_5$ radical at C-1 of the ligand to give $Mn(CO)_5(C_4H_6)$. This species then reacts with another $Mn(CO)_5$ radical to yield an intermediate $(CO)_5MnC_4H_6Mn-(CO)_5$, which loses CO and affords the product shown in Fig. 14a.

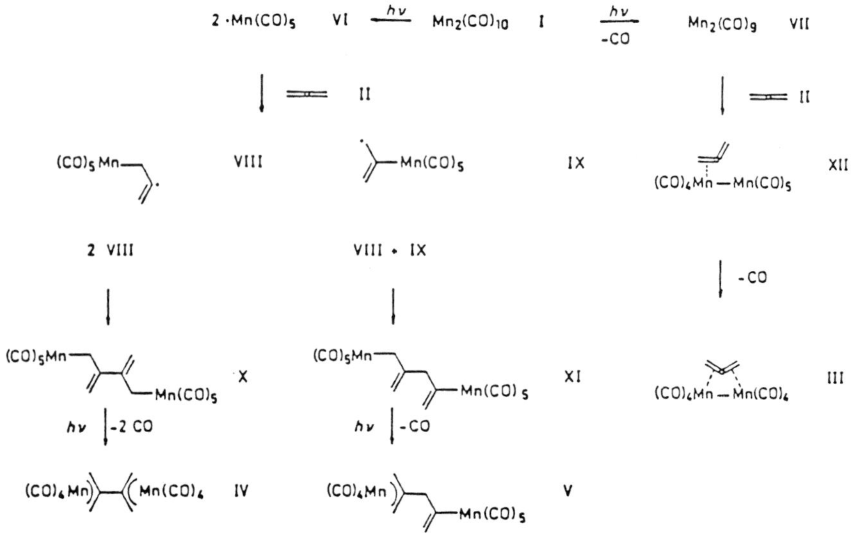

Fig. 15. Proposed mechanism for the photochemical reaction of $Mn_2(CO)_{10}$ with allene (reproduced with permission from ref. 139).

A related species $(CO)_5ReC_4H_6Re(CO)_5$ has been reported (ref. 135). Two other reaction products are formed by diene photosubstitution of CO in 14a, followed by diene insertion into the Mn-C bond. A minor by-product 14b is formed presumably by addition of the diene to the second primary photoproduct $Mn_2(CO)_9$ followed by thermal loss of CO.

The cumulated diene allene (C_3H_4) does not react photochemically at room temperature with $Mn_2(CO)_{10}$. However, when the reaction is performed at reduced temperature (233 K) three products can be isolated (III, IV and V in Fig. 15) (ref. 138,139). Product III is formed by reaction of allene with the primary photoproduct $Mn_2(CO)_9$ and further loss of CO. The other two complexes, IV and V, are thought to be formed by reaction of allene with the $Mn(CO)_5$ radicals.

$Re_2(CO)_{10}$ reacts differently with the dienes since it easily forms bridged hydride complexes. Thus, the main product of the reaction of $Re_2(CO)_{10}$ with 1,3-butadiene is $(\mu-H)(\mu-butadienyl)Re_2(CO)_8$ which is isostructural with the bridged species formed by reaction of $Re_2(CO)_{10}$ with monoolefins (Fig. 12) (ref. 140). Contrary to the products obtained from monoolefins, however, this compound is not very stable since it rearranges rapidly in polar solvents to the Re homolog of complex 14b. All reactions with dienes, trienes and tetraenes show the formation of the hydride, alkenyl- bridged species as the main reaction product.

Kreiter et al. also studied the photochemical reaction of $Re_2(CO)_{10}$ with aldehydes (ref. 141). Apart from $(\mu-H)Re_3(CO)_{14}$, $(\mu-H)(\mu-acyl)Re_2(CO)_8$ is formed in an oxidative addition reaction just as in the case of pyridine, water, mono- and diolefins. The acyl group R-C=O is σ bonded via C to one Re atom and via a lone pair of oxygen to the second one.

The corresponding $(\mu-hydrido)(\mu-alkynyl)$ complexes have been prepared by Franzreb and Kreiter upon photolysis of $Re_2(CO)_{10}$ with acetylene (ref. 142) and, thermally, by Nubel and Brown by reaction of $(\mu-H)(\mu-propenyl)-Re_2(CO)_8$ with phenylacetylene (ref. 125,127). The photochemical reaction affords a mixture of compounds from which $(\mu-H)(\mu-acetynyl)Re_2(CO)_8$ is the major product. The photochemical reaction of 1,2-eq,eq-$Re_2(CO)_8(L-L)$ (L-L=dppm, dmpm) with acetylene or phenylacetylene (ref. 143) does not lead to the formation of the $(\mu-hydrido)(\mu-alkynyl)$ complex as the dominant product in the early stage of the reaction; instead, it yields $(\mu-H)Re_2(CO)_7(L-L)(\eta'-C\equiv-CR')$ (see Fig. 16b). Prolonged irradiation affords the $(\mu-H)(\mu-C\equiv CR')$ complex (see Fig. 16c) and its photolysis products. This difference in behaviour between

$Re_2(CO)_{10}$ and $Re_2(CO)_8(L-L)$ during their photochemical reaction with alkynes has been attributed to different primary photoprocesses. The reaction of $Re_2(CO)_{10}$ is thought to proceed via a radical pathway; for the reaction of $Re_2(CO)_8(L-L)$, this pathway is presumably prevented by the fast recombination

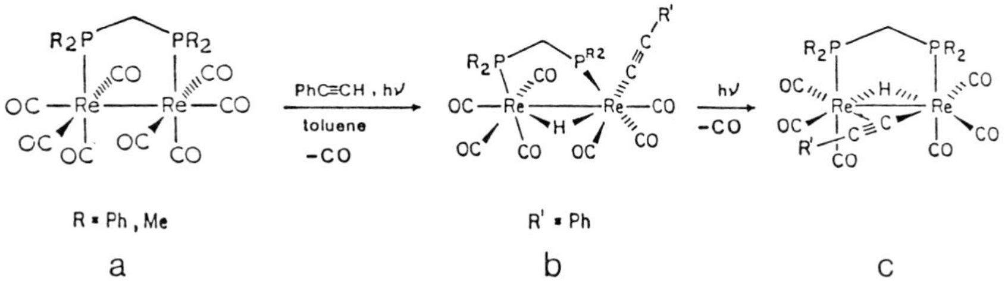

R = Ph, Me   R' = Ph

a   b   c

Fig. 16. Photochemical reaction of 1,2-eq,eq-$Re_2(CO)_8(L-L)$ with $PhC \equiv CH$ (reproduced with permission from ref. 143).

of the diradical compared to the slow reaction with the alkyne. The alkyne reaction then follows the slower CO dissociation pathway. This mechanism somewhat deviates from that proposed for the reaction with water and methanol (ref. 133) and may be open to discussion in view of recent flash photolysis results, which show that CO loss is an inefficient process for these 1,2-eq,eq-$Re_2(CO)_8(L-L)$ complexes (ref. 75).

4      $M_2(CO)_8(\alpha\text{-DIIMINE})(M=Mn,Re)$.

Derivatives whose ground- and excited state electronic structures deviate appreciably from those of $Mn_2(CO)_{10}$ are of special interest to photochemists. This holds e.g. for the complexes $(CO)_5MM'(CO)_3(\alpha\text{-diimine})$ (M,M'=Mn,Re; α-diimine=2,2'-bipyridine(bpy), 1,10-phenanthroline (phen), pyridine-2-carbalde-hyde-imine (R-PyCa), 1,4-diaza-1,3-butadiene(R-DAB)), which have the structure shown in Fig. 17 and which possess, apart from a $\sigma_b \rightarrow \sigma^*(M-M')$ transition at 340 nm, an intense metal to α-diimine charge transfer (MLCT) band at about 550 nm (Fig. 18) (ref. 30-32). Based on arguments presented in section 3.1, this low-energy band has been assigned to transitions from the $d\pi$ orbitals of M', not involved in the metal-metal bond, to the lowest π^* orbital of the α-diimine ligand (ref. 32).

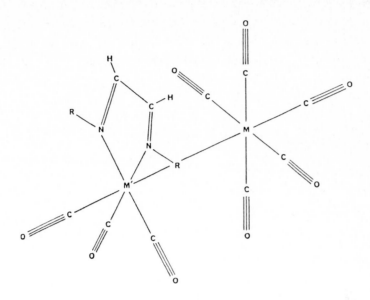

Fig. 17. Structure of $(CO)_5MM'(CO)_3(R\text{-}DAB)$ (M,M'=Mn,Re).

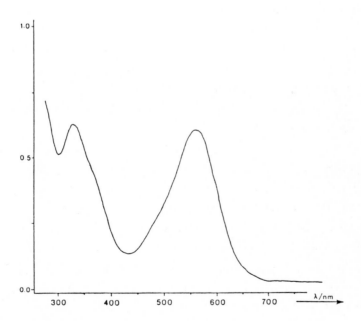

Fig. 18. Absorption spectrum of $(CO)_5MnMn(CO)_3(i\text{-}Pr\text{-}DAB)$ in n-pentane (reproduced with permission from ref. 36).

Irradiation into such a transition and population of a low-energy ^{3}MLCT ($^3d\pi\pi^*$) state is not expected to lead to an efficient photochemical reaction since these states are normally not reactive (ref. 144). Morse and Wrighton, however, observed the following reaction upon irradiation of $M_2(CO)_8$(phen) (M=Mn,Re) and $Re_2(CO)_8$(bquin) (bquin=2,2'-bquinoline) in CH_2Cl_2/CCl_4 (1:1) (ref. 30):

$$(CO)_5MM'(CO)_3L \xrightarrow[CH_2Cl_2/CCl_4 \ (1:1)]{h\upsilon} ClM(CO)_5 + ClM'(CO)_3L \quad (45)$$

M=M'=Mn,Re   L=phen,bquin

Photolysis of $M_2(CO)_8$(phen) (M=Mn,Re) in degassed benzene solution, afforded $M_2(CO)_{10}$ and $M_2(CO)_6$(phen)$_2$. The quantum yields of these reactions were practically wavelength independent and as high as 0.9 for $Mn_2(CO)_8$(phen).

In order to explain these high quantum yields and their wavelength independence, Meyer and Caspar (ref. 7) proposed a mechanism according to which the homolytic splitting of the metal-metal bond, observed by Morse and Wrighton, takes place from the dissociative $^3\sigma_b\sigma^*$ state just as in the case of $M_2(CO)_{10}$. This state is postulated to intersect with the ^{3}MLCT state, as visualized in the energy diagram of Fig. 19, which is just an extension of Fig. 5.

Irradiation into the ^{1}MLCT state will then lead to efficient cleavage of the metal-metal bond from the $^3\sigma_b\sigma^*$ state just as in the case of $Mn_2(CO)_{10}$. This energy diagram explains the high quantum yields even at low-energy excitation. This process of excited-state interconversion is an intramolecular sensitization in which energy is transferred from the metal-α-diimine MLCT state to the $^3\sigma_b\sigma^*$ state of the metal-metal bond. Related processes, involving sensitization of low-lying LF states following MLCT excitation, have been shown to be of importance for d^6 systems (ref. 144 and references therein).

This study of the photochemistry of $(CO)_5MM'(CO)_3$(α-diimine) complexes has been extended by Kokkes et al(ref. 36,119,145) to complexes of the α-diimine ligands R-DAB and R-PyCa, to reactions with nucleophiles and to measurements at lower temperatures and in rigid media. The complexes $(CO)_5MnMn(CO)_3$(α-diimine) were irradiated in 2-MeTHF at temperatures varying from 133 to 293 K(ref. 36). Above 200K the same homolysis reaction takes place as observed by Morse and Wrighton (ref. 30), viz.:

$$(CO)_5MnMn(CO)_3(\alpha\text{-diimne}) \xrightarrow{h\nu} Mn(CO)_5 + Mn(CO)_3(\alpha\text{-diimine}) \tag{46}$$

$$2\ Mn(CO)_5 \to Mn_2(CO)_{10} \tag{47}$$

$$2\ Mn(CO)_3(\alpha\text{-diimine}) \underset{\leftarrow}{\overset{\rightarrow}{\rightleftarrows}} Mn_2(CO)_6(\alpha\text{-diimine})_2 \tag{48}$$

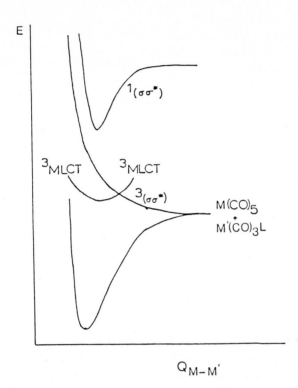

Fig. 19. Schematic energy level diagram for $(CO)_5MM'(CO)_3L$ ($L=\alpha$-diimine; M,M'=Mn,Re) (reproduced with permission from ref. 7).

Whereas the $Mn(CO)_5$ radicals are too short-lived to be detected, the $Mn(CO)_3$ (α-diimine) radicals could be studied with esr in the case of the bulky t-Bu-DAB ligand (see section 2.5). These latter radicals are in thermal equilibrium with the dimer and the position of the equilibrium depends on the temperature, the energy of the lowest π^* orbital of the α-diimine ligand and the steric requirements of this ligand (ref. 104).

When the photolysis reaction takes place at room temperature in the presence of a phosphine ligand, different reactions occur depending on the donor properties of the phosphine (ref. 119). Photolysis in the presence of PPh_3 affords $Mn_2(CO)_8(PPh_3)_2$ and $Mn_2(CO)_6$ (α-diimine) $(PPh_3)_2$; but, when the reaction takes place in the presence of the more basic phosphine $P(n-Bu)_3$, or a N-donor ligand, a completely different reaction is observed. The ions $[Mn(CO)_5]^-$ and $[Mn(CO)_3(α\text{-diimine})(L')]^+$ (L'=$P(n-Bu)_3$, N-donor ligand) are then formed as the result of a photocatalytic disproportionation reaction (eq 49-56) analogous to that of $Mn_2(CO)_{10}$ with pyridine and amines (eq 28-33)

$$(CO)_5MnMn(CO)_3L \quad \xrightarrow{h\nu} \quad Mn(CO)_5 + Mn(CO)_3L \qquad (49)$$

$$2\ Mn(CO)_5 \quad \rightarrow \quad Mn_2(CO)_{10} \qquad (50)$$

$$Mn(CO)_3L + L' \quad \rightarrow \quad Mn(CO)_3LL' \qquad (51)$$

$$Mn(CO)_3LL' + (CO)_5MnMn(CO)_3L \quad \rightarrow \quad [Mn(CO)_3LL']^+ + [(CO)_5MnMn(CO)_3L]^- \qquad (52)$$

$$[(CO)_5MnMn(CO)_3L]^- \quad \rightarrow \quad [Mn(CO)_5]^- + Mn(CO)_3L \qquad (53)$$

$$Mn(CO)_3LL' + Mn_2(CO)_{10} \quad \rightarrow \quad [Mn(CO)_3LL']^+ + [Mn_2(CO)_{10}]^- \qquad (54)$$

$$[Mn_2(CO)_{10}]^- \quad \rightarrow \quad Mn(CO)_5 + [Mn(CO)_5]^- \qquad (55)$$

$$Mn(CO)_3LL' + Mn(CO)_5 \quad \rightarrow \quad [Mn(CO)_3LL']^+ + [Mn(CO)_5]^- \qquad (56)$$

L=α-diimine

L'=$P(n-Bu)_3$,N-donor ligand

The quantum yield of this reaction is very high and can reach values of 100 and more by suitable choice of ligands. The highly reducing radicals $Mn(CO)_3$(α-diimine)L' can also catalyse substitution reactions. Thus, addition of a small amount of one of these complexes to a solution of $Ru_3(CO)_{12}$ and $P(n-Bu)_3$ starts, upon excitation with visible light, the catalytic substitution of CO in $Ru_3(CO)_{12}$ by $P(n-Bu)_3$ to give $Ru_3(CO)_{11}(P(n-Bu)_3)$. These complexes have the great advantage as photocatalysts over $Mn_2(CO)_{10}$ that the reducing radicals, $Mn(CO)_3$(α-diimine)L', are formed with high quantum yield by excitation with visible light and in the presence of low concentrations of monodentate ligands.

An analogous disproportionation product, $[Mn(CO)_3(α\text{-diimine})(S)]^+[Mn-(CO)_5]^-$, is obtained by irradiation of these complexes in 2-MeTHF or THF at temperatures at which they act as coordinating solvents (S). The above reactions are the results of a homolytic splitting of the metal-metal bond as the primary photoprocess.

There is, however, just in the case of $Mn_2(CO)_{10}$, evidence for the occurrence of a CO loss reaction as a second primary photoprocess. As mentioned above, the complexes $(CO)_5MnMn(CO)_3$(α-diimine)show homolytic splitting of the metal-metal bond and formation of $Mn_2(CO)_{10}$ and $Mn_2(CO)_6$(α-diimine)$_2$ in 2-MeTHF at temperatures above 200K. However, when the complexes are irradiated at lower temperatures, where the solvent becomes increasingly viscous, release of CO also occurs. This reaction becomes more and more important when the temperature is further lowered and it is the sole reaction at the melting point of the solvent. The CO-loss product has been characterized with ir spectroscopy (ref. 146).

Raising the temperature causes a backreaction of this photoproduct with CO to give the starting complex.

The photochemistry of these complexes is, however, still more complicated since a third reaction obtains for the R-DAB and R-PyCa complexes. Irradiation of these complexes not only yields $Mn_2(CO)_{10}$ and $Mn_2(CO)_6$(α-diimine)$_2$, but also compound 20a in the case of R-PyCa, and 20b for the R-DAB complexes. In complex 20a a CO ligand of the $Mn(CO)_5$ moiety is substituted by the C=N group of the R-PyCa ligand, in 20b two CO groups are substituted by the two C=N groups of the R-DAB ligand. For one of the R-PyCa complexes an X-ray structure determination has been carried out (ref. 146). It will be evident from Fig. 20 that no such products can be formed for the bpy and phen complexes.

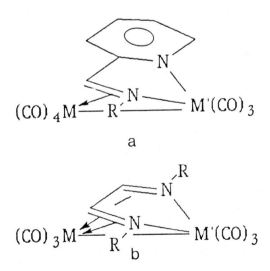

Fig. 20. Structure of $(CO)_4M(\sigma,\sigma,\eta^2$-R-PyCa)M'(CO)$_3$(a) and $(CO)_3M(\sigma,\sigma,\eta^4$-R-DAB)M'(CO)$_3$ (b) (M,M'=Mn,Re).

The quantum yield for the formation of these products increases upon going to higher energy excitation and so does addition of an extra amount of $Mn_2(CO)_{10}$. This means that the complexes are formed by attack of the R-DAB and R-PyCa ligands of the $Mn(CO)_3$ (α-diimine) radicals, via their C=N bonds, at the $Mn(CO)_5$ radicals, leading to substitution of CO. Irradiation at higher energy, or in the presence of excess $Mn_2(CO)_{10}$, increases only the concentration of the $Mn(CO)_5$ radicals, necessary for this reaction.

There is, however, evidence also for the formation of these complexes in non coordinating, viscous or rigid, media such as paraffin, Ar matrices or PVC films. The above complexes are then formed as the sole products.

5    $(\eta^5-C_5H_5)_2M_2(CO)_6$ (M=Mo,W)

These d^5-d^5 complexes were first prepared and described by Wilkinson (ref. 148) and the X-ray structure of $Cp_2Mo_2(CO)_6(Cp=\eta^5-C_5H_5)$ was determined by Wilson and Shoemaker (ref. 149). Just as with $M_2(CO)_{10}$ (M=Mn,Re) these complexes possess a direct metal-metal bond and no bridging carbonyls (see Fig. 21). The uv/vis spectra, shown in Fig. 22, exhibit, just as in the case of $M_2(CO)_{10}$ (see Fig. 3), two absorption bands. The high intensity band between 350 and 400 nm has again been assigned (ref. 6, 150) to the $\sigma_b\rightarrow\sigma^*$ (M-M) transition, the weak band at about 500 nm to $d\pi\rightarrow\sigma^*$ excitation.

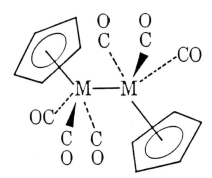

Fig. 21.   Structure of $Cp_2M_2(CO)_6$ (M=Mo,W).

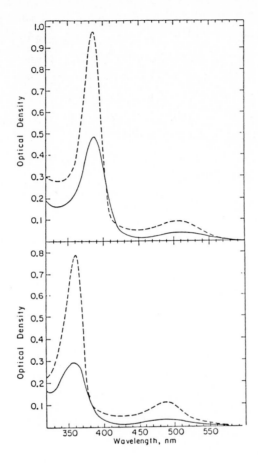

Fig. 22. Absorption spectra of $Cp_2Mo_2(CO)_6$(top) and $Cp_2W_2(CO)_6$(bottom) in EPA at 298 K (——) and 77 K (---) (reproduced with permission from ref. 150).

Photolysis in CCl_4 yielded two molecules $CpM(CO)_3Cl$ per molecule of the dimer, with quantum yields for the disappearance of the parent compound in the range 0.12-0.45, depending on M and on the wavelength of irradiation (ref. 150). Similar results were obtained by Giannotti and Merle(ref. 151) upon photolysis of $Cp_2Mo_2(CO)_6$ in neat CHX_3 and CX_4(X=Cl, Br).

Simultaneous excitation of $M'_2(CO)_{10}$ (M'=Mn, Re) and $Cp_2M_2(CO)_6$ affforded the mixed-metal complexes $(CO)_5M'M(CO)_3Cp$ in high yield (ref. 150). Based on these results, the primary photoprocess of these reactions was assumed to be homolysis of the metal-metal bond. The same mechanism was proposed for other reactions of these dimers, such as the formation of the photosubstitution products $Cp_2Mo_2(CO)_4(P(OPh)_3)_2$ (ref. 152) and the photochemical generation of the M-M triple- bonded dimers $Cp_2M_2(CO)_4$ (vide infra) (ref. 153, 154). Both products were thought to be formed after CO loss from the radicals $CpM(CO)_3$. These radicals were identified with ir by Rest and co-workers (ref. 155) who irradiated $CpM(CO)_3H$ in a CO matrix at 12 K. In contrast to the Mo and W radicals, $CpCr(CO)_3$ could be detected in solution (ref. 156, 157) and, recently, the X-ray structure of a stable derivative $CpCr(CO)_2(PPh_3)$ has been determined (ref. 158).

Nevertheless, by 1975, Meyer and co-workers (ref. 159) had already shown that homolysis of the metal-metal bond is not the only primary photoprocess of these $Cp_2M_2(CO)_6$ complexes. Using conventional flash photolysis of $Cp_2Mo_2(CO)_6$, they obtained evidence for the appearance of two distinct intermediates which differed appreciably in their lifetimes and spectral properties. The short-lived species did not absorb in the $\sigma_b \rightarrow \sigma^*$(M-M) spectral region and was therefore assumed to be the $CpMo(CO)_3$ radical. It reacted back to the dimer with a second-order rate constant of $10^9 M^{-1}s^{-1}$, which value approaches the diffusion-controlled limit. The long-lived intermediate absorbed strongly in the $\sigma_b \rightarrow \sigma^*$(M-M) spectral region and reacted back to the parent dimer with an estimated maximum rate constant of 10^7-$10^8 M^{-1}s^{-1}$ depending on the solvent. This process was thought to involve backreaction of CO with a second primary photoproduct $Cp_2Mo_2(CO)_5$. On the basis of these findings, the following two primary photoprocesses were proposed (eq. 57, 58):

$$Cp_2Mo_2(CO)_6 \xrightarrow{h\nu} \begin{cases} \longrightarrow & 2\ CpMo(CO)_3 \qquad\qquad (57) \\ \\ \longrightarrow & Cp_2Mo_2(CO)_5 + CO \qquad (58) \end{cases}$$

These dimers $Cp_2M_2(CO)_5$ were also indentified with ir and uv/vis by Hooker, Mahmoud and Rest, who photolyzed $Cp_2M_2(CO)_6$ (M=Mo,W) in frozen gas matrices at 12K and in polyvinyl chloride film (12-77K) (ref. 159). From the observation of free CO after photolysis, they concluded that the photoproducts are CO loss products. The presence of a metal-metal bond was evident from an intense uv absorption band belonging to the $\sigma_b \rightarrow \sigma^*$(M-M) transition. These authors proposed a structure, shown in Fig. 22, for this complex. The bridging carbonyl ligand, also present in $Mn_2(CO)_9$(vide supra), has a frequency of 1665 and 1635 cm^{-1} for the Mo- and W-complex, respectively. These low frequencies point to a four- electron donor bonding of this ligand. The same CO-bridged intermediate has been observed for $CpMo(CO)_3$, upon flash photolysis of $Cp_2Mo_2(CO)_6$, using ir detection (ref. 3).

Laine and Ford (ref. 160) studied the photochemical reaction of $Cp_2W_2(CO)_6$ with several chlorocarbon trapping agents. They observed a quantum yield dependence on lamp intensity, and chlorocarbon concentration, which they could explain by a kinetic scheme. In this scheme they introduced, apart from the $CpW(CO)_3$ radicals, a CO loss product which accounts for the photodecomposition of the dimer, in THF, in the absence of a trapping agent.

Turaki and Huggins (ref. 161) used the same competition between halogen atom abstraction and phosphine substitution, which had been exploited by Fox et al. for $Re(CO)_5$ (ref. 80) and by Herrinton and Brown for $Mn(CO)_5$ (ref. 81), to establish the substitution pathway of the $CpW(CO)_3$ radical. However, instead of photolyzing the dimer, they generated the $CpW(CO)_3$ radical by hydrogen abstraction from $CpW(CO)_3H$ using triphenylmethylradicals, $Ph_3C\cdot$. The subsequent competition between chlorine abstraction, from Ph_3CCl, and substitution of CO, by a phosphine ligand L, afforded $CpW(CO)_3Cl$ (2) and $CpW(CO)_2(L)Cl$ (3) as the only products. These reactions are schematically presented in Fig. 23.

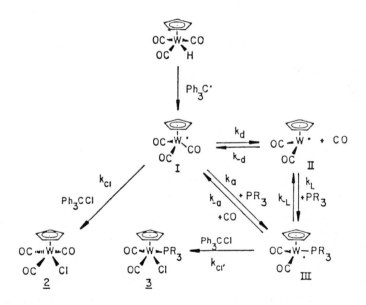

Fig. 23.   Proposed mechanism for the reaction between $CpW(CO)_3H$ and $Ph_3C\cdot$ (reproduced with permission from ref. 161).

The ratio of the concentrations of 3 and 2 showed a linear relationship with the concentration of L, with zero intercept, indicative of a dominant associative substitution process. Consistent with this associative nature of the reaction was the observed nucleophile reactivity order, which was determined both by steric and electronic effects of the ligands.

An interesting reaction, not observed for the $M'_2(CO)_{10}$ (M'=MnRe) complexes, is the formation of $Cp_2M_2(CO)_4$ ($(Cp(CO)_2M\equiv M(CO)_2Cp)$ by near uv or vis irradiation of $Cp_2M_2(CO)_6$(ref. 153,154). This reaction also proceeds thermally. The triple bonded complex $Cp_2Mo_2(CO)_4$(ref. 162) has a Mo-Mo bond which is about 0.8Å shorter than in the single bonded dimer (ref.163). As a result, the $\sigma_b \rightarrow \sigma^*$(M-M) transition is also at significantly lower energy in the latter complex (ref. 6).

Ginley et al. (153, 154) first investigated the photochemical conversion of the single bonded into the triple bonded complex and proposed a mechanism which involved cleavage of the metal-metal single bond, dissociative loss of CO from the $CpMo(CO)_3$ radicals formed, and combination of the $CpMo(CO)_2$ radicals to give the triple bonded dimer. This explanation does, however, not agree with the above mentioned results, which indicated an associative pathway for the substitution of CO in the $CpW(CO)_3$ radical without any contribution from a dissociative process (ref. 161). Turaki and Huggins (ref. 164), therefore, investigated the reactions of $MeCp(CO)_2Mo\equiv Mo(CO)_2Cp$ as well as of mixtures of $(MeCp)_2Mo_2(CO)_4$ and $Cp_2Mo_2(CO)_4$ with CO. The products were invariably the corresponding dimers without formation of any crossover product. Metal-metal bond cleavage was therefore excluded as the mechanism for the thermal and photochemical interconversion of $Cp_2Mo_2(CO)_4$ and $Co_2Mo_2(CO)_6$. Instead, the process shown in Fig. 24 was proposed in which CO directly adds to $Cp_2Mo_2(CO)_4$ or dissociates from $Cp_2Mo_2(CO)_6$.

Hepp and Wrighton (ref. 86) studied the photolysis of $Cp_2W_2(CO)_6$ in the presence of variable amounts of CCl_4 and one of several one-electron oxidants. They determined the ratios of the rate constants for oxidation (k_{ox}) and halogen abstraction (k_x). In the case of ferricenium as oxidant, this ratio appeared to increase from 13 in CH_2Cl_2 to 143 in CH_3CN. Apparently, coordination of CH_3CN to the $CpW(CO)_3$ radical affords the strongly reducing $19e^-$ radical $CpW(CO)_3(CH_3CN)$, which reacts with the oxidant, to give $CpW(CO)_3(CH_3CN)^+$. Special attention has been paid recently to the disproportionation reactions of the $Cp_2Mo_2(CO)_6$ complexes, reactions which are closely related to those of $Mn_2(CO)_{10}$ with O- and N-donor ligands (vide supra).

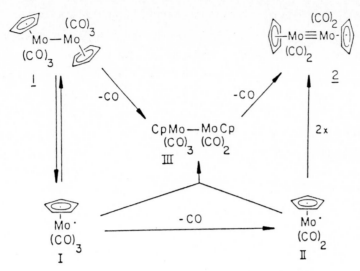

Fig. 24. Proposed mechanism for the photochemical interconvension of $Cp_2Mo_2(CO)_6$ and $Cp_2Mo_2(CO)_4$ (reproduced with permission from ref. 164)

Contrary to the results obtained with $Mn_2(CO)_{10}$, $Cp_2Mo_2(CO)_6$ dispropor-tionates in the presence of small phosphines and phosphites, according to reaction 59.

$$Cp_2Mo_2(CO)_6 + 2PR_3 \quad \xrightarrow{h\nu} \quad CpMo(CO)_2(PR_3)_2^+ + CpMo(CO)_3^- + CO \qquad (59)$$

For PPh_3, and larger ligands only, the monosubstituted cation forms according to reaction 60 (vide infra). Also, for smaller ligands, this reaction occurs when it is performed under an atmosphere of CO. (ref. 165)

$$Cp_2Mo_2(CO)_6 + PR_3 \quad \overset{h\nu}{\underset{\leftarrow}{\to}} \quad CpMo(CO)_3(PR_3)^+ + CpMo(CO)_3^- + \qquad (60)$$

Haines, Nyholm and Stiddard (ref. 166) were the first to establish that disproportionation of these dimers occurs in the presence of a number of phosphines and phosphites. King et al. (ref. 152) reported analogous reactions with bidentate phosphines and arsines. Haines et al. (ref. 166) claimed that disproportionation occurs photochemically as well as thermally but Stiegman et al. (ref. 167) showed that this pathway is always photochemical and that mono- and disubstituted complexes $Cp_2Mo_2(CO)_5(PR_3)$ and $Co_2Mo_2(CO)_4(PR_3)_2$ are formed in conjunction with the disproportionation products. From their results the authors concluded that the electronic and steric properties of the PR_3

ligand determine whether or not the dimer will disproportionate (ref. 167). They must be good electron donors and cannot be sterically bulky if disproportionation is to occur. If these two requirements are not met, then irradiation of the dimer in the presence of PR_3 leads to substitution products only. Disproportionation according to reaction 59 appeared to be limited to ligands with cone angles $\tau < 145°$. Since quantum yields in excess of unity were observed, a radical chain mechanism was proposed for this reaction (eq. 61-65) just as for the disproportionation of $Mn_2(CO)_{10}$ in pyridine (eq. 28-33).

$$Cp_2Mo_2(CO)_6 \; \overset{h\nu}{\underset{\rightleftarrows}{\rightleftarrows}} \; 2CpMo(CO)_3 \qquad (61)$$

$$CpMo(CO)_3 + L \; \rightleftarrows \; CpMo(CO)_2(L) + CO \qquad (62)$$

$$CpMo(CO)_2(L) + L \; \rightleftarrows \; CpMo(CO)_2(L)_2 \qquad (63)$$

$$CpMo(CO)_2(L)_2 + Cp_2Mo_2(CO)_6 \; \rightarrow \; CpMo(CO)_2(L)_2^+ + Cp_2Mo_2(CO)_6^- \qquad (64)$$

$$Cp_2Mo_2(CO)_6^- \; \rightarrow \; CpMo(CO)_3^- + CpMo(CO)_3 \qquad (65)$$

In this mechanism, the primary photoprocess is again homolytic cleavage of the metal-metal bond. After substitution (eq. 62) the 17-electron radical can recombine to form substituted dimers (eq. 66,67), or they can pick up an extra ligand L (eq. 63) to give the highly reducing 19e$^-$ radical, $CpMo(CO)_2(L)_2$, which

$$CpMo(CO)_3 + CpMo(CO)_2(L) \; \rightarrow \; Cp_2Mo_2(CO)_5(L) \qquad (66)$$

$$2CpMo(CO)_2(L) \; \rightarrow \; Cp_2Mo_2(CO)_4(L)_2 \qquad (67)$$

then undergoes an electron transfer reaction with the parent dimer. Kinetic evidence provided support for this 19e$^-$ intermediate. For example, high ligand concentrations, and bidentate ligands, increased the quantum yield of the disproportionation reaction. In the case of the bidentate phophine dppe (=bis(1,2-diphenylphosphino)ethane) disproportionation was the sole reaction with a quantum yield as high as 11.8 It must be recalled here that, in the case of $Mn_2(CO)_{10}$ especially, high quantum yields were obtained upon using tridentate ligands, a result which was ascribed to the formation of the 19e$^-$ intermediate, $Mn(CO)_3N_3$.

An interesting observation of Stiegman and Tyler (ref. 168) was the intensity dependence of the product distribution for the reaction of $(CH_3Cp)_2Mo_2(CO)_6$ with $P(OCH_3)_3$. Disproportionation appeared to dominate over substitution at low light intensities while, at higher intensities, more substitution products were formed. At low intensities, the concentration of the

radicals is low and no coupling reactions (eq. 66, 67) occur. Formation of the
19-electron intermediate is more efficient then, and relatively more
disproportionation occurs. At higher intensities, radical-radical coupling
reactions are more favourable. Further support for the above mechanism was
provided by generating the key intermediate $(CH_3Cp)_2MO_2(CO)_6^-$ by reduction
of $(CH_3Cp)_2Mo_2(CO)_6$ with Na (ref. 167). When the sodium dispersion was added
to a solution containing $(CH_3Cp)_2Mo_2(CO)_6$ only a very small amount of
$(CH_3Cp)Mo(CO)_3^-$ was detected. When the same experiment was performed in
the presence of $PPh_2(CH_3)$, complete disproportionation was observed within
two minutes of adding the sodium.

In the case of PPh_3, Stiegman and Tyler(ref. 169) observed a wavelength
dependent disproportionation reaction. Upon 290 nm irradiation, the dimers
$(RCp)_2Mo_2(CO)_6(R=H, CH_3)$ appeared to disproportionate in apolar solvents and,
in the presence of PPh_3, according to reaction 59. However, excitation with
405 nm led only to the formation of the monosubstitution product $(RCp)_2Mo_2-$
$(CO)_5(PPh_3)$. They later reinvestigated this reaction (ref. 165) and, first of all,
observed a solvent dependence of the disproportionation. Photolysis of
$(RCp)_2Mo_2(CO)_6$ with $\lambda > 525$ nm, in the presence of PPh_3, did, indeed, afford
no disproportionation products when this reaction was performed in nonpolar
solvents. In CH_2Cl_2, however, reaction 60 was observed in conjunction with
substitution. Upon standing in the dark for 1h, 90% of the ionic product had
reacted back to the parent dimer. Apparently, this backreaction is much faster
in nonpolar solvents, which explains the previous findings (ref. 169); i.e. that
these complexes do not disproportionate with PPh_3 upon near uv or vis
excitation in nonpolar solvents. The observation of reaction 59, upon 290 nm
excitation, was ascribed to a secondary photochemical reaction at this
wavelength of the initially formed $(RCp)Mo(CO)_3(PPh_3)^+$ (eq. 68). The
disubstituted cation and $(RCp)Mo(CO)_3^-$ do not back-react.

$$(RCp)Mo(CO)_3(PPh_3)^+ + PPh_3 \xrightarrow{\lambda=290 \text{ nm}} (RCp)Mo(CO)_2(PPh_3)_2^+ + CO \quad (68)$$

Whether the reaction proceeds via eq. 59 or 60 depends on the cone angle
of the phosphine and its electron donating ability. Ligands with cone angles
larger than 145° are too bulky to form a disubstituted cationic product, via
reactions 62-64, using low energy excitation.

Instead, they react according to eqs 69 and 70.

$$CpMo(CO)_3 + L \rightarrow CpMo(CO)_3(L) \qquad (69)$$
$$CpMo(CO)_3(L) + Cp_2Mo_2(CO)_6 \rightarrow CpMo(CO)_3(L)^+ + Cp_2Mo_2(CO)_6^- \qquad (70)$$

Also, in the case of smaller phosphines reaction 62 is prevented if it is performed in the presence of a CO atmosphere.

So, contrary to what had been proposed before (vide supra, ref. 167), steric effects are not important in determining whether or not dispropor-tionation of $Cp_2Mo_2(CO)_6$ will occur. They only determine whether the reaction proceeds via eqs 62-64 or via eqs 69 and 70. The same holds for the electronic effects. Poor electron donors, such as $P(OPh)_3$, do not react with $Cp_2Mo_2(CO)_6$ in benzene but do so in CH_2Cl_2. Again, this has been ascribed (ref. 165) to the inability of these ligands to form disubstituted cationic product and to the fast back-reaction in eq. 60.

Disproportionation occurs not only with phosphines but also in coordinating solvents in the absence of suitable ligands. Stiegman and Tyler (ref. 170) studied the disproportionation of $(CH_3Cp)_2Mo_2(CO)_6$ in acetone, CH_3CN and Me_2SO in the absence and presence of halides. The complex disproportionated in these solvents in the absence of halide. In the presence of a halide, (X^-) $(CH_3Cp)Mo(CO)_3X$ and $(CH_3Cp)Mo(CO)_3^-$ were formed according to reaction 71.

$$(CH_3Cp)_2Mo_2(CO)_6 + X^- \xrightarrow{h\nu} (CH_3Cp)Mo(CO)_3X + CH_3CpMo(CO)_3^- \qquad (71)$$

The quantum yield of this reaction was greater than one for low lamp intensity and appeared to be independent of the halide concentration. On the basis of these and other observations, a mechanism was proposed (eqs 72-77) in which the solvent(S) and not the halide bonds to the $(CH_3Cp)Mo(CO)_3$ radical. Reactions 73 and 74 are similar to eqs 69 and 70.

Just as for PPh_3 and other bulky phosphines, the halide complex $(CH_3Cp)Mo-(CO)_3X$ may react back with $(CH_3Cp)Mo(CO)_3^-$ to give the parent dimer and X^- in non-polar solvents. This is, in fact, the case since the above reaction is not observed in benzene (ref. 170).

$$(CH_3Cp)Mo_2(CO)_6 \xrightarrow[\leftarrow]{h\nu} 2(CH_3Cp)Mo(CO)_3 \qquad (72)$$

$$(CH_3Cp)Mo(CO)_3+S \rightleftarrows (CH_3Cp)Mo(CO)_3S \qquad (73)$$

$$(CH_3Cp)Mo(CO)_3S \rightarrow (CH_3Cp)Mo(CO)_3S^+$$
$$+(CH_3Cp)_2Mo_2(CO)_6 \qquad +(CH_3Cp)_2Mo_2(CO)_6^- \qquad (74)$$

$$(CH_3Cp)_2Mo_2(CO)_6^- \rightarrow (CH_3Cp)Mo(CO)_3^-$$
$$+(CH_3Cp)Mo(CO)_3 \qquad (75)$$

$$(CH_3Cp)Mo(CO)_3S^++X^- \rightarrow (CH_3Cp)Mo(CO)_3X+S \qquad (76)$$

$$(CH_3Cp)Mo(CO)_3S \rightleftarrows (CH_3Cp)Mo(CO)_3S^+$$
$$+(CH_3Cp)Mo(CO)_3 \qquad +(CH_3Cp)Mo(CO)_3^- \qquad (77)$$

6. $(\eta^5\text{-}C_5H_5)_2 Fe_2 (CO)_4$

The d^7-d^7 cyclopentadienyl complexes containing Fe, Ru, Os exist as dimeric species in different isomeric forms. The first row Fe complex has bridging CO groups (ref. 171) while Ru has both the bridged and the nonbridged forms and the Os complex is known only as the nonbridged structure. In solution, both the cis- and the trans-isomer of $Cp_2Fe_2(CO)_4$ exist and undergo a rapid interconversion. (ref. 172-177) (Fig. 25)

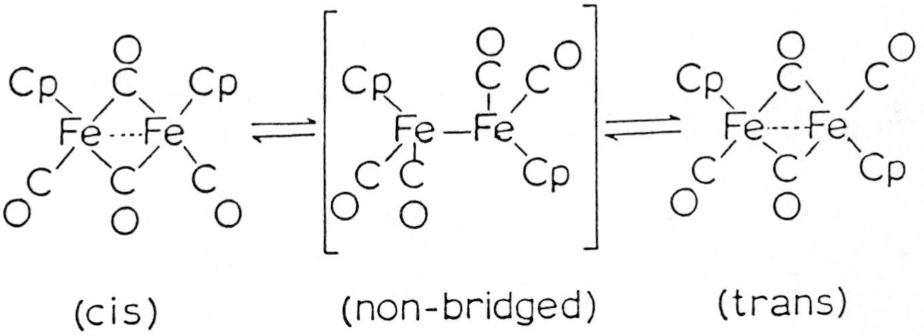

(cis)   (non-bridged)   (trans)

Fig. 25. Interconversion of the isomers of $Cp_2Fe_2(CO)_4$ (reproduced with permission from ref. 7).

The absorption spectrum of the Fe dimer, presented in Fig. 26 has an intense band at 346 nm (benzene solution at 298 K) which has been assigned by Harris and Gray (ref. 178) to a $\sigma_6 \rightarrow \sigma^*$ (M-M) transition in accordance with the nonbridged complexes $M_2(CO)_{10}$ (M=Mn, Re) and $Cp_2M_2(CO)_6$ (M=Mo,W) discussed before. However, in view of more recent low temperature X-ray (ref. 179) and molecular orbital (ref. 180,181) analyses, which showed that there is no net direct bonding between the carbonyl bridged iron atoms, this band is more appropriately assigned to a $\pi \rightarrow \pi^*$ transition of the $Fe_2(CO)_2$ bridge.

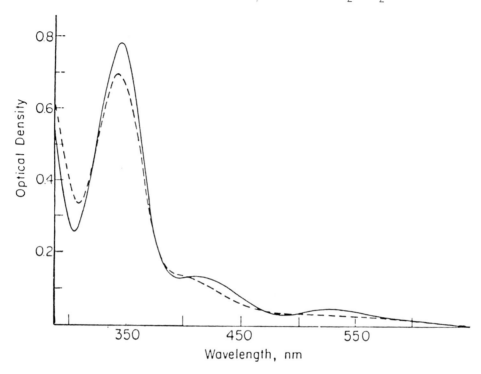

Fig. 26. Absorption spectra of $Cp_2Fe_2(CO)_4$ in isooctane (——) and CH_3CN (---) (reproduced with permission from ref. 177).

Abrahamson et al. (ref. 177) studied the photochemistry of $Cp_2M_2(CO)_4$ (M=Fe,Ru; Cp=η^5-C_5H_5) in CCl_4 at 298 K and observed the formation of $CpM(CO)_2Cl$ according to reaction 78.

$$Cp_2M_2(CO)_4 \xrightarrow[CCl_4]{h\upsilon} 2CpM(CO)_2Cl \tag{78}$$

M=Fe, Ru

Irradiation of $Cp_2Fe_2(CO)_4$ in the presence of PPh_3 (0,1M) resulted in the formation of the monosubstituted complex $Cp_2Fe_2(CO)_3(PPh_3)$. However, by using $P(OCH_3)_3$ instead of PPh_3 both the mono- and the disubstituted $(Cp_2Fe_2(CO)_2 (P(OCH)_3)_2)$ product were formed simultaneously. It was concluded that formation of the latter complex does not require irradiation of the monosubstituted species. The competition between halogen abstraction and substitution was investigated by irradiation of $Cp_2Fe_2(CO)_4$ in the presence of 0.1M PPh_3 and 0.1M CCl_4. The dominant product was $CpFe(CO)_2Cl$. Apparently, the substitution reaction is quenched by the halogen abstraction. The quenching appeared to be less effective when 0.1M of the less reactive radical trap $1-IC_5H_{11}$ was added to the solution, instead of CCl_4. In that case, the major product was still $Cp_2Fe_2(CO)_3(PPh_3)$ and $CpFe(CO)_2I$ was only formed to a small extent. From these results, it was concluded that substitution of PR_3 ligands takes place via substitution of CO from the radical rather than by dissociative loss of CO from the dimer. Similar results were obtained by Giannotti and Merle (ref. 151) upon irradiation of $Cp_2Fe_2(CO)_4$ in CHX_3 and CX_4 (X = Cl, Br).

Tyler et al. (ref. 182,183) reinvestigated these photoreactions of $Cp_2Fe_2(CO)_4$ with phosphines and chlorocarbons. For the reactions of CCl_4 an $CHCl_3$ they proposed, just as Wrighton and co-workers (ref. 177), a radical pathway since the chlorine atom abstraction was not inhibited by 6 atm. of CO. For the reactions with phosphines, however, they proposed a different mechanism based on the following observations. Irradiation of $Cp_2Fe_2(CO)_4$ in the presence of $P(O-i-Pr)_3$, PPh_3 or $P(n-Bu)_3$ afforded $Cp_2Fe_2(CO)_3(PR_3)$ as the sole product. These reactions were inhibited under 1 atm of CO. Only in the case of $P(OCH_3)_3$ was a disubstituted complex $Cp_2Fe_2(CO)_2(P(OCH_3)_3)_2$ obtained without any evidence for the intermediacy of the monosubstituted complex. Tyler et al. observed, however, that this latter complex reacts photochemically with $P(OCH_3)_3$ to give the disubstituted complex and that this photoreaction is much faster than the reaction of $P(OCH_3)_3$ with $Cp_2Fe_2(CO)_4$. The authors were able to stabilize an intermediate in the reaction with $P(O-i-Pr)_3$ by irradiation in ethylchloride solution at -78°C. Instead of the $Cp_2Fe_2(CO)_3(P-(O-i-Pr)_3)$ complex, a yellow intermediate was obtained which converted into the monosubstituted dimer by warming the solution to room temperature. The yellow intermediate had no intense absorption at about 350 nm, its ir spectrum contained a band at 1720 cm^{-1} belonging to a bridging carbonyl and it did not show any esr signal at temperatures as low as 15 K. Finally, this intermediate was only observed when the phosphite was present in the solution. On the basis of these results they proposed the reactions with phosphines and chlorocarbons to proceed by the pathways shown in Fig. 27.

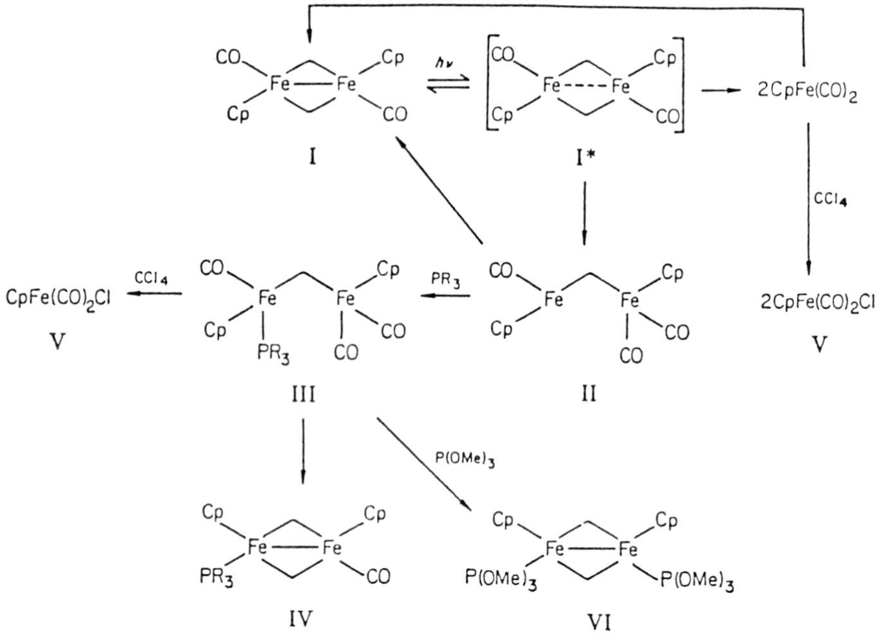

Fig. 27. Proposed mechanism for the photochemical reactions of $Cp_2Fe_2(CO)_4$ with CCl_4 and PR_3 (reproduced with permission from ref. 183).

The first step is the formation of the excited complex I*. From I* the 17-electron $CpFe(CO)_2$ radical and the binuclear intermediate II result as primary photoproducts. The radicals react with the chlorocarbons, whereas the reactions with phosphines and phosphites proceed mainly through II giving rise to the formation of the yellow intermediate III, which is apparently kinetically stable at -78°C. This reaction scheme agrees with the formation of only monosubstituted complexes of PPh_3 or $P(O-i-Pr)_3$. It might be argued that only monosubstituted products are formed if two substituted radicals $CpFe(CO)(PR_3)$ cannot combine to form a dimer for steric reasons. This alternative has, however, been ruled out by Tyler et al. (ref. 182) by using a Stern-Volmer plot for the reaction with PPh_3.

Caspar and Meyer (ref. 184) performed a conventional flash photolysis study on $Cp_2Fe_2(CO)_4$ both in the absence and presence of CCl_4 or PPh_3. Two distinct intermediates were observed which again differed appreciably in lifetimes. The short lived species absorbed only weakly in the vis and near uv and reacted back to the parent dimer via a second-order bimolecular reaction (eq. 79). It was proposed to be the $CpFe(CO)_2$ radical. The second, long lived, intermediate absorbed in the visible at 510 nm and reacted back thermally by a first-order process to the starting complex (eq. 80). It was therefore proposed to be the CO loss product $Cp_2Fe_2(CO)_3$.

$$Cp_2Fe_2(CO)_4 \xrightarrow{h\nu} \begin{cases} \longrightarrow & 2\ CpFe(CO)_2 & \qquad (79) \\ \longrightarrow & Cp_2Fe_2(CO)_3 + CO & \qquad (80) \end{cases}$$

By performing the flash photolysis experiment in the presence of CCl_4 and PPh_3, Caspar and Meyer could establish that both intermediates undergo reactions with CCl_4. For $Cp_2Fe_2(CO)_3$ the rate constant at 22°C was $4 \times 10^3 M^{-1} s^{-1}$, for the radical $CpFe(CO)_2$ the reaction was much faster with $k > 1.5 \times 10^5 M^{-1} s^{-1}$. The experiment with PPh_3 (0.1-0.001 M) did not show any reaction between the ligand and the radical. Under the same conditions the reaction between $Cp_2Fe_2(CO)_3$ and PPh_3 was first order in both reactants with $k(22°C) = 1.5 \times 10^5 M^{-1} s^{-1}$. These results agree with the conclusion of Tyler et al. (ref. 183) that phosphine ligands only react with the $Cp_2Fe_2(CO)_3$ intermediate.

The structure of the intermediate $Cp_2Fe_2(CO)_3$ could be determined spectroscopically by photolysis in different matrices (ref. 185,186) and by flash photolysis of $Cp_2Fe_2(CO)_4$ in room temperature solution using fast time-resolved infrared spectroscopy (ref. 187,188). Hooker et al. (ref. 185) were the first to stabilize and characterize $Cp_2Fe_2(CO)_3$ by photolyzing the dimer in a CH_4 matrix (12K) and PVC film (12-77K). In these rigid media only the CO loss product is formed due to the fast recombination of the radicals. The authors observed an ir band of free CO and a new CO vibration at 1812 cm^{-1}. At the same time, the bands of the trans-isomer of the parent dimer decreased in intensity, whereas those of the cis-isomer were not affected. Apparently, the CO loss product is only formed out of the trans-isomer of $Cp_2Fe_2(CO)_4$. From the observation of only one ir active bridging carbonyl vibration at 1812 cm^{-1}, Hooker et al. concluded that the new species has a symmetrical structure with three equivalent bridging CO ligands (D_{3h} local symmetry) just as $Fe_2(CO)_9$. The proposed structure is shown in Fig. 28.

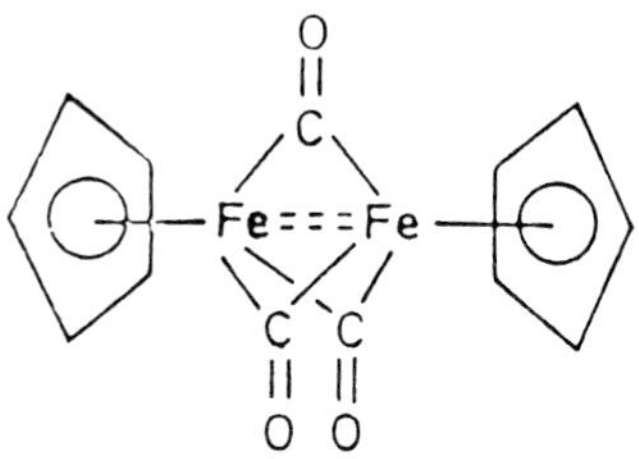

Fig. 28. Structure of $Cp_2Fe_2(\mu\text{-CO})_3$ (reproduced with permission from ref. 185).

Hepp et al. (ref. 186) photolyzed three complexes, $(\eta^5\text{-}C_5R_5)Fe_2(CO)_4$ (R=H,CH$_3$,CH$_2$Ph), in alkane and 2-MeTHF glasses, at 77K which differed in the relative importance of the cis- and trans-isomers. For R=H and CH$_3$, a significant fraction of the cis-isomer is present in solution, whereas for R=CH$_2$Ph only the trans-isomer is observed in both polar and nonpolar solvents. Also, in these matrices a triply CO-bridged dimer was formed, exclusively from the trans isomer. The difference in photosensitivity of the cis and trans isomer was ascribed to the fact that the trans isomer can form a symmetrical $(\mu\text{-CO})_3$ complex without gross movements of the C_5 ring.

Moore et al. (ref. 187,188) studied the flash photolysis of $Cp_2Fe_2(CO)_4$ in cyclohexane under 1 atm. pressure of CO with fast time-resolved ir spectroscopy (ref. 3). They observed the formation of two principal photoproducts. The first species had two CO vibrations which completely disappeared within 25 μs of the flash. The second product, having a single CO vibration at 1823 cm^{-1}, was much longer lived. The short lived photoproduct had its two vibrations almost identical to those of the matrix isolated $(\eta^4\text{-}C_4H_6)Fe(CO)_2$ and was therefore assumed to be the CpFe(CO)$_2$ radical. The long lived species had its ir band close to that of the dinuclear intrermediates observed in matrices and was therefore identified as $Cp_2Fe_2(CO)_3$.

Wrighton and co-workers (ref. 189) were able to isolate the complex $(\eta^5\text{-}C_5(CH_3)_5)Fe_2(CO)_3$ and to determine its X-ray structure. This structure indeed proved the highly symmetrical conformation of this complex. The Cp rings are pentahapto coordinated to the Fe atoms with their planes perpendicular to, and centered on, the Fe-Fe bond. The three CO ligands bridge symmetrically the two Fe-centers. The compound did not show a detectable ^{1}H NMR in solution since it is paramagnetic and the broadened resonance was not detected due to the low solubility of the complex. In view of these findings the reaction scheme of Fig. 27 has to be reconsidered. Instead of complex I, the triply CO-bridged complex of Fig. 28 will be formed as a primary photoproduct. This will then take up a ligand L and form the intermediate III which has been identified by Tyler et al. (ref. 182,183).

Contrary to the behaviour of the complexes $Mn_2(CO)_{10}$, $Co_2(CO)_8$ and $Cp_2Mo_2(CO)_6$ (M=Mo,W), $Cp_2Fe_2(CO)_4$ primarily shows substitution reactions in inert solvents and no disproportionation. Until recently, the only such reaction reported has been the thermal reaction with the bidentate ligand dmpe (=bis(dimethylphosphino)ethane) affording $[CpFe(CO)(dmpe)]^+[CpFe(CO)_2]^-$ (ref. 190). In the case of $Cp_2Mo_2(CO)_6$ Stiegman and Tyler (vide supra, ref. 165) observed that, although disproportionation of the dimer occurred in the presence of e.g. PPh_3, no photoproducts were obtained due to the fast backreaction of the ions. A detailed study by Goldman and Tyler (ref. 191) of the reactions of $Cp_2Fe_2(CO)_4$ showed that a similar process occurred for this dimer. Irradiation of $Cp_2Fe_2(CO)_4$ in various inert solvents, in the presence of one of a series of monodentate and (bulky)bidentate phosphines, did not lead to disproportionation. However, in CH_2Cl_2 the following reaction (eq. 81) occurred for L=P(OEt)$_3$ and L_2=dppe:

$$Cp_2Fe_2(CO)_4 + L_2 + CH_2Cl_2 \xrightarrow{h\nu} [CpFe(CO)L_2]^+Cl^- + CpFe(CO)_2(CH_2Cl) + CO \quad (81)$$

For this reaction Goldman and Tyler proposed the following mechanism (eq. 82-86).

$$Cp_2Fe_2(CO)_4 \xrightarrow{h\nu} 2CpFe(CO)_2 \quad (82)$$
$$CpFe(CO)_2 + L_2 \rightarrow CpFe(CO)L_2 + CO \quad (83)$$
$$CpFe(CO)L_2 + Cp_2Fe_2(CO)_4 \rightarrow CpFe(CO)L_2^+ + CpFe_2(CO)_4^- \quad (84)$$
$$CpFe_2(CO)_4^- \rightarrow CpFe(CO)_2^- + CpFe(CO)_2 \quad (85)$$
$$CpFe(CO)_2^- + CH_2Cl_2 \rightarrow CpFe(CO)_2(CH_2Cl) + Cl^- \quad (86)$$

Recall that reactions 82-85 are analogous to reactions 61-65 for $Cp_2Mo_2(CO)_6$. In CH_2Cl_2 the $CpFe(CO)_2^-$ ion is trapped and the backreaction (eq. 87) with CO and $CpFe(CO)L_2^+$, therefore inhibited. Since the backreaction is fast in inert solvents, no disproportionation is observed there.

$$Cp_2Fe_2(CO)_4 \xrightarrow{\text{h}\nu/\Delta(\text{Ar purge})} CpFe(CO)L_2^+ + CpFe(CO)_2^- + CO \quad (87)$$

Evidence for this backreaction was obtained by performing the reaction with dppe in Me_2SO. When the solution was purged with Ar, a complete disproportionation occurred thermally to give $[CpFe(CO)(dppe)]^+[CpFe(CO)_2]^-$. Bubbling CO through the solution then resulted in a backreaction of the ions. Apparently, disproportionation is not related to the polarity of the solvent. Rather, reaction 87 is reversible and the equilibrium lies far to the left. Only by removing one of the products, e.g. CO by purging with Ar, or $CpFe(CO)_2^-$ by reaction with CH_2Cl_2 (eq. 86) can the other products be observed. Disproportionation of $Cp_2Fe_2(CO)_4$, both thermal and photochemical, does not occur with bulky phosphine ligands such as PPh_3 or PCy_3 (Cy=cyclohexyl).

Goldman and Tyler (ref. 120) also studied the reducing ability of the $19e^-$ radical $CpFe(CO)(dppe)$, formed in reaction $83(L_2=dppe)$, using substrates other than $Cp_2Fe_2(CO)_4$ (eq. 88); for example

$$1/2Cp_2Fe_2(CO)_4 + dppe + S \xrightarrow{\text{h}\nu} [CpFe(CO)(dppe)]^+ + S^- \quad (88)$$

Many of such reactions were observed e.g. $CpMo(CO)_3Cl$ was reduced to $CpMo(CO)_3^-$ and $Mn_2(CO)_{10}$ to $Mn(CO)_5^-$. These reactions are analogous to the reduction of $Cp_2Mo_2(CO)_6$ in the presence of $Mn_2(CO)_{10}$ and diethylenetriamine, discussed before. A special property of the key intermediate $CpFe(CO)L_2$ is that it can act as initiator of electron-transfer-catalyzed (ETC) chain reactions (ref. 121, 192). Addition of a catalytic amount of $Cp_2Fe_2(CO)_4$ to a THF solution of $Ru_3(CO)_{12}$ and PMe_2Ph and irradiation with $\lambda > 525$ nm initiated the following ETC substitution of $Ru_3(CO)_{12}$ (eq. 89, ref. 191):

$$Ru_3(CO)_{12} + PMe_2Ph \xrightarrow[\text{h}\nu(\lambda > 525nm)]{Cp_2Fe_2(CO)_4} Ru_3(CO)_{11}(PMe_2Ph) \quad (89)$$

An interesting reaction occurred when an acetone solution of $Cp_2Fe_2(CO)_4$ and $[Cp_2Co]^+ [PF_6]^-$ was irradiated in the presence of dppe, $P(OEt)_3$ or PPh_3. With dppe and $P(OEt)_3$, the expected reduction of Cp_2Co^+ to cobaltocene was observed according to reaction 90:

$$1/2Cp_2Fe_2(CO)_4 + L_2 + Cp_2Co^+ \xrightarrow{h\nu} CpFe(CO)L_2^+ + Cp_2Co \qquad (90)$$

L_2 = dppe, 2 P(OEt)$_3$

However, upon addition of PPh$_3$, for which no disproportionation of $Cp_2Fe_2(CO)_4$ occurred (vide supra), leads to the reduction of Cp_2Co^+, although the rection only proceeded to ca 15% completion and $CpFe(CO)_2(PPh_3)^+$ was formed instead of $CpFe(CO)(PPh_3)_2^+$ (eq. 91):

$$1/2Cp_2Fe_2(CO)_4 + PPh_3 + Cp_2Co^+ \xrightarrow{h\nu} CpFe(CO)_2(PPh_3)^+ + Cp_2Co \qquad (91)$$

Apparently, the 19e$^-$ radical $CpFe(CO)_2(PPh_3)$, formed in the reaction of $Cp_2Fe_2(CO)_4$ with PPh$_3$ cannot reduce the dimer $Cp_2Fe_2(CO)_4$ but it can do so with Cp_2Co^+.

The question remains why reaction 90 is observed for P(OEt)$_3$ and 91 in the case of PPh$_3$. There are two important reasons for this different behaviour (ref. 191). First, PPh$_3$ cannot form a disubstituted radical $CpFe(CO)(PPh_3)_2$ because of steric reasons. Secondly, P(OEt)$_3$ is a poor donor ligand and the radical $CpFe(CO)_2(P(OEt)_3)$, formed by addition of P(OEt)$_3$ to the 17e$^-$ radical $CpFe(CO)_2$, before further reaction to give $CpFe(CO)(P(OEt)_3)_2$ (eq. 83), is apparently a bad reductor unable to reduce the parent dimer $Cp_2Fe_2(CO)_4$ or Cp_2Co^+. Only the more basic disubstituted radical $CpFe(CO)(P(OEt)_3)_2$ can do so (eq. 90). However, steric factors, as in the case of PPh$_3$, inhibit the formation of the disubstituted radical; but, in the case of the relatively small P(n-Bu)$_3$ ligand, the reaction proceeds according to 91. Only when the reaction neared completion was the formation of $CpFe(CO)(P(n-Bu)_3)_2^+$ observed. Goldman and Tyler explained this behaviour as follows. In the presence of sufficient Cp_2Co^+, formation of the highly reducing $CpFe(CO)_2(P(n-Bu)_3)$ radical is quickly followed by electron transfer to Cp_2Co^+ before disubstitution of the radical can take place (eq. 94). When this reaction is nearly completed, and much less Cp_2Co^+ is available, $CpFe(CO)_2(P(n-Bu)_3)$ can be further substituted by P(n-Bu)$_3$ to give $CpFe(CO)(P(n-Bu)_3)_2$ (eq. 95). This 19e$^-$ radical can, in turn, reduce $CpFe(CO)_2(P(n-Bu)_3)^+$ by an ETC pathway (eq. 95, 96). The complete reaction scheme for this photochemical reduction of Cp_2Co^+ in a solution of $Cp_2Fe_2(CO)_4$ and $P(n\text{-}Bu)_3$ is shown in eq. 92-96.

$$Cp_2Fe_2(CO)_4 \xrightarrow{h\nu} 2CpFe(CO)_2 \qquad (92)$$

$$CpFe(CO)_2 + L \rightarrow CpFe(CO)_2L \qquad (93)$$

$$CpFe(CO)_2L + Cp_2Co^+ \xrightarrow{fast} CpFe(CO)_2L^+ + Cp_2Co \qquad (94)$$

$$CpFe(CO)_2L + L \rightarrow CpFe(CO)L_2 + CO \qquad (95)$$

$$CpFe(CO)L_2 + CpFe(CO)_2L^+ \rightarrow CpFe(CO)L_2^+ + CpFe(CO)_2L \qquad (96)$$

$L = P(n\text{-}Bu)_3$

The above results clearly show that both steric and electronic effects of the phosphines are of crucial importance for the disproportionation and reduction reactions of $Cp_2Fe_2(CO)_4$. Moreover, disproportionation can only be observed when the backreaction of the ions, to give the parent dimer, is inhibited by removing one of the reaction products from the reaction mixture.

7   HETERODINUCLEAR METAL CARBONYLS

The photochemical studies of heterodinuclear metal-metal bonded carbonyls have mainly been confined to those complexes which form a combination of the radicals $M'(CO)_5$ ($M'=Mn,Re$), $CpM(CO)_3$ ($M=Mo,W$) and $CpFe(CO)_2$, discussed before.

Wrighton and co-workers studied in detail the spectroscopic properties and photochemical behaviour of the complexes $CpM(CO)_3M'(CO)_5$ ($M=Mo,W$; $M'=Mn,Re$), both in room temperature solution (ref. 194) and alkane glasses (ref. 197). Just as the homodinuclear analogues, these mixed-metal complexes possess an intense $\sigma_b \rightarrow \sigma^*(M-M')$ transition between 330 and 400 nm and a much weaker band at lower energy belonging to $d\pi \rightarrow \sigma^*(M-M')$ transitions.

Conventional flash photolysis of these complexes at room temperature afforded $M'_2(CO)_{10}$ and $Cp_2M_2(CO)_6$ in nearly a 1:1 ratio (eq. 97). Photolysis of the complexes in degassed CCl_4 yielded $CpM(CO)_3Cl$ and $M'(CO)_5Cl$ (eq. 98). The quantum yield of this latter reaction appeared to decrease substantially upon going from 366 to 436 nm excitation.

$$2 \ CpM(CO)_3M'(CO)_5 \ \underset{\Delta}{\overset{h\nu}{\rightleftarrows}} \ Cp_2M_2(CO)_6 + M'_2(CO)_{10} \qquad (97)$$

$$CpM(CO)_3M'(CO)_5 \ \underset{CCl_4}{\overset{h\nu}{\rightarrow}} \ CpM(CO)_3Cl + M'(CO)_5Cl \qquad (98)$$

M=Mo,W

M'=Mn,Re

Just as for the homodinuclear complexes, the occurrence of these cross-coupling and halogen atom abstraction reactions are clear evidence for a homolytic splitting of the metal-metal bond.

Quite recently, Pope and Wrighton (ref. 197) extended the photochemical studies on $CpMo(CO)_3Mn(CO)_5$ to rigid methylcyclohexane (MCH) at 93K. In this medium the products of homolytic cleavage of the metal-metal bond are normally not observed because of a fast recombination of the radicals (vide supra).

Irradiation of the complex with uv light (313 and 366 nm) resulted in the appearance of uncomplexed CO and in the formation of a CO loss product $CpMoMn(CO)_7$. The ir spectrum of this photoproduct shows the presence of two bridging carbonyls absorbing at 1760 and 1675 cm^{-1}, respectively. Melting of the glass (150 K) caused the disappearance of several ir bands of the terminal carbonyls and of the bridging CO vibration at 1760 cm^{-1}. Recooling the solution to 93K did not cause the reappearance of these bands. It was therefore concluded that two isomers of the photoproduct were formed in the MCH glass. For the stable isomer the structure of Fig. 29 was proposed, having a fully

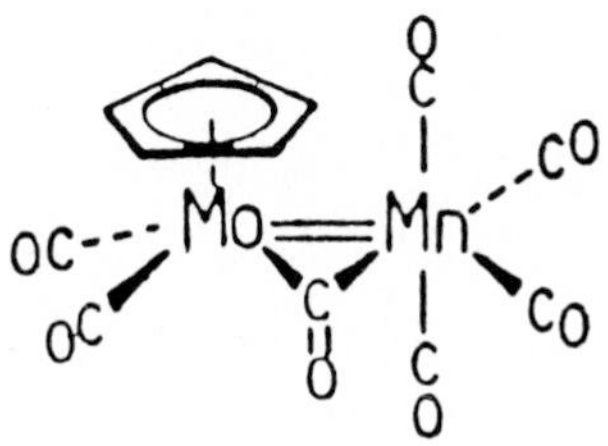

Fig. 29.   Proposed structure of $CpMoMn(CO)_7$ (reproduced with permission from ref. 197).

bridging CO associated with the 1680 cm^{-1} absorption. The less stable isomer was thought to have a semi-bridging CO ligand just as $Mn_2(CO)_9$ (Fig. 4). Unfortunately, ^{13}CO labeling did not establish from which metal atom the CO was photoejected because of rapid scrambling of the carbonyls.

The photoproduct reacted slowly with CO to the starting complex and this reaction could even be followed at room temperature. With PR_3 addition to the Mn atom occurred to give $CpMoMn(CO)_7(PR_3)$.

These results clearly show that these heterodinuclear complexes, just as the dimers, undergo two different primary photoprocesses viz. homolytic splitting of the metal-metal bond being the main reaction in fluid solution and loss of CO, observed as the sole photoprocess in rigid media. Heterolytic splitting of the metal-metal bond, which might be expected for such polar or even partly ionic species, was not obtained for the above complexes. It has, however, been proposed for the heterodinuclear complex $CpFe(CO)_2Co(CO)_4^-$ (vide infra).

Pope and Wrighton (ref. 198) also studied the photochemistry of the complexes $CpFe(CO)_2Mn(CO)_5$ and $Cp*Fe(CO)_2Mn(CO)_5$ $(Cp*=\eta^5-C_5Me_5)$ and, again, observed homolytic cleavage of the metal-metal bond in room temperature solution and loss of CO in alkane glasses. There is, however, a significant difference in structure between the CO loss products of these two systems which also obtains for the complexes themselves in low temperature solutions and alkane glasses. The CO-stretching region (ir) of $CpFe(CO)_2Mn(CO)_5$ hardly changes upon cooling the MCH solution to a glass and does not show a band between 1900-1500 cm^{-1}. $Cp*Fe(CO)_2Mn(CO)_5$, on the other hand, has a very weak band at 1796 cm^{-1} in room temperature solution which increases in intensity by a factor of 5 as the solution is cooled to 93 K. At the same time, the relative intensities of the other CO bands change appreciably. This effect was attributed to the formation of a structural isomer with a bridging carbonyl. This difference in structure between the two complexes in the MCH glass is also observed for their CO loss products. The photoproduct $CpFeMn(CO)_6$ has its lowest frequency CO band at 1914 cm^{-1} and therefore no bridging carbonyl, contrary to the behaviour of the photoproduct $CpMoMn(CO)_7$ discussed before. $Cp*FeMn(CO)_6$, on the other hand, has a bridging carbonyl absorbing at 1760 cm^{-1}.

Also, for these complexes ^{13}CO enrichement of one of the metal fragments did not reveal from which metal atom CO was photoejected because of rapid CO site exchange. Upon warming, both photoproducts reacted with CO to give the starting complex and with PR_3 to give the substitution products $CpFeMn(CO)_6(PR_3)$ and $Cp*FeMn(CO)_6(PR_3)$ by addition of PR_3 to the Mn-atom of the CO loss products.

The photochemistry of the complexes $CpM(CO)_3Fe(CO)_2Cp$ (M=Mo,W) has only been studied in room temperature solution (ref. 196). The flash photolysis of these complexes and their reactions with halocarbons proceeded according to eq. 97 and 98, respectively, which result was taken as evidence for homolysis of the metal-metal bond as the primary photoprocess.

Abrahamson and Wrighton (ref. 195) performed a comparative study of the photochemical reactions of the series of heterodinuclear complexes $CpM(CO)_3X$ (M=Mo,X=Co(CO)$_4$, $CpW(CO)_3$, $Mn(CO)_5$, $CpFe(CO)_2$; M=W, X=Co(CO)$_4$, $Mn(CO)_5$) and $MnRe(CO)_{10}$ with the radical scavenger 1-iodopentane. From the products formed, they derived the following order of radical reactivity towards halogen atom abstraction:

$$Re(CO)_5 > Mn(CO)_5 > CpW(CO)_3 > CpMo(CO)_3 > CpFe(CO)_2 > Co(CO)_4.$$

A similar ordering seemed to obtain for CCl_4.

The complexes $CpM(CO)_3Co(CO)_4$ (M=Mo,W) have only been studied thus far in room temperature solution in which homolytic cleavage of the metal-metal bond was observed (ref. 196). Fletcher et al. (ref. 69) studied the photolysis of the corresponding complex $CpFe(CO)_2Co(CO)_4$ in Ar, N_2 and CO-doped Ar matrices at 10 K. This complex has four terminal carbonyls and two CO-bridges. The Ar matrix irradiation with uv light afforded the CO loss product, $CpFeCo(CO)_5$. Attempts to establish the number of bridging carbonyls by ^{13}CO enrichement failed. In the N_2 matrix the CO substitution product $CpFe(CO)_5(N_2)$ was formed.

A very remarkable reaction was observed upon irradiation in Ar matrices doped with CO. Intense absorptions were obtained for the photoproduct between 1950 and 1850 cm^{-1}, where neutral unsubstituted dinuclear carbonyls do not normally absorb. These low frequency bands were close to those of the $Co(CO)_4^-$ anion in solution while the higher frequency bands were close to those of the cation $CpFe(CO)_3^+$. The reaction was therefore proposed to be the result of heterolysis of the metal-metal bond as shown in eq. 99. Up to now, no such reaction had been observed.

$$CpFe(CO)_2Co(CO)_4 + CO \xrightarrow{h\nu} CpFe(CO)_3^+ + Co(CO)_4^- \qquad (99)$$

Recently, van Dijk et al. (ref. 199) observed the formation of the ion pair $[Re(CO)_3(bpy)]^+[Co(CO)_4]^-$ upon irradiation of $(CO)_4CoRe(CO)_3(bpy)$ in toluene at temperatures below -50°C. This reaction was, however, completely quenched by radical scavengers, indicating a homolytic cleavage of the metal-metal bond as the primary photoprocess. Apparently, the radical $Re(CO)_3(bpy)$ formed by the photolysis reaction reduces the parent compound which then decomposes into $Co(CO)_4^-$ and a $Re(CO)_3(bpy)$ radical. Up to now such disproportionation reactions have only been observed in the presence of electron donor ligands and in coordinating solvents (ref. 9).

8   CONCLUDING REMARKS

The most important conclusion which can be drawn from the preceeding sections is that a close analogy exists between these homo- and heterodinuclear carbonyls with respect to their primary photoprocesses. All complexes show homolytic cleavage of the metal-metal bond and loss of CO. The major great significance of recent developments in uv/vis and ir fast time- resolved spectroscopy is that these primary photoprocesses can now be observed and that reactions of the primary photoproducts can be analyzed kinetically. At the same time, several of these primary photoproducts and reactive intermediates have been characterized spectroscopically by trapping them in rigid media.

The combined use of time-resolved and stabilization techniques is of crucial importance for a better understanding of the mechanisms of these photochemical reactions and both methods are now routinely used by a large number of research groups.

Contrary to LF photochemistry of coordination complexes, information about the properties of excited states and crossing of potential energy surfaces of these metal-metal bonded carbonyls is scarce. Much had to be done to improve the understanding in this field.

One of the most interesting properties of these complexes, discussed in detail in this article, is the formation of highly reducing 19e$^-$ radical species which are stable enough to start a catalytic disproportionation or substitution reaction. The formation of these radicals will certainly be of great importance for the application of these metal-metal bonded carbonyls in photocatalysis. It is to Tyler's merit that he developed this new research area in such a short period of time.

Although a few interesting results have already been obtained in this field (ref. 200-208), much has to be done to understand the photochemical reactions of metal clusters, which play an important role in catalysis.

REFERENCES

1.   A.W. Adamson, J. Phys. Chem., 71 (1967) 798.

2.   T. Kobayashi, K. Yasufuku, J. Iwai, H. Yesaka, H. Noda and H. Ohtani, Coord. Chem. Rev., 64 (1985) 1.

3.   H. Poliakoff and E. Weitz, Adv. Organomet. Chem., 25 (1986) 277.

4.   H. Hamaguchi, in "Vibrational Spectra and Structure" Vol. 16 p. 227; Ed. J.R. Durig; Elsevier Science Publishers, Amsterdam 1987.

5.   J.J. Turner, M.B. Simpson, M. Poliakoff and W.B. Maier II, J. Am. Chem. Soc., 105 (1983) 3898.

6.   G.L. Geoffroy and M.S. Wrighton, "Organometallic Photochemistry", Academic Press, New York, 1979.

7.   Th.J. Meyer and J.V. Caspar, Chem. Rev., 85 (1985) 187.

8.   A.E. Stiegman and D.R. Tyler, Acc. Chem. Res., 17 (1984) 61.

9.   A.E. Stiegman and D.R. Tyler, Coord. Chem. Rev., 63 (1985) 217.

10.   C.G. Kreiter, Adv. Organomet. Chem., 26 (1986) 297.

11.   E.O. Brimm, M.A. Lynch, Jr and W.J. Sesny, J. Am. Chem. Soc., 76 (1954) 3831.

12.   L.F. Dahl and R.E. Rundle, Acta Crystallogr., 16(1963) 419.

13.   M.F. Bailey and L.F. Dahl, Inorg. Chem., 4 (1965) 1140.

14.   L.F. Dahl, E. Ishishi and R.E. Rundle, J. Chem. Phys., 26 (1957) 1750.

15.   M.R. Churchill, K.N. Amoh and H.J. Wasserman, Inorg. Chem., 20 (1981) 1609.

16.   A.L. Rheingold, W.K. Meckstroth and D.P. Ridge, Inorg. Chem., 25 (1986) 3706.

17.   W.K. Meckstroth and D.P. Ridge, J. Am. Chem. Soc., 107 (1985) 2281.

18.   D.A. Brown, W.J. Chambers, N.J. Fitzpatrick and S.R.M. Rawlinson, J. Chem. Soc. (A), (1971) 720.

19.   S. Evans, J.C. Green, M.L.H. Green, A.F. Orchard and D.W. Turner, Disc. Faraday Soc., 47 (1969) 112.

20.   B.R. Higginson, D.R. Lloyd, S. Evans and A.F. Orchard, J. Chem. Soc. Faraday II, 71 (1975) 1913.

21.   M.B. Hall, J. Am. Chem. Soc., 97 (1975) 2057.

22.   R.R. Andrea, A. Terpstra, D.J. Stufkens and A. Oskam, Inorg. Chim. Acta, 96 (1984) L57.

23.   R.A. Levenson and H.B. Gray, J. Am. Chem. Soc., 97 (1975) 6042.

24.   W. Heijser, E.J. Baerends and P. Ros, Far. Symp. Chem. Soc., 14 (1980) 211.

25.   H.-J. Freund, B. Dick and G. Hohlneicher, Theoret. Chim. Acta, 57 (1980) 181.

26.   R.A. Levenson, H.B. Gray and G.P. Ceasar, J. Am. Chem. Soc., 92 (1970) 3653.

27.   M.S. Wrighton and D.S. Ginley, J. Am. Chem. Soc., 97 (1975) 2065.

28.   I.S. Butler, M.C. Barreto, N.J. Coville, P.D. Harvey and G.W. Harris, Inorg. Chem., in press.

29.   G.W. Harris, J.C.A. Boeijens and N.J. Coville, Organometallics, 4 (1985) 914.

30.   D.L. Morse and M.S. Wrighton, J. Am. Chem. Soc., 98 (1976) 3931.

31.   L.H. Staal, G. van Koten and K. Vrieze, J. Organomet. Chem., 175 (1979) 73.

32.   M.W. Kokkes, T.L. Snoeck, D.J. Stufkens, A. Oskam, M. Christophersen and C.H. Stam, J. Mol. Structure, 131 (1985) 11.

33.   W. Hieber and W. Schropp Jr, Z. Naturforsch., B15 (1960) 271.

34.   T. Kruck, M. Höfler and M. Noacke, Chem. Ber., 99 (1966) 1152.

35.   T. Kruck, M. Höfler and M. Noacke, Angew. Chem., 76 (1964) 786.

36.   M.W. Kokkes, D.J. Stufkens and A. Oskam, Inorg. Chem., 24 (1985) 2934.

37.   H. Saito, J. Fujita and K. Saito, Bull. Chem. Soc. Japan, 41 (1968) 359.

38.   H. Saito, J. Fujita and K. Saito, Bull. Chem. Soc. Japan, 41 (1968) 863.

39.   Y. Kaizi, I. Fujita and H. Kobayashi, Z. Phys. Chem., (Frankfurt/Main) 79 (1972) 298.

40   M.S. Wrighton and D.L. Morse, J. Organomet. Chem., 97 (1975) 405.

41   H.B. Abrahamson and M.S. Wrighton, Inorg. Chem., 17 (1978) 3385.

42   R.W. Balk, D.J. Stufkens and A. Oskam, Inorg. Chim. Acta, 34 (1979) 267.

43   R.W. Balk, D.J. Stufkens and A. Oskam, Inorg. Chem., 19 (1980) 3015.

44   R.W. Balk, D.J. Stufkens and A. Oskam, J. Chem. Soc. Dalton Trans., (1982) 275.

45   M.W. Kokkes, D.J. Stufkens and A. Oskam, J. Chem. Soc. Dalton Trans., (1983) 439.

46   M.S. Wrighton and D. Bredesen, J. Organomet. Chem., 50 (1973) C35.

47   D.R. Kidd and Th.L. Brown, J. Am. Chem. Soc., 100 (1978) 4095.

48   D.R. Kidd, C.R. Cheng and Th.L. Brown, J. Am. Chem. Soc., 100 (1978) 4103.

49   A. Freedman and R. Bersohn, J. Am. Chem. Soc., 100 (1978) 4116.

50   R.W. Wegman, R.J. Olsen, D.R. Gard, L.R. Faulkner and Th.L. Brown, J. Am. Chem. Soc., 103 (1981) 6089.

51   A. Hudson, M.F. Lappert and B.K. Nicholson, J. Chem. Soc. Dalton Trans., (1977) 551.

52   A.F. Hepp and M.S. Wrighton, J. Am. Chem. Soc., 105 (1983) 5934.

53   J.L. Hughey IV, C.P. Anderson and Th.J. Meyer, J. Organomet. Chem., 125 (1977) C49.

54   A. Fox and A. Poë, J. Am. Chem. Soc., 102 (1980) 2497.

55   H. Yesaka, T. Kobayashi, K. Yasufuku and S. Nagakura, J. Am. Chem. Soc., 105 (1983) 6249.

56   L.J. Rothberg, N.J. Cooper, K.S. Peters and V. Vaida, J. Am. Chem. Soc., 104 (1982) 3536.

57   W.L. Waltz, O. Hackelberg, L.M. Dorfman and A.J. Wojcicki, J. Am. Chem. Soc., 100 (1978) 7259.

58   S.P. Church, M. Poliakoff, J.A. Timney and J.J. Turner, J. Am. Chem. Soc., 103 (1981) 7515.

59   S.B. McCullen and Th.L. Brown, Inorg. Chem., 20 (1981) 3528.

60   J.M. Kelly, D.V. Bent, H. Hermann, D. Schulte-Frohlinde and E. Koerner v. Gustorf, J. Organomet. Chem., 69 (1974) 259.

61   J.M. Kelly, H. Hermann and E. Koerner v. Gustorf, J. Chem. Soc., Chem. Commun., (1973) 105.

62   D.G. Leopold and V. Vaida, J. Am. Chem. Soc., 106 (1984) 3720.

63   T.A. Seder, S.P. Church and E. Weitz, J. Am. Chem. Soc., 108 (1986) 7518.

64   S.P. Church, H. Hermann, F.-W. Grevels and K. Schaffner, J. Chem. Soc., Chem. Commun., (1984) 785.

65   I.R. Dunkin, P. Härter and C.J. Shields, J. Am. Chem. Soc., 106 (1984) 7248.

66   T. Kobayashi, H. Ohtani, H . Noda, S. Teratani, H. Yamazaki and K. Yasufuku, Organometallics, 5 (1986) 110.

67   K. Yasufuku, H. Noda, J. Iwai, H. Ohtani, M. Hoshino and T. Kobayashi, Organometallics, 4 (1985) 2174.

68   C.A. Coulson and I. Fischer, Philos. Mag., 40 (1949) 386.

69   S.C. Fletcher, M. Poliakoff and J.J. Turner, J. Organomet. Chem., 268 (1984) 259.

70   St. Firth, W.E. Klotzbücher, M. Poliakoff and J.J. Turner, Inorg. Chem., in press.

71 H. Huber, E.P. Kundig and G.A. Ozin, J. Am. Chem. Soc., 96 (1974) 5585.

72 H.W. Walker, R.S. Herrick, R.J. Olsen and Th.L. Brown, Inorg. Chem., 23 (1984) 3748.

73 A. Hudson, M.F. Lappert, P.W. Lednov and B.K. Nicholson, J. Chem. Soc., Chem. Commun., (1974) 966.

74 L.S. Benner and A.L. Balch, J. Organomet. Chem., 134 (1977) 121.

75 K-W Lee, J.M. Hanckel and Th.L. Brown, J. Am. Chem. Soc., 108 (1986) 2266.

76 R.S. Herrick and Th.L. Brown, Inorg. Chem., 23 (1984) 4550.

77 U. Koelle, J. Organomet. Chem., 155 (1978) 53.

78 M.S. Ziegler, H. Haas and R.K. Sheline, Chem. Ber., 98 (1965) 2454.

79 B.H. Byers and Th.L. Brown, J. Am. Chem. Soc., 99 (1977) 2527.

80 A. Fox, J. Malito and A. Poë, J. Chem. Soc., Chem. Commun., (1981) 1052.

81 Th.R. Herrinton and Th.L. Brown, J. Am. Chem. Soc., 107 (1985) 5700.

82 Q-Z. Shi, Th.G. Richmond, W.C. Trogler and F. Basolo, J. Am. Chem. Soc., 106 (1984) 71.

83 J.L. Hughey, C.B. Bock and T.J. Meyer, J. Am. Chem. Soc., 97 (1975) 4440.

84 W.K. Meckstroth, R.T. Walters, W.L. Waltz, A. Wojcicki and L.M. Dorfman, J. Am. Chem. Soc., 104 (1982) 1842.

85 R.S. Herrick, Th.R. Herrinton, H.W. Walker and Th.L. Brown, Organometallics, 4 (1985) 42.

86 A.F. Hepp and M.S. Wrighton, J. Am. Chem. Soc., 103 (1981) 1258.

87 J.M. Hanckel, K-W. Lee, P. Rushman and Th.L. Brown, Inorg. Chem., 25 (1986) 1852.

88 S.B. McCullen and Th.L. Brown, J. Am. Chem. Soc., 104 (1982) 7496.

89 W. Kaim, Coord. Chem. Rev., 76 (1987) 187.

90 K.A.M. Creber, K.S. Chen and J.K.S. Wan, Revs. Chem. Intermed., 5 (1984) 37.

91 A. Alberti and C.M. Camaggi, J. Organomet. Chem., 161 (1978) C63.

92 A. Alberti and C.M. Camaggi, J. Organomet. Chem., 181 (1979) 355.

93 L. Pasimeni, P.L. Zanonato and C. Corvaja, Inorg. Chim. Acta, 37 (1979) 241.

94 T. Foster, K.S. Chen and J.K.S. Wan, J. Organomet. Chem., 184 (1980) 113.

95 K.A.M. Creber and J.K.S. Wan, J. Am. Chem. Soc., 103 (1981) 2101.

96   K.A.M. Creber and J.K.S. Wan, Chem. Phys. Lett., 81 (1981) 453.

97   T.-I. Ho, K.A.M. Creber and J.K.S. Wan, J. Am. Chem. Soc., 103 (1981) 6524.

98   K.A.M. Creber, T.-I. Ho, M.C. Depen, D. Weir and J.K.S. Wan, Can. J. Chem., 60 (1982) 1504.

99   W.G. McGimpsey and J.K.S. Wan, J. Photochem., 22 (1983) 87.

100   K.A.M. Creber and J.K.S. Wan, Can. J. Chem., 61 (1983) 1017.

101   K.A.M. Creber and J.K.S. Wan, Trans. Met. Chem., 8 (1983) 253.

102   A. Alberti and A. Hudson, J. Organomet. Chem., 241 (1983) 313.

103   A. Alberti and A. Hudson, J. Organomet. Chem., 248 (1983) 199.

104   R.R. Andréa, W.G.J. de Lange, T. van der Graaf, M. Rijkhoff, D.J. Stufkens and A. Oskam, Organometallics, in press.

105   W.G. McGimpsey, M.C. Depew and J.K.S. Wan, Organometallics, 3 (1984) 1684.

106   A. Alberti, M.C. Depew, A. Hudon, W.G. McGimpsey and J.K.S. Wan, J. Organomet. Chem., 280 (1985) C21.

107   D. Fenske, Chem. Ber., 112 (1979) 363.

108   P.J. Krusic, H. Stoklosa, L.E. Manzer and P. Meakin, J. Am. Chem. Soc., 97 (1975) 667.

109   T. van der Graaf, D.J. Stufkens and A. Vlcek, Jr, unpublished results.

110   H.K. van Dijk, D.J. Stufkens and A. Oskam, to be published.

111   P.C. Servaas, D.J. Stufkens and A. Oskam, to be published.

112   R.R. Andréa, W.G.J. de Lange, D.J. Stufkens and A. Oskam, to be published.

113   W. Hieber and W. Schropp, Z. Naturforsch., B 15 (1960) 271.

114   D.M. Allen, A. Cox, T.J. Kemp, Q. Sultana and R.B. Pitts, J. Chem. Soc., Dalton Trans., (1976) 1189.

115   S.B. McCullen and Th.L. Brown, Inorg. Chem., 20 (1981) 3528.

116   A.E. Stiegman and D.R. Tyler, Inorg. Chem., 23 (1984) 527.

117   A.E. Stiegman, A.S. Goldman, C.E. Philbin and D.R. Tyler, Inorg. Chem., 25 (1986) 2976.

118   A.E. Stiegman and D.R. Tyler, J. Am. Chem. Soc., 105 (1983) 6032.

119   M.W. Kokkes, W.G.J. de Lange, D.J. Stufkens and A. Oskam, J. Organomet. Chem., 294 (1985) 59.

120   A.E. Stiegman, A.S. Goldman, D.B. Leslie and D.R. Tyler, J. Chem. Soc., Chem. Commun., (1984) 632.

121   J.K. Kochi, J. Organomet. Chem., 300 (1986) 139.

122 D.R. Gard and Th.L. Brown, J. Am. Chem. Soc., 104 (1982) 6340.

123 D.R. Gard and Th.L. Brown, Organometallics, 1 (1982) 1143.

124 P.O. Nubel, S.R. Wilson and Th.L. Brown, Organometallics, 2 (1983) 515.

125 P.O. Nubel and Th.L. Brown, J. Am. Chem. Soc., 104 (1982) 4955.

126 P.O. Nubel and Th.L. Brown, J. Am. Chem. Soc., 106 (1984) 644.

127 P.O. Nubel and Th.L. Brown, J,. Am. Chem. Soc., 106 (1984) 3474.

128 K.-W. Lee and Th.L. Brown, Organometallics, 4 (1985) 1030.

129 B.F.G. Johnson and J. Lewis, J. Chem. Soc. A, (1968) 2859.

130 D.F. Jones, P.H. Dixneuf, A. Benoit and J.-Y. Le Marouille, Inorg. Chem., 22 (1983) 29.

131 S.W. Lee, L.F. Wang and C.P. Cheng, J. Organomet. Chem., 248 (1983) 189.

132 C.S. Young, S.W. Lee and C.P. Cheng, J. Organomet. Chem., 282 (1985) 85.

133 K.-W. Lee, W.T. Pennington, A.W. Cordes and Th.L. Brown, Organometallics, 3 (1984) 404.

134 K.-W. Lee, W.T. Pennington, A.W. Cordes and Th.L. Brown, J. Am. Chem. Soc., 107 (1985) 631.

135 W. Beck, K. Raab, U. Nagel and W. Sacher, Angew. Chem., 97 (1985) 498

136 C.G. Kreiter and W. Lipps, Angew. Chem., Int. Ed. Engl., 20 (1981) 201.

137 C.G. Kreiter and W. Lipps, Chem. Ber., 115 (1982) 973.

138 M. Leyendecker and C.G. Kreiter, J. Organomet. Chem., 260 (1984) C67.

139 C.G. Kreiter, M. Leyendecker and W.S. Sheldrick, J. Organomet. Chem., 302 (1986) 35.

140 K.H. Franzreb and C.G. Kreiter, Z. Naturforsch, 37B (1982) 1058.

141 C.G. Kreiter, K.H. Franzreb and W.S. Sheldrick, J. Organomet. Chem., 270 (1984) 71.

142 K.H. Franzreb and C.G. Kreiter, Z. Naturforsch., 39B (1984) 81.

143 K.-W. Lee, W.T. Pennington, A.W. Cordes and Th.L. Brown, J. Am. Chem. Soc., 107 (1985) 631.

144 Th.J. Meyer, Pure and Appl. Chem., 58 (1986) 1193.

145 M.W. Kokkes, D.J. Stufkens and A. Oskam, Inorg. Chem., 24 (1985) 4411.

146 T. van der Graaf, D.J. Stufkens and A. Oskam, unpublished results.

147 P.O. Nubel and Th.L. Brown, Organometallics, 3 (1984) 29.

148 G. Wilkinson, J. Am. Chem. Soc., 76 (1954) 209.

149 F.C. Wilson and D.P. Schoemaker, J. Chem. Phys., 27 (1957) 809.

150 M.S. Wrighton and D.S. Ginley, J. Am. Chem. Soc., 97 (1975) 4246.

151 C. Giannotti and G. Merle, J. Organomet. Chem., 105 (1976) 97.

152 R.B. King and K.H. Pannell, Inorg. Chem., 7 (1968) 2356.

153 D.S. Ginley and M.S. Wrighton, J. Am. Chem. Soc., 97 (1975) 3533.

154 D.S. Ginley, C.R. Bock and M.S. Wrighton, Inorg. Chim. Acta, 23 (1977) 85.

155 K.A. Mahmoud, A.J. Rest and H.G. Alt, J. Organomet. Chem., 246 (1983) C37.

156 Th. Madach and H. Vahrenkamp, Z. Naturforsch, 33B (1978) 1301.

157 Th. Madach and H. Vahrenkamp, Z. Naturforsch, 34B (1979) 573.

158 N.A. Cooley, K.A. Watson, S. Fortier and M.C. Baird, Organometallics, 5 (1986) 2563.

159 R.H. Hooker, K.A. Mahmoud and A.J. Rest, J. Organomet. Chem., 254 (1983) C25.

160 R.M. Laine and P.C. Ford, Inorg. Chem., 16 (1977) 388.

161 N.N. Turaki and J.M. Huggins, Organometallics, 5 (1986) 1703.

162 R. Klingler, W. Butler and M.D. Curtis, J. Am. Chem. Soc., 97 (1975) 3535.

163 R.D. Adams, D.M. Collins and F.A. Cotton, Inorg. Chem., 13 (1974) 1086.

164 N.N. Turaki and J.M. Huggins, Organometallics, 4 (1985) 1766.

165 C.E. Philbin, A.S. Goldman and D.R. Tyler, Inorg. Chem., 25 (1986) 4434.

166 R.J. Haines, R.S. Nyholm and M.H.B. Stiddard, J. Chem. Soc. A, (1968) 43.

167 A.E. Stiegman, M. Stieglitz and D.R. Tyler, J. Am. Chem. Soc., 105 (1983) 6032.

168 A.E. Stiegman and D.R. Tyler, J. Photochem., 24 (1984) 311.

169 A.E. Stiegman and D.R. Tyler, J. Am. Chem. Soc., 104 (1982) 2944.

170 A.E. Stiegman and D.R. Tyler, J. Am. Chem. Soc., 107 (1985) 967.

171 O.S. Mills, Acta Crystallogr., 11 (1958) 620.

172 J.G. Bullitt, F.A. Cotton and T.J. Marks, Inorg. Chem., 11 (1972) 671.

173 J.G. Bullitt, F.A. Cotton and T.J. Marks, J. Am. Chem. Soc., 92 (1970) 2155.

174 R.D. Fischer, A. Vogler and K. Noack, J. Organomet. Chem., 7 (1967) 135.

175 K. Noack, J. Organomet. Chem., 7 (1967) 151.

176 P. McArdle and A.R. Manning, J. Chem. Soc. A, (1970) 2120.

177 H.B. Abrahamson, M.C. Palazotto, C.L. Reichel and M.S. Wrighton, J. Am. Chem. Soc., 101 (1979) 4123.

178 D.C. Harris and H.B. Gray, Inorg. Chem., 14 (1975) 1215.

179   A. Mitschler, B. Rees and M.J. Lehman, J. Am. Chem. Soc., 100 (1978) 3390.

180   E.D. Temmis, A.R. Pinhos and R. Hoffmann, J. Am. Chem. Soc., 100 (1978) 7259.

181   M. Bernard, Inorg. Chem., 18 (1979) 2782.

182   D.R. Tyler, M.A. Schmidt and H.B. Gray, J. Am. Chem. Soc., 101 (1979) 2753.

183   D.R. Tyler, M.A. Schmidt and H.B. Gray, J. Am. Chem. Soc., 105 (1983) 6018.

184   J.V. Caspar and Th.J. Meyer, J. Am. Chem. Soc., 102 (1980) 7794.

185   R.H. Hooker, H.A. Mahmoud and A.J. Rest, J. Chem. Soc., Chem. Commun., (1983) 1022.

186   A.F. Hepp, J.P . Blaha, C. Lewis and M.S. Wrighton, Organometallics, 3 (1984) 174.

187   B.D. Moore, M.B. Simpson, M. Poliakoff and J.J. Turner, J. Chem. Soc., Chem. Commun., (1984) 972.

188   B.D. Moore, M. Poliakoff, M.B. Simpson and J.J. Turner, J. Phys. Chem., 89 (1985) 850.

189   J.P. Blaha, B.E. Bursten, J.C. Dewan, R.B. Frankel, C.L. Randolph, B.A. Wilson and M.S. Wrighton, J. Am. Chem. Soc., 107 (1985) 4561.

190   R.B. King, K.H. Pannell, C.A. Eggers and L.W. Houk, Inorg. Chem., 7 (1968) 2353.

191   A.S. Goldman, and D.R. Tyler, Inorg. Chem., 26 (1987) 253.

192   M. Julliard and M. Chanon, Chem. Rev., 83 (1983) 425.

193   F.R. Lemke and C.P. Kubiak, Inorg. Chim. Acta, 113 (1986) 125.

194   D.S. Ginley and M.S. Wrighton, J. Am. Chem. Soc., 97 (1975) 4908.

195   H.B. Abrahamson and M.S. Wrighton, J. Am. Chem. Soc., 99 (1977) 5510.

196   H.B. Abrahamson and M.S. Wrighton, Inorg. Chem., 17 (1978) 1003.

197   K.R. Pope and M.S. Wrighton, Inorg. Chem., 26 (1987) 2321.

198   K.R. Pope and M.S. Wrighton, J. Am. Chem. Soc., 109 (1987) 4545.

199   H.K. van Dijk, D.J. Stufkens and A. Oskam, unpublished results.

200   M.F. Desrosiers and P.C. Ford, Organometallics, 1 (1982) 1715.

201   M.F. Desrosiers, D.A. Wink and P.C. Ford, Inorg. Chem., 24 (1985) 1.

202   M.F. Desrosiers, D.A. Wink, R. Trautman, A.E. Friedman and P.C. Ford, J. Am. Chem. Soc., 108 (1987) 1917.

203   J. Malito, S. Markiewicz and A.J. Poë, Inorg. Chem., 21 (1982) 4335.

204   A.J. Poë and C.V. Sekhar, J. Am. Chem. Soc., 108 (1986) 3673.

205   J.L. Graff and M.S. Wrighton, J. Am. Chem. Soc., 102 (1980) 2123.

206   J.G. Bentsen and M.S. Wrighton, J. Am. Chem. Soc., 106 (1984) 4041.

207   R.A. Epstein, T.R. Gaffney, G.L. Geoffroy, W.L. Gladfelter and R.S. Henderson, J. Am. Chem. Soc., 101 (1979) 3847.

208   H.C. Foley and G.L. Geoffroy, J. Am. Chem. Soc., 103 (1981) 7176.

209   B.L. Tumanskij, K. Sarbasov, S.P. Solodovnikov, N.N. Bubnov, A.I. Prokofev and M.I. Kabachnik, Dokl. Akad. Nauk USSR, 259 (1981) 611.

210   A. Vlček, Jr., J. Organomet. Chem., 306 (1986) 63.

STEREOCHEMICAL ASPECTS OF ORGANOMETALLIC CLUSTERS. A VIEW OF THE POLYHEDRAL SKELETAL ELECTRON PAIR THEORY

D. Osella and P.R. Raithby

Stereochemical Aspects of Organometallic Clusters. A View of the Polyhedral Skeletal Electron Pair Theory.

Domenico Osella[1] and P.R. Raithby[2]

1) Dipartimento di Chimica Inorganica, Chimica Fisica e Chimica dei Materiali, Università di Torino, Via P. Giuria 7, 10125 Torino, Italy and
2) University Chemical Laboratory, Lensfield Road, Cambridge CB2 1EW, United Kingdom

304

1. INTRODUCTION

Over the last two decades organometallic cluster complexes have been widely studied for a number of different reasons. They represent good molecular models for chemisorbed species on metallic surfaces. Potentially, they are active homogeneous catalysts, or at least catalytic precursors. They serve as useful stoichiometric reagents in organic chemistry.

In particular, the Cluster Surface Analogy [1] represents an important methodological approach to a deeper understanding of the chemisorption phenomenon, and its relationship to heterogeneous catalysis. It is now relatively simple, in a majority of cases, to determine the structures and stereochemistry of molecular organometallic clusters. This contrasts the situation in the chemistry of heterogeneous catalysts, where, despite recent developments in surface spectroscopic techniques, it remains something of a problem to unambiguously determine the structure of a species chemisorbed onto a surface. However, a comparison of IR, Raman, and ^{13}C NMR spectroscopic data, obtained on the surface species with that obtained from a molecular cluster, where the geometry is accurately known, can provide useful information as to the structure of the former [2]. Thus, a detailed knowledge of the stoichiometric chemistry of organometallic clusters should aid our understanding of the chemistry of organic molecules chemisorbed onto metallic surfaces.

Transtion Metal Carbonyl Clusters (*TMCC's*) represent the largest group of polymetallic complexes with a formal oxidation state of *zero*, which should suitably model metal surfaces. A variety of unstaturated hydrocarbons are capable of reacting with *TMCC's* , and of these, alkynes [3] display the highest reactivity and the widest range of coordination modes. Thus, an investigation of cluster - alkyne chemistry should represent a good general model for metal - hydrocarbon interactions, since several of the alkyne bonding modes are identical to those found in analogous dienic derivatives, obtained through metal-assisted double bond transposition, and olefinic derivatives, obtained *via* metal-assisted dehydrogenation. It is, therefore, the aim of this review to describe the chemistry of alkyne substituted transition metal clusters, and related complexes, in terms of their synthesis, structure, reactivity and dynamics, and to systematize the structural types observed using the Polyhedral Skeletal Electron Pair approach. This review does not set out to be comprehensive but is intended to illustrate the structural features observed in hydrocarbon - substituted *TMCC's* and to indicate the general trends in the chemistry of these complexes. Several other review articles describing various aspects of the chemistry of organic groups directly coordinated to metal clusters have appeared previously [3,4].

2. Synthesis

The reaction between an alkyne and a *TMCC* depends on the experimental conditions, the substituents on the alkyne, and the nature of the metal cluster framework itself. These factors are interrelated, and the influence of a particular factor may vary from reaction to reaction. However, some general trends are set out in the following sections.

2.1. Ligand substitution

Generally the addition of an alkyne to a *TMCC* occurs *via* substitution of one or more carbonyl ligands on the cluster. The M-CO bond cleavage is achieved by thermal, photochemical or chemical activation. The synthesis of the $Co_2(CO)_6(RCCR)$ series from the binary carbonyl $Co_2(CO)_8$ at moderate temperatures, in apolar solvents, represents a classical example [5].

$$(CO)_3Co \overset{\triangle}{} Co(CO)_3 \quad \xrightarrow[-2CO]{+RC \equiv CR} \quad (CO)_3Co \overset{R-C \equiv C-R}{} Co(CO)_3$$

Photochemical activation, particularly at the wavelength corresponding to a metal to ligand charge transfer process [6], can favour carbonyl loss, even before the coordination of the incoming alkyne takes place [7].

$$(CO)_3CpMo \text{---} MoCp(CO)_3 \quad \xrightarrow[-2CO]{h\nu} \quad (CO)_2CpMo \equiv MoCp(CO)_2$$

$$\xrightarrow{+RC \equiv CR}$$

Chemical activation is generally carried out by using Me_3NO as the reactant [8]. This reagent removes coordinated carbon monoxide by oxidation to CO_2, while the reactive intermediate is stabilized by donor solvents such as MeCN or THF. The reaction is generally applicable to both low and high nuclearity cluster complexes [9].

$$\xrightarrow[NCMe]{+2 Me_3NO} \quad +2CO_2 \; + 2 Me_3N$$

2.2. Metal-metal edge cleavage

In some cases, the addition of an alkyne to a *TMCC* occurs with concomitant cleavage of an edge of the metal framework. This is frequently observed for first row transition metal clusters where the metal-metal bonding is relatively weak, and also occurs in photochemical activation reactions where the wavelength used corresponds to the $\sigma \rightarrow \sigma^*$ transition [6]. For example, in the reaction of $Co_4(CO)_{12}$ with alkynes, the tetrahedral Co_4 framework of the parent carbonyl undergoes edge cleavage to give the "butterfly" metal conformation in the resultant Co_4C_2 arrangement [10]. With the isoelectronic and isostructural heterometallic clusters $Co_2Rh_2(CO)_{12}$ [11], $Co_2Ru_2(CO)_{13}$ [12], and $[Co_3Ru(CO)_{12}]^-$ [13] the same reaction path is followed, but with specific rupture of the weaker Co-Co bonds.

2.3. Carbon-carbon bond rupture

Under forcing reaction conditions of high temperature, pressure, and long reaction times, cleavage of the carbon-carbon triple bond of the alkylenic unit may occur. This results in the formation of two carbynic species which may themselves coordinate to the cluster [14].

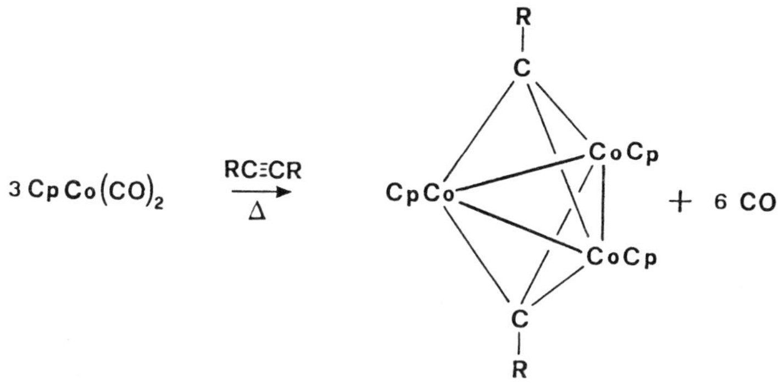

$$3\,CpCo(CO)_2 \xrightarrow[\Delta]{RC{\equiv}CR} \quad \text{[cluster structure]} \quad +\ 6\ CO$$

The reverse reaction, involving the coupling of two carbyne fragments which were originally coordinated to a single metal centre, is also possible [15]. These products are of particular interest because of the analogy with surface carbynes which are postulated as intermediates in Fischer - Tropsch catalysis [16].

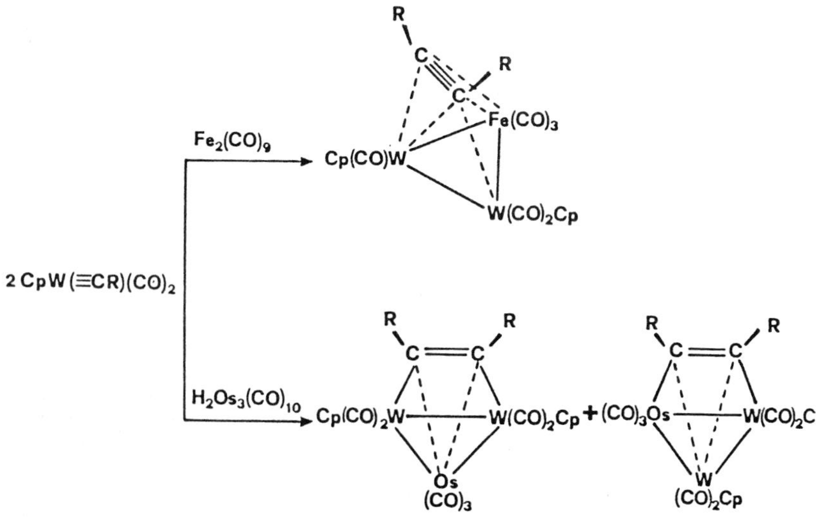

$$2\,CpW(\equiv CR)(CO)_2$$

Fe$_2$(CO)$_9$

H$_2$Os$_3$(CO)$_{10}$

2.4. Syntheses involving diaryl alkynes

Diaryl alkynes, such as PhC≡CPh, usually coordinate to a *TMCC* without major structural modification, adopting a μ-η^2 bonding mode (see also Section 2.5). With excess of the alkyne there is a tendency, particularly with first and second row transition metals, for metal assisted alkyne oligomerization to occur. This sometimes involves CO insertion and results in the formation of metalla-cyclopentadiene and metalla-cyclohexadienone rings [5].

The use of the heterometallic clusters $Fe_2Ru(CO)_9(\mu_3-\eta^2-\bot$ -$PhC_2Ph)$, obtained through the photolysis of $Fe_3(CO)_{12}$ and $Ru_3(CO)_{12}$ in the presence of PhC_2Ph, has confirmed the above mechanism, indicating that once multicentred alkyne coordination to the metal triangle has occurred, there is no intermolecular exchange as the reaction proceeds [17].

By way of contrast, when excess alkyne is reacted with a third row transition metal carbonyl, or its activated derivative, where the formal metal-metal bond strength is greater and the complex less kinetically labile, alkyne oligomerization does not always take place, and two alkynes may coordinate to the cluster independently [18,19].

However, when more forcing conditions are used with the osmium systems, in reactions to coordinate three akynes to a triosmium cluster unit, metal-metal edge cleavage and partial alkyne oligomerization occurs [20].

2.5. Orthometallation reactions

For the heavier metals, especially the third row elements, oxidative addition, including orthometallation of a phenyl ring, is a common process. If a C-H bond is forced to approach closely to a metal centre by coordination of the adjacent triple bond, the C-H breaking process, which is now kinetically possible, is favoured by the subsequent formation of stable metal-hydrogen and metal-carbon bonds [21]. The right hand side of the scheme below shows an example of orthometallation in the reaction of $Os_3(CO)_{12}$ with PhC_2Ph to afford the highly reactive cluster $Os_3H(CO)_8\{(PhC_2Ph)(PhC_2C_6H_4)\}$. It is of interest that the oxidative addition can be reversed by bubbling CO through a solution of this product [22].

312

2.6. Syntheses involving dialkyl alkynes with active β hydrogens

When disubstituted alkynes, having a hydrogen atom in the position α to the triple bond, such as $RCH_2C{\equiv}CR'$, are employed, oxidative - addition reactions with second and third row transition metal clusters such as $Ru_3(CO)_{12}$ [23] and $Os_3(CO)_{12}$ [24] are possible. These reactions afford the kinetically favoured "allenic" isomer, which, on heating, rearranges to the thermodynamically stable "allylic" isomer. These isomers are also the major products of the reactions of $Ru_3(CO)_{12}$ with dienes [25] and alkenes [26].

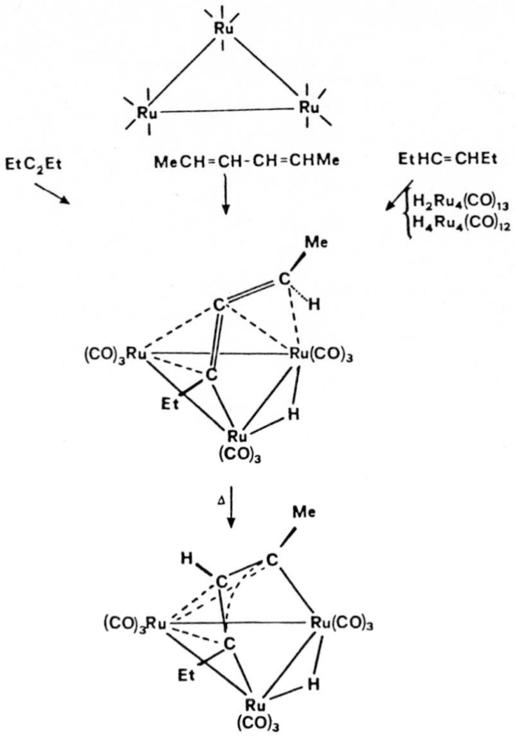
Ru
Ru
Ru
EtC$_2$Et
MeCH=CH-CH=CHMe
EtHC=CHEt
H$_2$Ru$_4$(CO)$_{13}$
H$_4$Ru$_4$(CO)$_{12}$
Me
C
C
H
(CO)$_3$Ru
Ru(CO)$_3$
C
Et
H
Ru
(CO)$_3$
Δ
Me
H
C
C
(CO)$_3$Ru
Ru(CO)$_3$
C
Et
H
Ru
(CO)$_3$

2.7. *Syntheses involving terminal alkynes*

In the presence of the clusters $M_3(CO)_{12}$ (M = Ru, Os), terminal alkynes of the type RC≡CH undergo specific cleavage of the reactive C_{sp}-H bond to give μ_2-hydrido and μ_3-acetylido ligands [27]. Haloalkynes [28] and phosphinoalkynes [29] show similar behaviour.

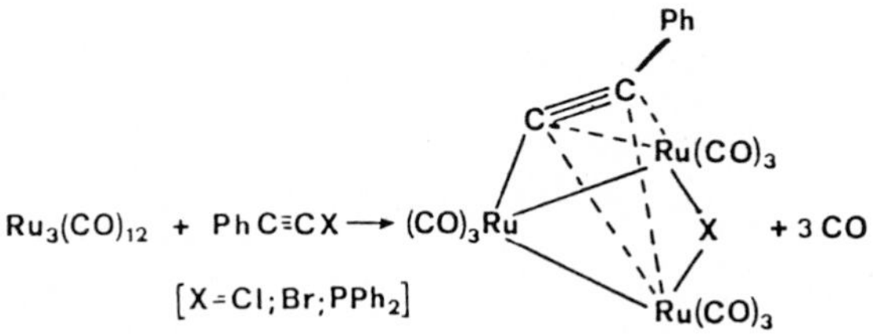
Ph
C
C
Ru(CO)$_3$
Ru$_3$(CO)$_{12}$ + PhC≡CX ⟶ (CO)$_3$Ru
X
+ 3 CO
[X=Cl; Br; PPh$_2$]
Ru(CO)$_3$

The reaction of the activated hexanuclear cluster $Os_6(CO)_{17}(NCMe)$ with terminal alkynes also affords a product in which alkylenic C-H bond cleavage has occurred to give $Os_6(\mu-H)(CO)_{17}(\mu_4-\eta-CCEt)$, in which the acetylido ligand coordinates to four metal atoms [30].

$$Os_6(CO)_{17}(NCMe) \xrightarrow{\text{HCCR}}$$

2.8. Syntheses involving functionalised alkynes

Functionalised alkynes react in a manner similar to that observed for alkyl and aryl - substituted alkynes. The presence of oxygenated functionalities, which can be transformed into good leaving groups, make possible the cleavage of a C-C bond α to the carbon-carbon triple bond [31].

The presence in the organic chain of a strong electron - releasing substituent, such as $-NMe_2$, can modify the ground state structure of the products, which are more correctly viewed as zwitterionic complexes, the reaction pattern being similar [32].

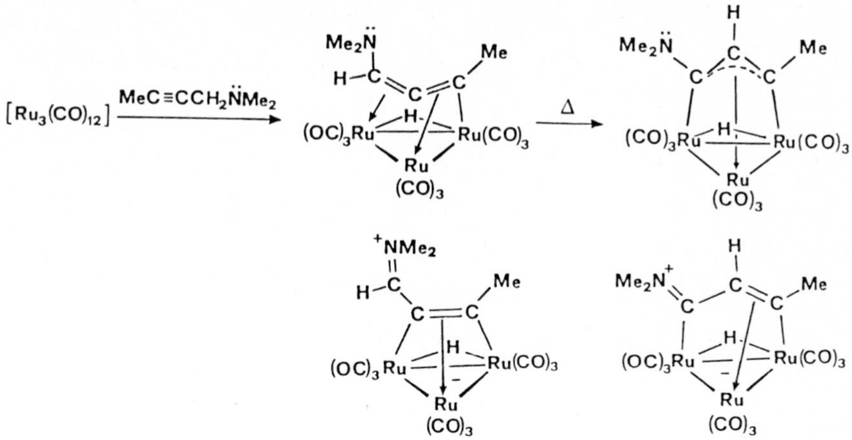

$$[Ru_3(CO)_{12}] \xrightarrow{MeC\equiv CCH_2\ddot{N}Me_2}$$

Me₂N̈\
H–C=C–C–Me\
H–\
(OC)₃Ru⟵⟶Ru(CO)₃\
Ru\
(CO)₃

$$\xrightarrow{\Delta}$$

H\
Me₂N̈–C–Me\
C---C\
(CO)₃Ru⟵H⟶Ru(CO)₃\
Ru\
(CO)₃

⁺NMe₂\
H–C–Me\
C=C\
(OC)₃Ru⟵H⟶Ru(CO)₃\
Ru\
(CO)₃

H\
Me₂N⁺=C–C–Me\
H·\
(OC)₃Ru⟵⟶Ru(CO)₃\
Ru\
(CO)₃

The zwitterionic character of these complexes is further corroborated by their electrochemical behaviour in non-aqueous solvents when compared with that of allenic and allylic isomers, bearing aliphatic substituents only. The presence of the dimethylamino group lowers the oxidation potentials and increases their reduction potential, thus indicating that there is a higher electron density on the metal triangle [33].

2.9. Cluster build up

For first and second row transition metals, where the metal-metal bonding is relatively weak, condensation of metal carbonyl fragments around the alkyne species often occurs, resulting in the formation of higher nuclearity clusters. The strong coordination of the alkylenic carbon-carbon bond to several metallic units requires that short intermetallic distances are maintained. This promotes metal-metal bond formation. For example, the reaction of $Ru_3(CO)_{12}$ with PhC_2Ph yields the tetrameric "butterfly" cluster $Ru_4(CO)_{12}(PhC_2Ph)$ [34]. Similar products may be obtained by the reaction with styrene [35], through a dehydrogenation pathway, ethylene [36], through an oligomerisation and dehydrogenation pathway, and cyclohexadiene [37], via a concerted double bond and hydrogen shift mechanism.

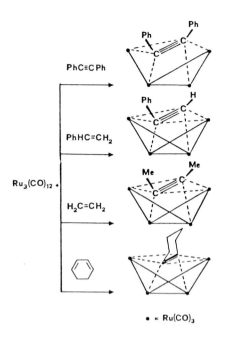

Ph C≡C Ph

Ph HC=CH₂

Ru₃(CO)₁₂ +

H₂C=CH₂

• = Ru(CO)₃

2.10. Syntheses involving the use of unsaturated or lightly ligated clusters

The use of unsaturated clusters, which are very reactive towards nucleophiles, or those which are lightly ligated by the presence of coordinated cyclooctene, cyclohexa-1,3-diene or acetonitrile, allows an easy addition of alkynes to the metal framework. In these reactions it has been possible to isolate intermediates which could not be detected by the conventional synthetic routes. This is the case in the reaction of the unsaturated cluster $Os_3H_2(CO)_{10}$ (a 46 electron system) [38], or the labilised cluster $Os_3(CO)_{10}(NCMe)_2$ [39], with PhC_2Ph occurs under milder conditions than a similar reaction with the parent carbonyl, $Os_3(CO)_{12}$. The scheme in *Section 2.5* gives a comparison of the reaction sequences using saturated and unsaturated clusters as starting materials, and shows that the final product is the same as for the alkyne oligomerisation process [40].

3. Structures of and Rationalisation of Bonding in
Alkyne-substituted Clusters

In the area of organometallic chemistry, as elsewhere in chemistry, the number of valence electrons associated with the interacting atoms is closely related to the three dimensional structure which is adopted by the molecule. A number of "electron counting rules" have been developed which help the chemist rationalise, and to an extent predict, the structure of the the complexes under consideration [41]. The most widely used of these "counting rules" will be discussed in this section with particular reference to the structures of and bonding in alkyne - substituted transition metal clusters.

3.1. Polyhedral structures

The framework geometries adopted by the majority of transition metal carbonyl clusters, and their organometallic derivatives, are based on deltahedral structures (closed polyhedra with triangulated faces). Particular deltahedra are associated with specific electron counts, and the number of valence electrons may be said to characterise these geometries. With particular regard to clusters which contain alkyne moieties, these fragments may be treated as ligands, which can donate between two and six electrons to the cluster, depending on the mode of coordination. Alternatively, the alkylenic carbon atoms may be treated as part of the cluster framework, in which a "CR" fragment can contribute a specific number of electrons and orbitals to cluster bonding in a similar manner to a "$M(CO)_3$" fragment.

48 electrons   60 electrons

72 electrons

3.2. The Eighteen Electron Rule and the Effective Atomic Number Rule

The simplest of these counting schemes is the "18-Electron Rule" or the "Effective Atomic Number Rule" (E.A.N.) which assumes that the skeletal cluster framework is held together by a network of localised two-centre two-electron bonds. The remaining valence orbitals, which are of low energy, are used either for metal - ligand bonding or contain non-bonding electrons. Since each transition metal has 9 valence orbitals ($5(n-1)d$, $1(n)s$, and $3(n)p$), in order to completely fill these and achieve the inert-gas configuration it requires a total of 18 electrons [42]. When this is applied to cluster systems the expected Total Valence Electron (T.V.E.) count for the various nuclearity clusters is as follows :-

$$\text{TVE} \left\{ \bullet\!\!-\!\!\bullet \right\} = 2(M)\cdot 18 - 2(M\text{-}M)\cdot 2 = 34$$

$$\text{TVE} \left\{ \triangle \right\} = 3(M)\cdot 18 - 3(M\text{-}M)\cdot 2 = 48$$

$$\text{TVE} \left\{ \diamondsuit \right\} = 4(M)\cdot 18 - 6(M\text{-}M)\cdot 2 = 60$$

The great majority of triangular carbonyl clusters possess 48 electrons as predicted by the the E.A.N. Rule. The most striking exception is the cluster $Os_3H_2(CO)_{10}$ (46 T.V.E.) which is considered to be unsaturated, and may in some ways be viewed as if it contained a localised Os=Os double bond. Although this view is consistent with its reactivity since, under mild reaction conditions, it reacts with Lewis bases (L) to afford adducts $Os_3H_2(CO)_{10}L$ [43], the bonding within this molecule is better viewed as containing a four-centre delocalised "OsH_2Os" unit [44]. However, Molecular Orbital treatments of this type indicate that the addition of an electron pair to a *TMCC*, whether it be by increasing the net negative charge on the cluster by two units or by the addition of a two-electron donor ligand, should be accompanied by the cleavage of an edge of the metal framework.

$$\underset{\substack{H_2Os_3(CO)_{10} \\ 46\,e^-}}{} \quad \overset{+CO}{\underset{-CO}{\rightleftharpoons}} \quad \underset{\substack{H_2Os_3(CO)_{11} \\ 48\,e^-}}{} \quad \overset{+CO}{\underset{-CO}{\rightleftharpoons}} \quad \underset{\substack{H_2Os_3(CO)_{12} \\ 50\,e^-}}{}$$

In electron counting procedures based on the E.A.N. Rule, a coordinated hydrocarbon is regarded as an external ligand, which simply donates a number of electrons to the metal framework. As shown in the Scheme below, an alkyne can act as a two- (mode a) or a four-electron donor (modes c, e, f, h, and l) when coordinated without oxidative addition, and as a one- (mode b), or a three- (mode d) or a five- electron donor (modes g, i, m) when coordinated as an acetylido fragment, or overall as a six- electron donor when two independent "CR" carbyne units are incorporated into the cluster (mode n).

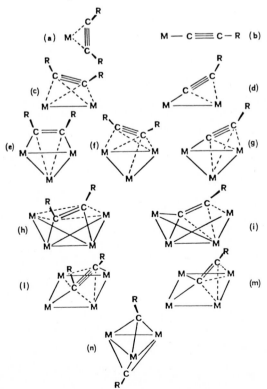

Unfortunately, the E.A.N. Rule frequently breaks down for higher nuclearity cluster systems, as well as for metal π-hydrocarbon complexes, in which several metal-carbon interactions are present.

3.3. The Polyhedral Skeletal Electron Pair (P.S.E.P.) Approach

This approach, which was developed by Wade and extended by Mingos [45], rationalises the link which exists between the electrons available for delocalised skeletal bonds and the shape adopted by clusters by using the stringent analogy with boranes. Boranes and carboranes tend to adopt structures based on triangular - faced polyhedra (deltahedra), and belong to three structural classes, *closo* (closed), *nido* (nest-like), and *arachno* (cobweb). If each BH or CH^+ unit is regarded as *sp* hybridised, and one *sp* lobe and one electron is used in forming a two - centre two - electron bond with the hydrogen atom, then the remaining *sp* hybrid lobe and the two orthogonal *p* atomic orbitals (A.O.s), together with the other two electrons, are available for skeletal bonding.

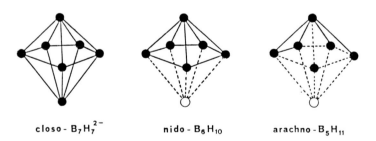

$closo - B_7H_7^{2-}$ $nido - B_6H_{10}$ $arachno - B_5H_{11}$

Quantum chemical calculations on *closo*-boranes $B_nH_n^{2-}$ (n = 6 - 12) have shown that these polyhedra have the appropriate symmetry to generate (n + 1) bonding molecular orbitals (M.O.s) from the 3n A.O.s present in the n skeletal atoms. Thus the complete filling of all available bonding M.O.s corresponds to a Skeletal Electron Pair number, $S = n+1$. The same polyhedra form the basis for the *nido* and *arachno* boranes, in which n skeletal atoms are held together by (n+2) or (n+3) skeletal electron pairs, so that one and two vertices are left vacant in the parent polyhedron, respectively.

The link between boranes and *TMCCs* is provided by the "isolobal" analogy developed by Hoffmann [46]. In this analogy, if a metal carbonyl fragment, $M(CO)_3$, is considered to be d^2sp^3 hybridized, three hybrid orbitals are used to accommodate the σ - electron pairs from each CO and three pure d orbitals (the t_{2g} set in O_h symmetry) are used for the π - back-bonding interaction to the same ligands. This leaves three hybrid orbitals for cage bonding. Thus a d^8 fragment, such as $Fe(CO)_3$, possesses three mutually orthogonal orbitals and two electrons for skeletal bonding, which is the identical frontier orbital situation to that of BH or CH^+.

In the case described above the BH and $M(CO)_3$ fragments have frontier orbitals which are similar in number, symmetry, and energy, and have the same number of electrons, and are, therefore, defined as being "isolobal".

It is perhaps surprising that the apparently electron - rich *TMCCs* can be rationalised by a bonding scheme which was originally developed for electron deficient compounds such as boranes. However, the increasing number of metallo - boranes and metallo - carboranes so far synthesized (in which BH units have been replaced by isolobal $Fe(CO)_3$ or $Co(\eta\text{-}C_5H_5)$ units) gives support to the isolobal connection between transition metal and borane clusters.

$$Fe(CO)_3 \leftrightarrow BH \leftrightarrow CH^+ \leftrightarrow Co(C_5H_5)$$

In the case of organometallic clusters, the Polyhedral Skeletal Electron Count can be obtained easily as follows :

A main group element E provides three A.O.s for cluster bonding and uses the remaining valence shell A.O.s to bond to external ligands (generally H) or to accommodate a lone pair of electrons. For the EH_x fragment, the skeletal electron pair number (S) is :

$$S = 0.5 \, (v + x - 2)$$

Similarly, for a transition metal fragment ML_z, having five more (n - 1)d A.O.s, and the S.E.P. number (S) is :

$$S = 0.5 \, (v + x - 12)$$

where v = number of valence electrons for E or M, and x = number of electrons from the external ligands.

The relationship between the Skeletal Electon Pair number (S), as derived above, and the shape of some selected organometallic clusters is illustrated   below.

S = 6, fundamental polyhedron : Trigonal bipyramid

Closo -species : **n = 5**

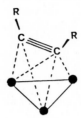 

Some examples (references are quoted in the text) :

$Fe_3(CO)_9(RC_2R)$, $Fe_2Ru(CO)_9(RC_2R)$, $Ru_3H(CO)_9(C_2R)$, $Co_3(C_5H_5)_3(CR)_2$,

$Fe_3(CO)_9(CR)(COR)$

Nido -species : **n = 4**

(tetrahedron)

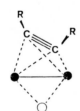 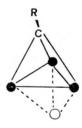

$Mo_2(CO)_4(C_5H_5)_2(RC_2R)$, $Co_2(CO)_6(RC_2R)$, $Ni_2(C_5H_5)_2(RC_2R)$, $Co_3(CO)_9(CR)$

$Co_2ML_n(CO)_6(CR)$ / $ML_n = Mo(CO)_2(C_5H_5)$; $Fe(CO)(C_5H_5)$; $Ni(C_5H_5)$/

S = 7, fundamental polyhedron ; Octahderon

Closo - species : **n = 6**

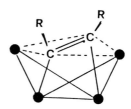

$Co_4(CO)_{10}(RC_2R)$, $Ru_4(CO)_{12}(RC_2R)$, $Co_2Ru_2(CO)_{11}(RC_2R)$.

Nido -species : **n = 5**

(Square-pyramid)

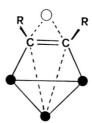

$M_3(H)_2(CO)_9(RC_2R)$ /M=Ru,Os/, $Os_3(CO)_{10}(RC_2R)$, $FeCo_2(CO)_9(RC_2R)$,
$Os_3H(CO)_9(RC_2PR_3)$

S = 8, fundamental polyhedron : Pentagonal Bipyramid

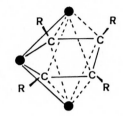

Closo - species : **n = 7**

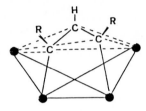

$M_3(CO)_8(RC_2R)_2$ /M=Fe,Ru/, $Ru_3Fe(CO)_9(C_5H_5)(RC-CH-CR)$,

$Ru_3Ni(CO)_8(C_5H_5)(RC-CH-CR)$.

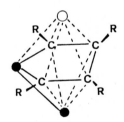

Nido -species : **n = 6**

(Pentagonal pyramid)

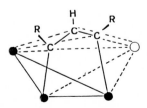

$Fe_2(CO)_6(RC_2R)_2$, $Ru_2(CO)_6(MeC=CH-CH=CH)$, $Ru_3H(CO)_9(RC-CH-CR)$

S = 9, fundamental polyhedron : Dodecahedron

Closo -species : **n = 8**

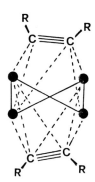

$Fe_4(CO)_{11}(HC_2Et)_2$, $Ru_4(CO)_{11}(RC_2R)_2$, $Ir_4(CO)_8(MeCO_2C_2CO_2Me)_4$

In the PSEP approach the hydrocarbon unit is considered as a source of "CR" or "CR$_2$" skeletal units rather than as an external ligand, so that the pair of electrons used to form the σ C-C bond, within the alkylenic unit, is included in the skeletal count. The geometries described below are in a sense idealised, because of the severe distortions which occur as a result of the difference in size between the carbon atoms and the transition metal atoms forming the cluster framework. It is interesting to note that, generally, in organometallic clusters the carbon atoms occupy the vertices of lower connectivity.

3.4. The Capping Principle

The PSEP approach, as described in the previous section, does not allow the rationalisation of the structures of all *TMCC* 's. As the nuclearity of the clusters increases so does the variety and complexity of the metal framework geometry [47]. A situation which occurs in *TMCC's* but does not occur in borane clusters is that of capping framework atoms; for example, where a metal atom sits over a triangle of metals, capping them, to form a tetrahedron. Using the PSEP approach to count electrons in capped systems it is found that a *mono* -capped cluster with n framework atoms is held together by n skeletal electron pairs, so that, formally a capping group adds no electron pairs to the framework, and the structure is based on an n-1 vertex polyhedron. This rule may be extended to systems with two and three caps [48] which are held together by n-1 and n-2 skeletal electron pairs, respectively. The number of Polyhedral Skeletal Electron Pairs (S) may be obtained using the same equations as discussed previously, namely :

$$S = 0.5[PEC - 12n] \text{ for transition metals}$$

$$S = 0.5[PEC - 2n] \text{ for main group elements}$$

$$PEC = \text{polyhedral electron count.}$$

3.5. Condensed polyhedra

Mingos has recently pointed out that this "Capping Principle" is merely a subset of the generalised condensation principle [49], which may be used to rationalise the structures of larger clusters in terms of fused polyhedra which share faces or edges.

With the introduction of capping atoms into the metal framework an ambiguity arises since it is possible to have more than one geometry with the same electron count. This feature is best illustrated using an example from hexaosmium cluster chemistry [50]. The structure, determined by X-ray diffraction, for $Os_6H_2(CO)_{18}$ is based on a capped *nido* -octahedral framework (a capped square-based pyramid), and thus has a different structure from that of the *closo* -octahedral $[Os_6(CO)_{18}]^{2-}$ dianion although both clusters have 7 skeletal electron pairs for framework bonding. Both these clusters have an electron pair more than the parent binary carbonyl, $Os_6(CO)_{18}$, for which S = 6, and the structure is based on a *closo* - five vertex polyhedron with one cap, a capped trigonal bipyramid.

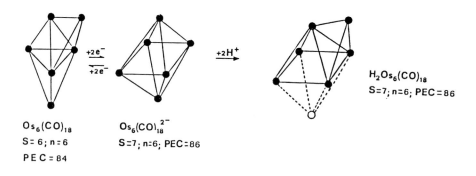

$Os_6(CO)_{18}$
S= 6; n=6
PEC = 84

$Os_6(CO)_{18}^{2-}$
S=7; n=6; PEC=86

$H_2Os_6(CO)_{18}$
S=7; n=6; PEC=86

Capped structures are not restricted to all metal frameworks. There are a number of carbyne or alkylidyne substituted clusters in which 'CR' fragments are condensed over a metallic surface. Activation of the heterometallic cluster $Os_3WH(CO)_{12}(C_5H_5)$ by Me_3NO in the presence of RC≡CR (R = Ph or C_6H_4Me) affords the alkyne substituted complex $Os_3WH(CO)_{10}(C_5H_5)(RC_2R)$ which then forms the bis-alkylidyne $Os_3WH(CO)_9(C_5H_5)(\mu_3-CR)_2$ under vacuum pyrolysis

[51]. This sequence is analogous to the Os_6 cluster interconversion described above, and represents a further example of $C\equiv C$ bond scission on a cluster surface.

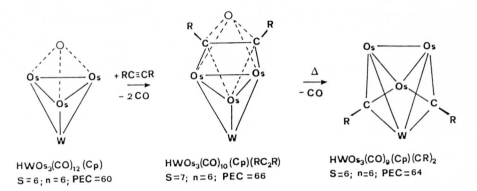

$HWOs_3(CO)_{12}(Cp)$
$S=6; n=6; PEC=60$

$HWOs_3(CO)_{10}(Cp)(RC_2R)$
$S=7; n=6; PEC=66$

$HWOs_3(CO)_9(Cp)(CR)_2$
$S=6; n=6; PEC=64$

The "Capping Principle" may also be used to rationalise the occurrence of two isomers of $Fe_3(CO)_8(RC_2R)_2$ [5]. The thermodynamically stable black isomer may be described in terms of the PSEP approach as a *closo* -pentagonal bipyramid. The violet isomer, however, can be viewed as the result of the condensation of two $Fe_3(\mu_3-\eta^2-\text{II}-RC_2R)$ polyhedra sharing the common Fe_3 face. Then, invoking the Mingos condensation principle [49], the PEC for this cluster will be 54 + 54 - 48 = 60 as expected for a cluster with the formula $Fe_3(CO)_8(RC_2R)_2$.

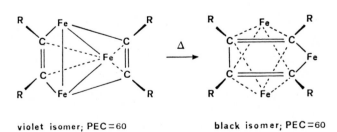

violet isomer; PEC=60

black isomer; PEC=60

3.6. Mechanisms for framework rearrangements

It has been stated above that the addition of an electron pair to a cluster is often concommitant with the cleavage of a framework edge. This suggests a mechanism by which framework geometries may interconvert, by going through an intermediate which as one less framework edge, and an electron count two in excess of that of the original cluster. This cleaved edge is then reformed between other atoms in the framework resulting in an interconversion of the cluster. Johnson [52] has recently proposed a mechanism for octahedral clusters *via* capped trigonal bipyramidal intermediates, as illustrated below. Johnson suggests that the shape adopted by a cluster is strongly dependent on attractive forces between the constituent framework atoms. A measure of the sum of these forces is the cohesive energy (C.E.) which for a transition metal is proportional to the square root of the connectivity Z. Substantial differences in C.E. are found between apparently closely related polyhedra, such as an octahedron compared to a trigonal prism. These differences arise from changes in connectivity within the cluster core. As a consequence, for any polyhedral arrangement within a cluster core, it might be expected that a pathway in which the change in cohesive energy is kept to a minimum would be preferred. In other words, the rearrangement is expected to occur by cleavage of only one framework edge at a time.

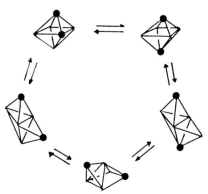

4. REACTIVITY

The PSEP approach not only provides a reasonable explanation of the geometries adopted by organo-substitued *TMCC's*, but in some cases it may be used to predict rational synthetic routes to new cluster systems.

4.1. Cluster Expansion through insertion of metallic units

The insertion of new metallic fragments into a cluster is possible when a *nido* or *arachno* complex is treated with a precursor, usually a bimetallic complex, which is able to generate reactive fragments *in situ* by thermal or photochemical cleavage of metal - metal bonds. If the incoming ML_x fragment can supply three empty AOs for cluster bonding, eg. $Cr(CO)_3$, $Mn(C_5H_5)$, or $Fe(CO)_2$, a clean expansion occurs with the new fragment inserting into the vacant apex.

arachno[n] CLUSTER $\xrightarrow{MLx(S=0,n=1)}$ nido[n+1] CLUSTER $\xrightarrow{MLx}$ closo[n+2] CLUSTER

It is known that the black isomer $Fe_3(CO)_8(RC_2R)_2$ degrades on heating to the "ferrole" $Fe_2(CO)_6(RC_2R)_2$ [5]. This reaction can be reversed in the presence of $Fe_2(CO)_9$, which acts as a source of "$Fe(CO)_x$" fragments. The yields depend on the steric bulk of the alkyne substituents and drops to zero for phenyl substituents [53]. The X-ray structure of $Fe_2(CO)_6(PhC_2Ph)_2$ [54] reveals that the phenyl rings bend towards the vacant apex, which hinders the insertion of the incoming "$Fe(CO)_2$" fragment.

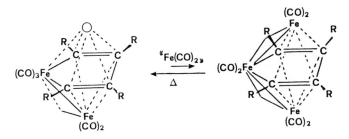

Yield(→) : Me > CO₂Me ≫ Ph

When the precursor is not able to generate a zero electron species which can occupy the vacant vertex, the original cluster may rearrange in order to maintain the correct number of skeletal electron pairs by eliminating external ligands. For example, when the *nido* - pentagonal bipyramidal cluster $Ru_3H(CO)_9(MeC-CH-CMe)$ is heated with $[Fe(CO)_2(C_5H_5)]_2$, the insertion of a "$Fe(C_5H_5)$" fragment (a 1 electron donor) is concomitant with the ejection of the hydride ligand. Similarly the addition of a "$Ni(C_5H_5)$" fragment (a 3 electron donor), to the same cluster causes the elimination of the hydride and one carbonyl ligand [55].

It is important to note that the PSEP Approach can rationalise the shape of the cluster framework obtained by the expansion process, but it cannot predict the position that is occupied by the incoming metal fragment or the distribution of the ligands. In the case of the cluster expansion of $Ru_3H(CO)_9(MeC-CH-CMe)$ with the iron and nickel fragments, under the same experimental conditions, the Fe atom occupies an apical position, and the nickel an equatorial one. In the latter case bridging carbonyls are present, and these probably help to alleviate the electron imbalance within the cluster core.

It is often difficult to generate zero-electron fragments from the precursor *in situ*, which possess few surrounding ligands and are quite unstable as a result. When the *nido* -cluster $Ru_3H(CO)_9(MeC-CH-CMe)$ reacts with $Ru_3(CO)_{12}$, instead of the zero electron fragment "$Ru(CO)_2$", the two electron fragment "$Ru(CO)_3$" is incorporated [48]. At the same time the organic chain rearranges, through a hydrogen shift, to give a *closo* -octahedral Ru_4C_2 core, with the formal reduction in polyhedral skeletal electron pairs from 8 to 7.

A similar hydrogen shift occurs when the *nido* -pentagonal bipyramidal cluster $Ru_3H(CO)_9(MeC-CH-CMe)$ is treated with alkyne. The change in connectivity between the Ru and C atoms give rise to a Ru_3C_4 *closo* -pentagonal bipyramidal core geometry [23]. This rearrangement is of interest because of its relevance to the hydrocarbon metathesis process, indeed similar compounds have been used as stoichiometric intermediates in olefin and acetylene metathesis reactions [56].

If a metal fragment ML_x, which can supply only one skeletal electron pair, i.e. $Fe(CO)_3$, $Co(C_5H_5)$, $Ni(PR_3)_2$, is incorporated into an existing cluster, the original polyhedron expands to a new arrangement with one additional vertex.

closo				closo		
nido	[n] CLUSTER	$\xrightarrow{ML_x \ (S=1, \ n=1)}$		nido		[n+1] CLUSTER
arachno				arachno		

The following example is taken from the extensively studied organometallic chemistry of the mixed Ni-Fe systems [3]. The first and second steps illustrated below represent the insertion of a two electron donor "$Fe(CO)_3$" fragment and the expansion from a (n)-*nido* to a (n+1)-*nido* polyhedron. The third and final step corresponds to the closing up of the *nido* species to give the *closo* -pentagonal bipyramid. An analogous Ni-Ru pentanuclear cluster, $Ru_3Ni_2(CO)_8(C_5H_5)_2(PhC_2Ph)$, has been obtained recently by an alternative route [57].

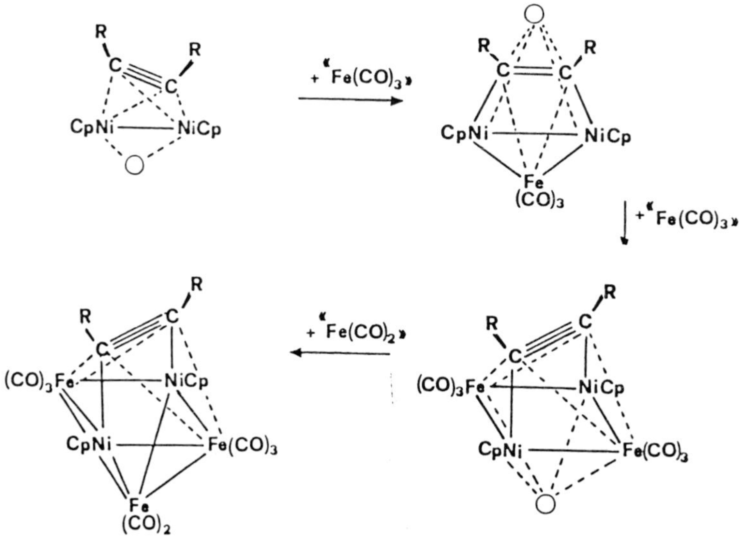

A systematic study of cluster expansion reactions from *nido* -trigonal bipyramidal (i.e. tetrahedral) to *nido* -octahedral (i.e. square-pyramidal) clusters has been carried out by Jaouen *et. al* [58].

Again the PSEP Approach rationalises the shape of the cluster produced by the expansion reaction, but does not predict the distribution of the metal atoms within the polyhedron. In $FeCoNi(CO)_6(C_5H_5)(RC_2R)$ and $FeNi_2(CO)_3(C_5H_5)_2(RC_2R)$ the added "$Fe(CO)_3$" fragment occupies an apical position, but in $FeNiMo(CO)_5(C_5H_5)_2(RC_2R)$ the iron atom occupies an equatorial site. Surprisingly, the structurally and electronically related $Co_2(CO)_6(RC_2R)$ complex does not undergo the expansion reaction, but the mixed metal clusters $FeCo_2(CO)_9(RC_2R)$ and $RuCo_2(CO)_9(RC_2R)$ have been obtained by a different route starting from $FeCo_3H(CO)_{12}$ [59], and $[RuCo_3(CO)_{12}]^-$ [13], respectively.

In the context of the mechanisms of these cluster expansion reactions, it is interesting to note that in a study of the stoichiometric hydrogenation of $Co_2(CO)_6(RC_2R')$ complexes, the kinetic data does not show that Co-CO bond cleavage is the rate determining step, as would be expected for a saturated species by the E.A.N. formalism, and an insertion of dihydrogen into a vacant apex has been postulated [60].

336

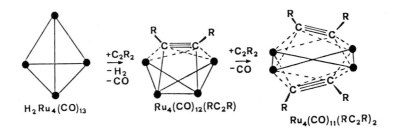

4.2. Cluster expansion through incorporation of organic ligands

Cluster expansion may also be obtained by the incorporation of additional organic fragments into the core polyhedron. This occurs in the stepwise synthesis of $Ru_4(CO)_{12}(RC_2R)$ and $Ru_4(CO)_{11}(RC_2R)_2$ [61].

According to the 18-electron formalism, both the alkyne-substituted clusters appear to be two electrons short of the required count, if the μ_4-η^2 alkyne groups act as four electron donors. In fact the metal framework should be associated with the following T.V.E. numbers, but the structures can be easily rationalised by using the PSEP Approach.

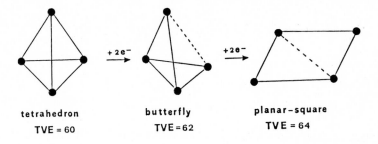

tetrahedron
TVE = 60

butterfly
TVE = 62

planar–square
TVE = 64

The related 62 electron cluster, $Fe_4(CO)_{11}(\mu_4\text{-}PR)_2$, exhibits Lewis acidity, giving rise to 1:1 adducts when stirred at room temperature with PR_3, $P(OR)_3$, or CNR.[61] By contrast, neither $Ru_4(CO)_{12}(RC_2R)$ nor $Ru_4(CO)_{11}(RC_2R)_2$ react under the same experimental conditions. However, at high temperature or under photochemical stimuli clean stoichiometric substitution of carbonyls by Lewis bases occurs [63].

It is worth pointing out again that the 18-electron Rule does not correctly rationalise the structure and chemical behaviour of organometallic clusters having significant metal - carbon π-interactions.

Another interesting example of cluster expansion through alkyne insertion is observed in the trapping of the very reactive imido cluster with diphenylacetylene which is formed during the protonation of the $[Ru_4(CO)_{12}(\mu_4\text{-}N)]^-$ anion [64].

Looking at the metal framework in the above conversion (the dotted lines in the Scheme below), the initial butterfly geometry is transformed into a near planar arrangement with the fifth Ru-Ru edge being broken as a consequence of the formal addition of two electrons to the cluster. A similar type of insertion occurs in the phosphinidene cluster $Ru_4(CO)_{11}(PPh)(PhC_2Ph)$ [65].

A remarkable example of cluster expansion occurs via the coupling of a coordinated carbyne and a free alkyne in the reaction of $Ru_3H_3(CO)_9CX$ with alkynes, which affords 1,3-dimetallo-allyl derivatives [66].

The complementary reaction, the coupling of a coordinated alkyne and a free carbyne has also been reported, giving rise to similar allyl derivatives [67].

Finally, the coupling between two alkylidyne units already coordinated to a trinuclear cluster and a free alkyne molecule has been reported [68]. The resultant complex contains a dimetallacyclopentadienyl ring.

These derivatives, in which facile C-C bond formation occurs, represent good models for the Fischer-Tropsch chain growth over a triangular face of a metallic surface [69].

4.3. Cluster Contraction

The reverse reaction to cluster expansion, cluster contraction, can occur in organometallic chemistry, and is generally favoured by the addition of nucleophilic reagents such as CO, PR_3 or H_2.

The two following examples are taken from the study of the photochemical behaviour of organo-ruthenium clusters in the presence of H_2, and give a clean conversion from a *closo* to a *nido* polyhedron. In the first case, this occurs via the complete hydrogenation of the β alkyne carbon [70], and in the second case, by the elimination of "$Ru(CO)_3$" fragments, which recombine to give the stable hydrido cluster $Ru_4H_4(CO)_{12}$ [63].

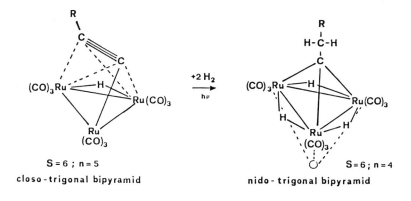

$S = 6 \ ; \ n = 5$

closo-trigonal bipyramid

$S = 6 \ ; \ n = 4$

nido-trigonal bipyramid

$S = 7 ; n = 6$

closo-octahedron

$S = 7 ; n = 5$

nido-octahedron

Another type of cluster contraction involves the extrusion of a Ni atom, as $[Ni(C_5H_5)(PMe_3)_2]^+$, and gives the *nido*-octahedral cluster $[Fe_2Ni(CO)_6(C_5H_5)(RC_2R)]^-$ anion, which is isoelectronic with $FeNi_2(CO)_3(C_5H_5)_2(RC_2R)$ (*vide infra*) [71].

A particular cluster contraction reaction is represented by the electrochemical reduction of the *nido*-pentagonal bipyramidal cluster HRu$_3$(CO)$_9$(MeC-CH-CMe) and subsequent chemical oxidation by air (O$_2$) in acidic solution of the short-lived radical anion [33]. This process affords the complex Ru$_2$(CO)$_6$(MeC=CH-CH=CH) where the elimination of a metallic unit is compensated by an extensive rearrangement of the organic chain with concomitant elimination of a molecule of hydrogen.The same product can be obtained by pyrolysis of the parent cluster [55].

4.4. Metal exchange processes

A related set of reactions is represented by a metal exchange process [72]. This can be viewed as the result of a sequence of cluster expansion and cluster contraction reactions, which preserve the overall nuclearity but modify the metal composition. Vahrenkamp has shown that the most important factor is the presence in the cluster of capping main group elements such as μ$_3$-CR, μ$_3$-GeR, μ$_3$-PR, or μ$_3$-S, which are able to clasp the metallic units [73], during the reaction sequence in which several metal - metal edges are broken. The following example shows the stepwise substitution of the good leaving groups "Co(CO)$_3$" in Co$_3$(CO)$_9$(μ$_3$-CR) by the isolobal fragments such as Mo(CO)$_2$(C$_5$H$_5$), W(CO)$_2$(C$_5$H$_5$), Fe(CO)(C$_5$H$_5$), or Ni(C$_5$H$_5$), generated *in situ* from the corresponding binuclear precursors by thermal or photochemical activation.

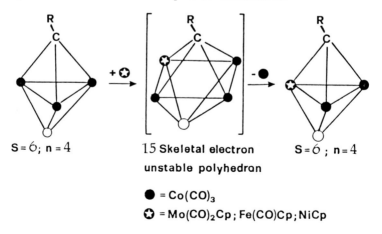

$ML_m, M'L_m = Mo(CO)_3Cp, Fe(CO)_2Cp, Ni(CO)Cp$

$ML_n, M'L_n = Mo(CO)_2Cp, Fe(CO)Cp, NiCp$

This reaction sequence, as well as synthesising new heterometallic clusters, affords chiral tetrahedral complexes, which can be separated as pure enantiomers, and tested as catalysts in asymmetric reactions [74], in the hope that the chirality of the metal framework is not lost during the catalytic sequence.

Since all the incoming fragments are three electron donors, it may be assumed that the elusive expansion product, which may be a *nido*-octahedron, has one more skeletal bonding electron than expected from the PSEP Approach. This extra electron would lie in an antibonding orbital, so that the whole cluster would be relatively unstable and would tend to eject either the incoming group (an unproductive reaction), or a three electron donor "Co(CO)$_3$" fragment (an exchange reaction), to reform an electron precise tetrahedron.

$S = 6; n = 4$ 15 Skeletal electron unstable polyhedron $S = 6; n = 4$

● = $Co(CO)_3$

✪ = $Mo(CO)_2Cp; Fe(CO)Cp; NiCp$

4.5. Nucleophilic attack on a coordinated ligand

An unusual cluster geometry modification can be obtained by the nucleophilic attack of PR_3 or CNR molecules on the α - carbon atom of the coordinated acetylido ligand in $M_3H(CO)_9(\mu_3\text{-}C{\equiv}CR)$ [M = Ru, Os] [75]. Indeed, CNDO calculations predict that in these systems the α - carbon atoms possess very low electron densities and that they are susceptible to nucleophilic attack [76]. The structural and spectroscopic data for the adducts is consistent with a zwitterionic formulation having negative charge localised in the metal framework.

A very similar nucleophilic attack of H^- on the isoelectronic and isostructural cluster $Fe_3(CO)_9(CNR)$ gives a similar adduct, which can be easily protonated using H_3PO_4. This sequence represents the ionic alternative route to the direct hydrogenation [77].

4.6. Redox processes

The main assumption in the PSEP Approach is that the highest occupied (HOMO) and the lowest unoccupied molecular orbital (LUMO) are cluster bonding and antibonding in character, respectively. From this view, the addition of electrons should cause cluster opening and vice-versa .

closo[n] CLUSTER $\xrightarrow{+2e^-}$ nido[n] CLUSTER $\xrightarrow{+2e^-}$ arachno[n] CLUSTER

The simplest way to induce this behaviour is a redox process which adds or removes electrons. The process occurs at an electrode, and organo-substituted *TMCCs* seem to be particularly suitable for electrochemical studies since the metallic framework can serve as an electron "resevoir" [78]. While binary *TMCCs* do not generally exhibit reversible electrochemical processes, the capping effect of a main group element or of an organic chain, which is able to hold the metal units together, may increase the lifetimes of the electrogenerated redox species. This is the case in the series of complexes $Co_3(CO)_9(\mu_3\text{-}ER)$ [E = C, Ge, P] and their related heterometallic derivatives, and their radical anions have been studied extensively [79].

A remarkable change in cluster geometry induced by an electrochemical redox process has been observed for alkyne - substituted trimetallic clusters [80]. Two coordination modes have been found in these trimetallic systems. In $Fe_3(CO)_9(RC_2R)$ [3] the C-C triple bond is orthogonal to an Fe-Fe vector, while in $M_3H_2(CO)_9(RC_2R)$ [M = Ru, Os] [3] and in $Os_3(CO)_{10}(RC_2R)$ the the C-C triple bond is parallel to a metal-metal bond, denoted $\mu_3\text{-}\eta^2\text{-}\perp$ and $\mu_3\text{-}\eta^2\text{-}\|$, respectively, according to the Muetterties' notation [81].

By the EAN rule,the $Fe_3(CO)_9(RC_2R)$ series represents an unsaturated 46 e- system, but no evidence of Lewis acidity has been found by chemical tests [82] . On the contrary the $\mu_3\text{-}\eta^2\text{-}\|$ complexes are saturated 48 e^- systems. In a detailed Extended Huckel Molecular Orbital (EHMO) study, Schilling and Hoffmann [83] have explained the preference for the perpendicular orientation in the model complex $Fe_3(CO)_9(HC_2H)$ as a consequence of the symmetry properties of the frontier orbitals : addition of two electron is expected to reverse the orientation preference from perpendicular to parallel. In the PSEP approach, $Fe_3(CO)_9(RC_2R)$ is a six skeletal electron pair system and therefore assumes a *closo*-trigonal bipyramidal shape with carbon atoms in equatorial and apical position. On the contrary $M_3H_2(CO)_9(RC_2R)$ and $Os_3(CO)_{10}(RC_2R)$, having S=7, assume a *nido*-octahedral geometry.

This change in orientation has been verified experimentally as two "Fe(CO)$_3$" fragments (two electron donors) are formally replaced by two "Co(CO)$_3$" [58] or "Ni(C$_5$H$_5$)" [84] fragments (three electron donors) or by one each of these fragments [85].

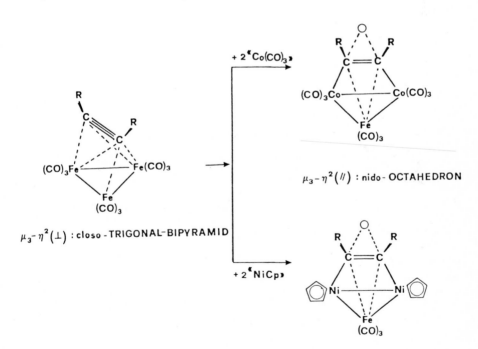

Direct evidence for this reorientation comes from the electrochemical behaviour of $Fe_3(CO)_9(RC_2R)$. Two distinct, subsequent cathodic processes (A,B) are observed in the CV response for these complexes. Analysis of these CV parameters, with scan rate and controlled potential coulometric tests, indicate that each step corresponds to a one electron reversible process [80].

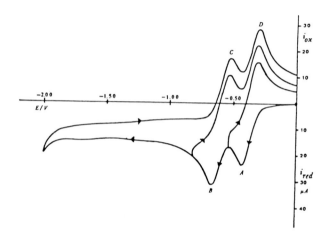

[1]H and [13]C n.m.r. studies on the relatively stable anion $[Fe_3(CO)_9(EtC_2Et)]^{2-}$ show that it is diamagnetic, with both electrons being added to the same molecular orbital (a singlet spin state). In the neutral parent complex $Fe_3(CO)_9(EtC_2Et)$, n.m.r. data shows a large spread of chemical shifts for the alkylenic C atoms and ethyl carbons but, in the dianion, the two alkylenic carbons and the ethyl substituents are equivalent at room temperature. The most reasonable explanation for this is that the alkyne has changed coordination mode, and the C≡C bond is now parallel to an Fe-Fe edge as expected on the basis of PSEP approach.

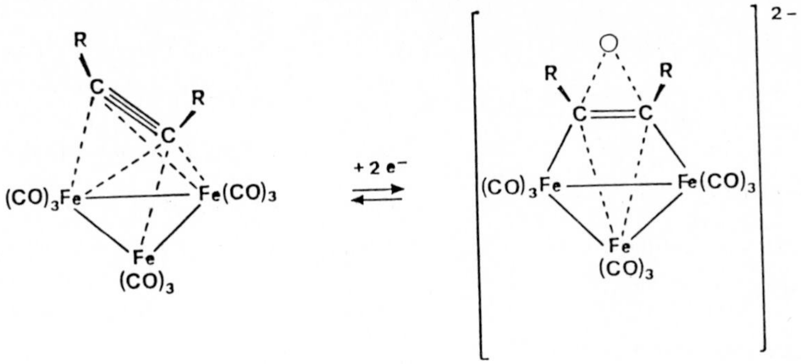

R
C
C R
(CO)$_3$Fe — Fe(CO)$_3$
Fe
(CO)$_3$

+2 e$^-$ ⇌

$\left[\begin{array}{c} \end{array} \right]^{2-}$

O
R R
C═C
(CO)$_3$Fe — Fe(CO)$_3$
Fe
(CO)$_3$

5. DYNAMICS

5.1. Fluxionality in binary transition metal carbonyls

Stereochemical non-rigidity or fluxionality of TMCC's has been a focus of interest over the past two decades. Several reviews have dealt with fluxionality behaviour of binary TMCC's [86]. Briefly, two main mechanisms have been found for carbonyl scrambling processes : axial-equatorial CO exchange localized at each metal centre and concerted bridging-terminal CO exchange delocalized over the whole metal framework. An alternative explanation of the fluxionality of binary TMCC's has been proposed by Johnson [87] : the ligand-ligand van der Waals' interactions are emphasized and the shape of the ligand envelope plays an important role in determining the overall cluster geometry and its scrambling pathways. The various kinds of CO exchange processes are then interpreted as the result of the rotation of the metal skeleton within the carbonyl envelope,which may remain fixed or move to an alternative polyhedral rearrangement in the transition state. Variable temperature (V.T.) nmr studies of TMCC's in the solid state have lent support to this proposal : the rotation of the Fe$_3$ triangle in Fe$_3$ (CO)$_{12}$ [88] or the Co$_4$ tetrahedron in Co$_4$(CO)$_{12}$ [89] occurs within the icosahedron defined by the twelve carbonyl ligands, which is static because of lattice constaints.

5.2. Fluxionality in organometallic clusters

As far as organometallic clusters are concerned, the mobility of the co-ordinated organic moiety has to be considered too. In several organo-clusters the motion of the organic chain has been proved to generate mirror planes and then to equilibrate selected sets of CO ligands. A low energy process, involving the exchange of the vinyl group, which oscillates between two osmiun centres in a "windscreen wiper" style, has been reported for $Os_3H(CO)_{10}(HC=CHR)$ compounds [90], a similar pathway has been proposed for the fluxionality of $Os_3H(CO)_9[P(Et)_2-C=CH_2]$ cluster [91].

A remarkable "wagging" motion of the allenyl chain coupled with the "hopping" motion of the hydride has been found to be operative in $M_3H(CO)_9(MeC=C=CMe_2)$ (M=Ru, Os) [92].

A change in" helicity" of the dienone ligand, able to interconvert diastereomeric forms of the flyover bridge complexes $M_2(CO)_6[(RC_2R')_2CO]$ (M=Ru, Fe) has been verified by ^{13}C-VT-nmr study of Group 15 ligand substituted derivatives [93]. The molecule traverses the symmetrical ring-opened transition state structure, originally proposed by Hoffmann *et al.* [94] on the basis of EHMO calculations of the homologous "fulvene" complex.

On the contrary, in several organo-clusters, the organic moiety is found to be static over a wide range of temperatures and only carbonyl scrambling (localized or delocalized) occurs.

To some extent, the PSEP approach can contribute to a rationalisation of the different dynamic behaviour of organo-clusters. In this view, the polytopal rearrangement of the entire cluster has to be considered. Suggestions for this methodologic approach come again from the borane-*TMCC*'s analogy. Polytopal motion of the cage atoms in boranes has been indeed detected by ^{11}B-VT-nmr spectroscopy [95]. Moreover, thermal isomerisation of metallocarboranes (in the slower chemical time-scale) through rearrangement of the polyhedral atoms is a well documented phenomenon [96].

$Co_2Cp_2C_2B_3H_5$: ⊘,Co ; ●,CH ,○,BH.

Jaouen and co-workers [97] developed this analogy and suggested that *TMCC's* having a vacant apex on the polyhedral surface might exhibit easy structural flexibility. The fluxionality in *nido* (or *arachno*) clusters can be then envisaged as the result of subsequent migration steps of the vacant apex over the polyhedron; this migration would not involve formal scission of chemical bonds and therefore would represent a low activation energy pathway. In this context, the rapid racemisation [98] of the ferrole complex, $Fe_2(CO)_6(ROC=CBut-CH=COR)$, a *nido*-pentagonal bipyramid, can be accounted for by the occurrence of an "open"-*nido*-transition state, where a mirror plane is generated by the presence of the $Fe(CO)_3$ moieties both in apical positions. A similar explanation can be forwarded for the dynamic behaviour of $Co_2Cp_2(C_4H_4)$ complex [99], this process actually occurs at the highest temperature limit of the VT-nmr technique.

This approach can be quite useful for the rationalisation of the dynamic behaviour of the organo-*TMCC*'s so far considered. The *nido* species should exhibit facile fluxionality of the organic chain as a consequence of a rearrangement of the entire polyhedron, on the contrary, in the closed species, the organic moiety would be static and only scrambling of the external ligands (carbonyls, hydrides) would be operative in a localised at the single metal center or delocalized fashion.

Jaouen *et al.* has elegantly verified the "hidden" fluxionality of the *nido-* trigonal bipyramidal (tetrahedral) $Co_2(CO)_6(RC_2R)$ complex through the formal reduction of symmetry from C_{2v} to C_s by using different alkyne substituents, and to C_1 by using different metallic unit [a $Co(CO)_3$ unit is replaced by an isoelectronic NiCp fragment] [100].

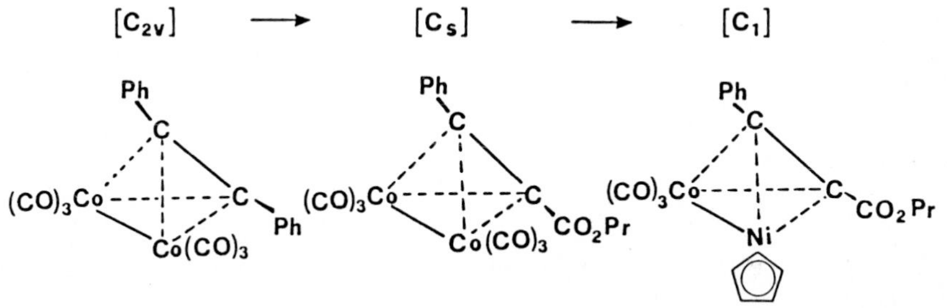

The resulting complex, $Co(CO)_3NiCp(PhC_2COOPr^i)$ is now chiral and the racemisation process, which corresponds to interchange of polyhedral vertices, can be detected by ^{1}H-VT-nmr spectroscopy by looking at the diastereotopic methyl groups of the isopropyl ester funtion. At room temperature they appear as a pair of partially overlapped doublets, at higher temperature these signals coalesce with an estimated activation energy of about 21 kcal/mol [100].

The following scheme shows the formal migration of the vacant site over the *nido*-trigonal bipyramid structure of the parent complex.

A similar migration scheme can be invoked to explain the high fluxionality of the organic ligand over a metallic triangle in the *nido*-octahedral compounds, homometallic as $Rh_3Cp_3(CO)(RC_2R)$ [101], $Os_3(CO)_7(ER_2)(C_6H_4)$ $(ER_2=PPh_2,PMe_2,AsMe_2)$ [102], $M_3H_2(CO)_9(RC_2R)$ (M=Ru,Os) [103], $Fe_3(CO)_9(EtC_2Et)^{2-}$ [80] or hererometallic as $FeCo_2(CO)_9(EtC_2Et)$ [59], $NiCpCo(CO)_3M(CO)_3(RC_2R)$ (M=Fe,Ru,Os) [104] and $NiCpFe(CO)_3M'(RC_2R)$, $[M=NiCp,Co(CO)_3,Mo(CO)_2Cp]$ [105].

On the other hand, in the case of a *closo*-polyhedron the organic chain remains static. This is the case in the "butterfly" clusters $Co_4(CO)_{10}(RC_2R)$ [106] and $Ru_4(CO)_{12}(RC_2R)$ [34, 35]. In both scrambling of external CO ligands occurs while the whole polyhedron mantains its stereochemical integrity. In the cobalt *closo*-octahedron, the presence of bridging CO's allows delocalised scrambling, in the isoelectronic ruthenium compound only localised exchange at each

Ru(CO)$_3$ moiety takes place with different activation energy depending on the alkyne substituents.

Interestingly, the analogous mixed-metal FeRu$_3$(CO)$_{12}$(PhC$_2$Ph) derivative exhibits thermal interconversion between the two possible isomers, having the iron atom in hinge or wingtip position respectively, only on the slow chemical time-scale. This indicates that a considerable energy barrier has to be surmounted when no vacant site is available [107].

In the closo-trigonal bipyramidal systems Fe$_3$(CO)$_9$(RC$_2$R) [108] and Fe$_2$Ru(CO)$_9$(RC$_2$R) [17] localised scrambling at the two structurally different M(CO)$_3$ set takes place.

The same behaviour is shown by the *closo*-dodecahedron Ru$_4$(CO)$_{11}$(RC$_2$R)$_2$, the almost planar disposition of the four metal atoms permits a merry-go-round process to occur involving all the equatorial CO's.

In the case of $Fe_3(CO)_8(RC_2R)_2$ derivatives (*closo*-pentagonal bipyramidal geometry), there is an exchange process between COa and COb sets through the opening of the longer bond of the asymmetric bridging carbonyls and rotation of the resulting apical $Fe'(CO)_3$ units. The equatorial $Fe(CO)_2$ moiety, as well as the ferracyclopentadienyl ring, remain rigid as unambiguously demonstrated by the VT-nmr study of the mixed-alkyne derivative $Fe_3(CO)_8(RC_2R)(HC_2H)$ [109], The equatorial carbonyl are no longer equivalent and show a singlet (COc) and a doublet (COd) by virtue of the *trans*-relationship with the CH unit of the ring : this pattern remains unchanged up to the complex decomposition temperature.

6. CONCLUSION

It is hoped that the data presented here has indicated that the PSEP Approach to cluster bonding has made a contribution to the understanding of not only the structures of organo-substituted transition metal clusters but also of their reaction chemistry and dynamic behaviour in solution. Using this approach it is possible to progress a stage further than the serendipitous pyrolysis reactions previously used to prepare cluster complexes. It is now possible to rationalise the structures of a majority of cluster species, predict the sites of nucleophilic or electrophilic attack, and so set out on the path to logical and planned synthetic routes which will lead to high yields of the desired product.

7. ACKNOWLEDGEMENTS

We thank the European Economic Community (EEC) for the support within the Stimulation Action Project (Contract No.ST2J-O214-1), and the Royal Society (London) - Accademia dei Lincei (Rome) for study grant (to D. O.).

References

1. R. Ugo, *Catal Rev.,* 1975, **11**, 225; E.L. Muetterties, *Science,* 1977, **196**, 839; E.L. Muetterties, *Bull. Soc. Chim. Belg.,* 1975, **84**, 959.

2. I.A. Oxton, *Rev. Inorg. Chem.,* 1982, **4**, 1; J. Evans and G.S. McNulty, *J. Chem. Soc., Dalton Trans.,* 1983, 639; *ibid,* 1984, 79.

3. E. Sappa, A. Tiripicchio, and P. Braunstein, *Chem. Rev.,* 1983, **83**, 203; E. Sappa, A. Tiripicchio, and P. Braunstein, *Coord. Chem. Rev.,* 1985, **65**, 219; K. Burgess, *Polyhedron,* 1984, **3**, 1175; J. Halton, M.F. Lappert, R. Pearce, and P.I.W. Yarrow, *Chem. Rev.,* 1983, **83**, 135; B.F.G. Johnson and J. Lewis, *Adv. Inorg. Chem. Radiochem.,* 1981, **24**, 225; H. Vahrenkamp, *Adv. Organomet. Chem.,* 1983, **22**, 169; A.J. Deeming in "Transition Metal Clusters", Ed. B.F.G. Johnson, J. Wiley, New York, 1980; R.D. Adams and I.T. Horvath, *Prog. Inorg. Chem.,* 1985, **33**, 127.

4. P.R. Raithby and M.J. Rosales, *Adv. Inorg. Chem. Radiochem.,* 1985, **29**, 169.

5. W. Hubel in "Organic Syntheses via Metal Carbonyls", Ed. F. Wender and P. Pino, J. Wiley, New York, 1968.

6. G.L. Geoffroy and M.S. Wrighton in "Organometallic Photochemistry", Academic Press, New York, 1978.

7. D.S. Ginley, C.R. Bock, and M.S. Wrighton, *Inorg. Chim. Acta.,* 1977, **23**, 85.

8. B.F.G. Johnson, J. Lewis, and D. Pippard, *J. Organomet.Chem.,* 1978, **160**, 263.

9. B.F.G. Johnson, R.A. Kamarudin, F.J. Lahoz, J. Lewis, and P.R. Raithby, *J. Chem. Soc., Dalton Trans.,* in the press.

10. L.F. Dahl and D.L. Smith, *J. Am. Chem. Soc.,* 1962, **84**, 2450; G. Gervasio, R. Rossetti, and P.L. Stanghellini, *Organometallics,* 1985, **4**, 1612.

11. I.T. Horvath, L. Zsolnai, and G. Huttner, *Organometallics,* 1986, **5**, 180.

12. E. Roland and H. Varhenkamp, *Organometallics,* 1983, **2**, 183.

13. P. Braunstein, J. Rose, and O. Bars, *J. Organomet. Chem.,* 1983, **252**,

C101.

14. J.R. Fritch and K.P.C. Vollhardt, *Angew. Chem., Int. Ed. Engl.,* 1980, **19**, 559.

15. D. Nuel, F. Dahan, and R. Mathieu, *Organometallics,* 1985, **4**, 1436.

16. E.L. Muetterties and J. Stein, *Chem. Rev.,* 1979, **79**, 479.

17. V. Busetti, G. Granozzi, S. Aime, R. Gobetto, and D. Osella, *Organometallics,* 1984, **3**, 1510.

18. B.F.G. Johnson, R. Khattar, F.J. Lahoz, J. Lewis, and P.R. Raithby, *J. Organomet. Chem.,* 1987, **319**, C51.

19. R.A. Kamarudin, Ph. D. Thesis, University of Cambridge, 1987.

20. B.F.G. Johnson, R. Khattar, J. Lewis, and P.R. Raithby, *J. Organomet. Chem.,* in the press.

21. E.L. Muetterties, *Chem. Soc. Rev.,* 1982, **12**, 283; M.L.H. Green and D. O'Hare, *Pure & Appl. Chem.,* 1985, **57**, 1897.

22. O. Gambino, G.A. Vaglio, R.P. Ferrari, and G. Cetini, *J. Organomet. Chem.,* 1971, **30**, 381; R.P. Ferrari, G.A. Vaglio, O. Gambino, M. Valle, and G. Cetini, *J. Chem. Soc., Dalton Trans.,* 1972, 1998.

23. S. Aime, L. Milone, D. Osella, and M. Valle, *J. Chem. Research (S),* 1978, 77; *(M)* 1978, 0785-0797.

24. A.J. Deeming, S. Hasso, and M. Underhill, *J. Chem. Soc., Dalton Trans.,* 1975, 1614.

25. O. Gambino, M. Valle, S. Aime, and G.A. Vaglio, *Inorg. Chim. Acta.,* 1974, **8**, 71.

26. M. Castiglioni, L. Milone, D. Osella, G.A. Vaglio, and M. Valle, *Inorg. Chem.,* 1976, **15**, 394.

27. E. Sappa, O. Gambino, L. Milone, and G. Cetini, *J. Organomet. Chem.,* 1972, **39**, 169.

28. S. Aime, D. Osella, A. Tiripicchio, A.M. Manotti - Lanfredi, and A.J. Deeming, *J. Organomet. Chem.,* 1983, **244**, C47.

29. A.J. Carty, S.A. McLaughlin, and N.J. Taylor, *J. Organomet. Chem.,* 1981,

204, C27.

30. M.P. Gomez-Sal, B.F.G. Johnson, R.A. Kamarudin, J. Lewis, and P.R. Raithby, *J. Chem. Soc., Chem. Commun.,* 1985, 1622.

31. S. Aime, L. Milone, and A.J. Deeming, *J. Chem. Soc., Chem. Commun.,* 1980, 1168.

32. S. Aime, D. Osella, A.J. Deeming, A.J. Arce, M.B. Hursthouse, and H.M. Dawes, *J. Chem. Soc., Dalton Trans.,* in the press.

33. P. Zanello, S. Aime, and D. Osella, *Organometallics,* 1984, **3**, 1374.

34. B.F.G. Johnson, J. Lewis, and K.T. Schorpp, *J. Organomet. Chem.,* 1975, **91**, C13; B.F.G. Johnson, J. Lewis, B.E. Reichert, K.T. Schorpp, and G.M. Sheldrick, *J. Chem. Soc., Dalton Trans.,* 1977, 1417.

35. B.F.G. Johnson, J. Lewis, S. Aime, L. Milone, and D. Osella, *J. Organomet. Chem.,* 1982, **233**, 247.

36. P.F. Jackson, B.F.G. Johnson, J. Lewis, P.R. Raithby, G.J. Will, M. McPartlin, and W.J.H. Nelson, *J. Chem. Soc., Chem. Commun.,* 1980, 1190.

37. S. Aime, L. Milone, D. Osella, G.A. Vaglio, M. Valle, A. Tiripicchio, and M. Tiripicchio - Camellini, *Inorg. Chim. Acta.,* 1979, **34**, 49.

38. M. Tachikawa, J.R. Shapley, and C.G. Pierpont, *J. Am. Chem. Soc.,* 1975, 7172; W.G. Jackson, B.F.G. Johnson, J.W. Kelland, J. Lewis, and K.T. Schorpp, *J. Organomet. Chem.,* 1975, **87**, C27.

39. P.A. Dawson, B.F.G. Johnson, J. Lewis, J. Puga, P.R. Raithby, and M.J. Rosales, *J. Chem. Soc., Dalton Trans.,* 1982, 233.

40. G.A. Vaglio, O. Gambino, R.P. Ferrari, and G. Cetini, *Inorg. Chim. Acta.,* 1973, **7**, 193.

41. S.M. Owen, *Polyhedron,* in the press.

42. N.V. Sidgwick and R.W. Bailey, *Proc. Soc. London, Ser. A.,* 1934, **144**, 521.

43. J.R. Shapley, J.B. Keister, M.R. Churchill, and B.G. DeBoer, *J. Am. Chem. Soc.,* 1975, **97**, 4145; A.J. Deeming and S. Hasso, *J. Organomet. Chem.,* 1975, **88**, C21.

44. D.E. Sherwood Jr. and M.B. Hall, *Inorg. Chem.,* 1982, **21**, 3458.

45. K. Wade, *Adv. Inorg. Chem. Radiochem.,* 1976, **18**, 1; K. Wade in "Transition Metal Clusters", Ed. B.F.G. Johnson, J. Wiley & Sons, New York, 1980; C.E. Housecroft and K. Wade, *Gazz. Chim. Ital.,* 1980, **110**, 87; D.M.P. Mingos, *Acc. Chem. Res.,* 1984, **17**, 311; D.M.P. Mingos, *Chem. Soc. Rev.,* 1986, **15**, 31.

46. R. Hoffmann, *Angew. Chem., Int. Ed. Engl.,* 1982, **21**, 711.

47. P.R. Raithby in "Transition Metal Clusters", Ed. B.F.G. Johnson, J. Wiley & Sons, New York, 1980.

48. C.R. Eady, B.F.G. Johnson, and J. Lewis, *J. Chem. Soc., Dalton Trans.,* 1975, 2006.

49. D.M.P. Mingos, *J. Chem. Soc., Chem. Commun.,* 1983, 706.

50. M. McPartlin, *Polyhedron,* 1984, **3**, 1279.

51. J.T. Park, J.R. Shapley, M.R. Churchill, and C. Bueno, *J. Am. Chem. Soc.,* 1983, **105**, 6182.

52. B.F.G. Johnson, *J. Chem. Soc., Chem. Commun,* 1986, 27.

53. D. Osella, private communication ; Interestingly, Hubel [5] reported that by refluxing $Fe_2(CO)_6(HC_2CO_2Me)_2$ with $Fe_3(CO)_{12}$ in ether little amount of the corresponding trinuclear compound $Fe_3(CO)_8(HC_2CO_2Me)_2$ could be obtained. Obviously at that time this reaction was not interpreted as a cluster expansion process.

54. S.R. Prince, *Cryst. Struct. Comm.,* 1976, **5**, 451.

55. S. Aime and D. Osella,*Inorg. Chim. Acta.,* 1982, **57**, 207; D. Osella, E. Sappa, A Tiripicchio, and M. Tiripicchio - Camellini, *ibid.,* 1980, **42**, 183.

56. M. Castiglioni, R. Giordano, and E. Sappa, *J. Organomet. Chem.,* 1984, **275**, 119.

57. A. Tiripicchio, M. Tiripicchio - Camellini, and E. Sappa, *J. Chem. Soc., Dalton Trans.,* 1984, 627.

58. G. Jaouen, A. Marinetti, B. Mentzen, R. Murtin, J.-Y. Saillard, B.G. Sayer, and M. McGlinchey, *ibid.,* 1982, **1**, 753; M. Miekuz, P. Bougard, B.G. Sayer,

S. Peng, M.J. McGlinchey, A. Marinetti, J.-Y. Saillard, J.B. Naceur, B. Mentzen, and G. Jaouen, *ibid.,* 1985, **4**, 1123.

59. S. Aime, L. Milone, D. Osella, A. Tiripicchio, and A.M. Manotti - Lanfredi, *Inorg. Chem.,* 1982, **21**, 501.

60. D. Osella, S. Aime, D. Boccardo, M. Castiglioni, and L. Milone, *Inorg. Chim. Acta.,* 1985, **100**, 97.

61. S. Aime, G. Nicola, D. Osella, A.M. Manotti - Lanfredi, and A. Tiripicchio, *Inorg. Chim. Acta.,* 1984, **85**, 161.

62. T. Jaeger, S. Aime, and H. Vahrenkamp, *Organometallics,* 1986, **5**, 245; H. Vahrenkamp and D. Woelters, *Organometallics,* 1982, **1**, 874.

63. D. Osella, S. Aime, G.Nicola, L. Milone, R. Amadelli, and V. Carassiti, *J. Chem. Soc., Dalton Trans.,* in the press.

64. M.L. Blohm and W.L. Gladfelter, *Organometallics,* 1986, **5**, 1049.

65. J. Lunniss, S.A. MacLaughlin, N.J. Taylor, A.J. Carty, and E. Sappa, *Organometallics,* 1985, **4**, 2066.

66. L.B. Beanon and J.B. Keister, *Organometallics,* 1985, **4**, 1713.

67 A.D. Clauss, J.R. Shapley, and S.R. Wilson, *J. Am. Chem. Soc.,* 1981, **103**, 7387.

68. D. Nuel, F. Dahan, and R. Mathieu, *J. Am. Chem. Soc.,* 1985, **107**, 1658.

69. W.A. Herrman, *Adv. Organomet. Chem.,* 1982, **20**, 160.

70. R. Amadelli, C. Bartocci, V. Carassiti, S. Aime, D. Osella, and L. Milone, *Gazz. Chim. Ital.,* 1985, **115**, 337.

71. M.I. Bruce, J.R. Rodgers, M.R. Snow, and F.S. Wong, *J. Chem. Soc., Chem. Commun.,* 1980, 1285; *J. Organomet. Chem.,* 1982, **240**, 299.

72. H. Beurich and H. Vahrenkamp, *Angew. Chem., Int. Ed. Engl.,* 1981, **20**, 98.

73. H. Vahrenkamp, *Comments Inorg. Chem. ,* 1985, **4**, 253.

74. C.U. Pittman Jr., M.G. Richmond, M. Absi - Halabi, H. Beurich, F. Richter, and H. Vahrenkamp, *Angew. Chem., Int. Ed. Engl.,* 1982, **21**, 786.

75. K. Henrick, M. McPartlin, A.J. Deeming, S. Hasso, and P. Manning, *J.*

Chem. Soc., Dalton Trans., 1982, 899; S.A. MacLaughlin, J. P. Johnson, N.J. Taylor, A.J. Carty, and E. Sappa, *Organometallics,* 1985, **2**, 352.

76. G. Granozzi, E. Tondello, R.Bertoncello, S. Aime, and D. Osella, *Inorg. Chem.,* 1985, **24**, 570.

77. M.I. Bruce, T.W. Hambley, and B.K. Nicholson, *J. Chem. Soc., Chem. Commun.,* 1982, 353.

78. P. Lemoine, *Coord. Chem. Rev.,* 1982, **47**, 55; W.E. Geiger and N.G. Connelly, *Adv. Organomet. Chem.,* 1985, **24**, 87.

79. P.N. Lindsay, B.M. Peaks, B.H. Robinson, J. Simpson, U. Honrath, H. Vahrenkamp, and A.M. Bond, *Organometallics,* 1984, **3**, 413 and references therein.

80. D.Osella, R. Gobetto, P. Montangeno, P. Zanello, and A. Cinquantini, *Organometallics,* 1986, **5**, 1247.

81. M.G. Thomas, E.L. Muetterties, R.O. Day, and V.W. Day, *J. Am. Chem. Soc.,* 1976, **98**, 4645.

82. E.Sappa, *J. Organomet. Chem.,* 1987, **323**, 83.

83. B.E.R.Shilling and R.Hoffmann, *J. Am. Chem. Soc.,* 1979, **101**, 3456 ; J.-F. Halet, J.-Y. Saillard, R.Lissillour, M.McGlinchey and G.Jaouen, *Inorg.Chem.,* 1985, **24**, 218.

84. E. Sappa, A.M. Manotti - Lanfredi, and A. Tiripicchio, *J. Organomet. Chem.* 1981, **221**, 93.

85. F.W.B. Einstein, K.G. Tyers, A.S. Tracey, and D. Sutton, *Inorg. Chem.,* 1986, **25**, 1631 and references therein.

86. F.A.Cotton,and B.E.Hanson in "Rearrangements in Ground and Exited State", Academic Press, New York , 1980 ; E.Band and E.L.Muetterties, *Chem.Rev.,* 1978, **78**, 639 ; J.Evans, *Adv. Organomet. Chem.,* 1977, **16**, 319 ; S.Aime and L.Milone, *Prog.NMR Spectosc.,* 1977, **11**, 183 ; B.E.Mann in "Comprehensive Organometallic Chemistry" G.Wilkinson, F.G.A.Stone,and E.W.Abel Eds.,Pergamon Press,Oxford 1983.

87. B.F.G.Johnson, *J. Chem. Soc., Chem. Commun.,* 1976, 703 ; B.F.G.Johnson

and R.E.Benfield, *J. Chem. Soc., Dalton Trans.*, 1978, 1554; *ibid.*, 1980,

1743 ; R.E.Benfield and B.F.G.Johnson, *Transition Met.Chem.*, 1981, **6**, 131.

88. H.Dorn, B.E.Hanson, E.Motell, *Inorg. Chim. Acta*, 1981, **54**, L71 ;

B.E.Hanson, E.C.Lisic, J.T.Petty, and G.A.Iannaconne, *Inorg. Chem.*,1986,

25, 4062.

89. B.E.Hanson, and E.C.Lisic, *Inorg. Chem.,* 1986, **25**, 716.

90. J.R.Shapley, S.I.Richter, M.Tachikawa, and J.B.Keister, *J. Organomet.*

Chem., 1975, **94**, C43 ;

S.Aime, R.Gobetto, D.Osella, L.Milone, E.Rosenberg, and E.V.Anslyn, *Inorg.*

Chim. Acta, 1986, **111**, 95.

91. A.J.Deeming, *J. Organomet. Chem.* 1977, **128**, 63.

92. S.Aime, R.Gobetto, D.Osella, L.Milone, and E.Rosenberg, *Organometallics*,

1982, **1**, 640.

93. S.Aime, R.Gobetto, G.Nicola, D.Osella, L.Milone, and E.Rosenberg,

Organometallics, 1986, **5**, 1829.

94. D.L.Thorn, and R.Hoffmann, *Inorg. Chem.*, 1978, **17**, 126.

95. E.L.Muetterties, E.L.Hoel, C.G.Salentine, and M.F.Hawthorne, *Inorg.Chem.*,

1975, **14**, 950.

96. K.P.Callahan, and M.F.Hawthorne, *Adv. Organomet. Chem.*, 1976, **14**,

145.

97. M.J.McGlinchey, M.Mlekuz, P.Bougeard, B.G.Sayer, A.Marinetti, J.-Y.

Saillard, and G.Jaouen, *Can. J. Chem.,* 1983, **61**, 1319 ; J.-F.Halet, G.Jaouen,

M.McGlinchey, and J.-Y.Saillard, *L'Actualite' Chimique,* 1985, Avril, 23.

98. R.Case, E.R.H.Jones, N.V.Schwartz, and M.C.Whiting, *Proc. Chem. Soc.*,

1962, 256.

99. M.Rosenblum, B.North, D.Wells, and W.P.Giering, *J. Am. Chem. Soc.,* 1972,

94, 1239.

100. G.Jaouen, A.Marinetti, J.-Y.Saillard, B.G.Sayer, and M.J.McGlinchey,

Organometallics, 1982, **1**, 225.

101. T.Yamamoto, A.R.Garden, G.M.Bodner, L.J.Todd, M.D.Rausch, and

362

S.A.Gardner, *J. Organomet. Chem.*, 1973, **56,** C23.

102. A.J.Deeming, R.S.Nyholm, and M.Underhill, *J. Chem. Soc. Chem. Commun.,* 1972, 224; A.J.Deeming, R.E.Kimher, and M.Underhill, *J. Chem. Soc. Dalton Trans.* 1973, 2589.

103. A.J.Canty, A.J.P.Domingos, B.F.G.Johnson, and J.Lewis, *J. Chem. Soc. Dalton Trans.*, 1973, 2056; J.Evans, B.F.G.Johnson, J.Lewis, and T.Matheson, *J. Organomet. Chem.*, 1975, **97,** C16; A.J.Deeming, *J. Organomet. Chem.* 1978, **150,** 123 ; S.Aime, R.Bertoncello, V.Busetti, R.Gobetto G.Granozzi, D.Osella, *Inorg. Chem.,* 1986, **25,** 4004.

104. F.W.B.Einstein, K.G.Tyers, A.S.Tracey, and D.Sutton, *Inorg. Chem.,* 1986, **25,** 1631.

105. G.Jaouen, A.Marinetti, B.Mentzen, R.Mutin, J.-Y.Saillard, B.G.Sayer, M.McGlinchey, *Organometallics,* 1982, **1,** 75; M.Mlekuz, P.Bougeard, B.G.Sayer, S.Peng, M.J.McGlinchey, A.Marinetti, J.-Y.Saillard, J.B.Naceur, B.Mentren, and G.Jaouen, *Organometallics*, 1985, **4,** 1123.

106. J.Evans, B.F,.G.Johnson, J.Lewis, and T.W.Mathenson, *J. Am. Chem. Soc.,* 1975, **97,** 1245; S.Aime, L.Milone, and E..Sappa, *Inorg. Chim. Acta.*, 1976, **16,** L7.

107. J.R. Fox, W.L. Gladfelter, G.L. Geoffroy, I. Tavanaiepour, S. Abdel-Mequid, and V.W. Day, *Inorg. Chem.* 1981, **20,** 3230.

108. G. Granozzi, E. Tondello, M. Casarin, S. Aime, and D. Osella, *Organometallics,* 1983, **2,** 430.

109. D. Osella, private communication.

THE STEREOCHEMISTRY OF THE SAKURAI REACTION

Yoshinori Yamamoto and Nobuki Sasaki

THE STEREOCHEMISTRY OF THE SAKURAI REACTION

Yoshinori Yamamoto and Nobuki Sasaki

Department of Chemistry, Faculty of Science, Tohoku
University, Sendai, Japan 980

6. CONCLUDING REMARKS

7. REFERENCES

8. ADDED IN PROOF

ABBREVIATIONS

L.A	Lewis Acid
TMS	trimethysilyl
TFA	trifluoroacetic acid
OTf	trifluoromethanesulfonate
MCPBA	meta-chloroperbenzoic acid
TBAF	tetrabutylammonium fluoride

1 INTRODUCTION

1.1 HISTORICAL BACKGROUND

In 1974, Calas et al. reported that silylated hydrocarbons having an activated Si-C bond, such as allyl-, ethynyl-, vinyl-, propargyl-, and phenyl-silanes, added to chloral or chloroacetone in the presence of a Lewis acid, and gave after hydrolysis the corresponding α-trichloromethylated carbinols (ref. 1).

Calas, Dunoguès

In 1975, Abel and Rowley reported that hexafluoroacetone reacted with allyltrimethylsilane in the presence of a Lewis acid catalyst to give the allylated alcohol along with an ene product and a Friedel-Crafts type adduct (ref. 2). The initial scope of the allylation via allyltrimethylsilane was limited to the activated aldehydes and ketones such as chloral, chloroacetone, and perfluoroacetone, and thus did not provide a strong impact upon the field of organic synthesis.

In 1976, Hosomi and Sakurai found that the allylation of a wide variety of aliphatic, alicyclic and aromatic carbonyl compounds took place with allyltrimethylsilane in the presence of $TiCl_4$ (ref. 3). The reaction proceeded rapidly even at -78°C with high yield, and the regioselective carbon-carbon bond formation occurred exclusively at the γ-position of allylic silanes (1, 2).

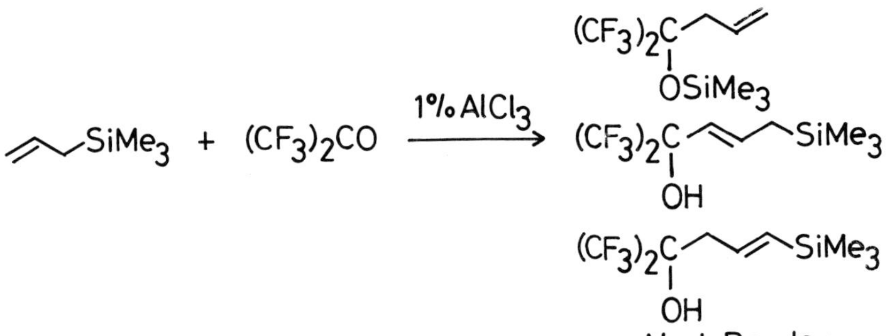

Abel, Rowley

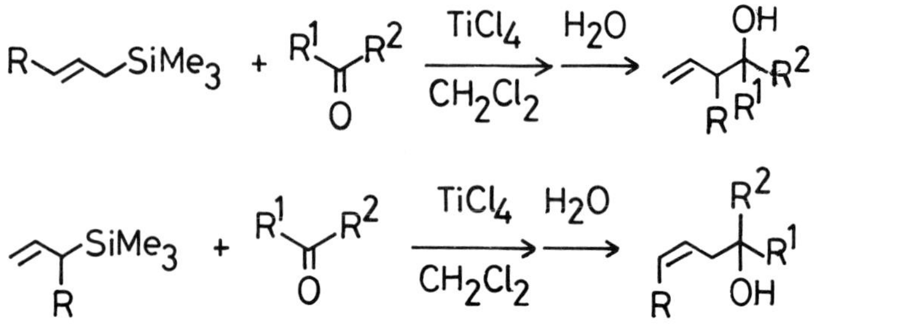

Hosomi, Sakurai

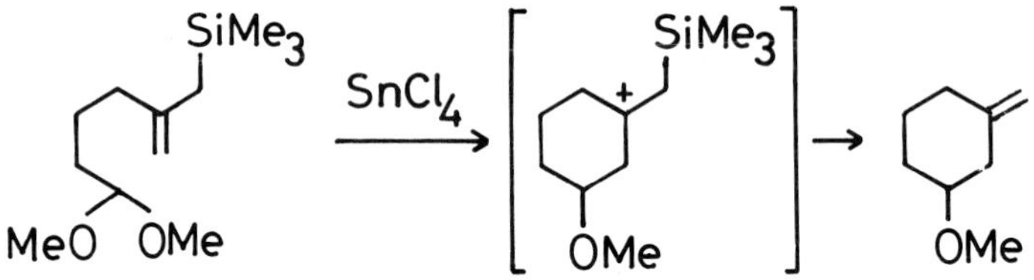

Fleming

This discovery greatly increased the synthetic utility of allyl-silane derivatives.   It should be noted that the $TiCl_4$-mediated reaction of allylsilanes (now well known as the Sakurai reaction) is structually in the same category as the $TiCl_4$-promoted reaction of silyl enol ethers which nowadays is given the name, Mukaiyama reaction (ref. 4).   Both in the Sakurai reaction as well as in the Mukaiyama reaction, $TiCl_4$ is added in order to increase the reactivity of carbon electrophiles toward the only weakly nucleophilic silyl reagents.

Another important finding was made by Fleming in 1976 (ref. 10b). The intramolecular reaction of allylsilane with acetal in the presence of $SnCl_4$ gave the cyclization product in which the double bond was formed by loss of the trimethylsilyl group. This finding also triggered the organic synthesis via allyl-silanes.

Since the discovery made by Hosomi and Sakurai, the Lewis acid mediated reactions of allylic silanes with various electro-philes have found frequent use in modern organic syntheses. Allylic silanes are thermally stable and storable without any special cautions, and can be handled under open atmosphere. In addition, the reagent can react smoothly with various electrophiles in the presence of Lewis acids. These characteri-stics are in marked contrast to those of the previous allylating reagents such as allyl-magnesium halides, -lithiums, -coppers, and -titanium compounds, which are thermally unstable and should be treated under nitrogen atmosphere and under moisture free conditions. For these reasons, allylic silanes have been used extensively for the selective carbon-carbon bond formation.

1.2 MECHANISM AND SCOPE

The facile and regioselective reaction of allylsilanes with electrophiles (E-N) can be explained by the β-effect of the silyl substituent (ref. 5). The electrophile attacks at the γ-carbon of the allyl system to generate a carbocation which is stabilized by a β-trimethylsilyl group through σ-π conjugation (3) (ref. 6). This σ-π conjugation may make the carbon-silicon bond weak and thus assist the cleavage process. In fact, the silicon-carbon bond of allylsilanes undergoes a facile cleavage with proton and some heteroatom electrophilic reagents such as hydrogen chloride, bromine, iodine and sulfuric acid (ref. 7). These reactions produce propene and allyl halides. Synthetically, the carbon-carbon bond formation is obviously more important than the heterolysis reaction. A

Me₃Si + E—N ⟶ E⁺ ...SiMe₃ / N ⟶ E + Me₃Si—N   3

number of carbon electrophiles can react with allylsilanes in
the presence of Lewis acids (**4**).   Ordinary aldehydes and
ketones (ref. 3), acid halides (ref. 8), epoxides (ref. 9),
acetals (ref. 10), iminium salts (ref. 11), t-alkyl halides
(ref. 9), allylic halides and ethers (ref. 12), allylic acetates
(ref. 13), α-halo-ethers (ref. 14), α-chlorosulfides (ref. 15),
monothioacetals (ref. 16), [(propargyl)dicobalt hexacarbonyl]
cations (ref. 17), and tricarbonylcyclohexadienyliumiron
cations (ref. 18) undergo a facile allylation with high yields.
Conjugate allylation takes place with Michael acceptors, such
as α,β-unsaturated carbonyl compounds (ref. 19) and nitroolefins
(**5**) (ref. 20).

$$Me_3Si\diagdown\!\diagup\diagdown + E-N \xrightarrow{LA} \diagup\diagdown\!\diagup E + Me_3Si-N \qquad 4$$

$$E-N: R^1R^2CO,\ RCOCl,\ \diagup\!\!\bigwedge\!\!O\!\!\diagdown\!\!\diagup,\ R^1R^2C(OR)_2,$$

$$[R^1R^2C=N^+R^3R^4]X^-,\ RX,\ \diagdown\!\!\diagup\!\!\diagdown X(X=halides,\ OR,\ OAc),$$

$$R^1R^2CXY(X;Y=OR;\ halides,\ SR;\ halides,\ SR;\ OR)$$

$$\left[\begin{matrix}\equiv\!\!\!-\!\!\!\!\overset{+}{\diagup}\!\!\overset{R^1}{\underset{R^2}{\diagdown}}\\ Co_2(CO)_6\end{matrix}\right]X^-\ ,\quad \left[\!\!\begin{matrix}\bigcirc\!\!\!+\end{matrix}\!\!-Fe(CO)_3\right]X^-$$

$$Me_3Si\diagdown\!\diagup\diagdown + \diagup\!\!\diagdown\!\!\overset{}{\underset{EWG}{}} \xrightarrow{LA} \diagup\!\!\diagdown\!\!\diagup\!\!\overset{}{\underset{EWG}{}} \qquad 5$$

$$EWG: COR,\ NO_2$$

Normally, titanium tetrachloride is used as a Lewis acid.
Other Lewis acids such as $SnCl_4$, $AlCl_3$, BF_3 OEt_2, Me_3SiOTf,
Me_3SiI, and $Me_3O^+BF_4^-$ are also utilized frequently.   Although
the Lewis acid is utilized commonly in stoichiometric amounts,
catalytic amounts of Me_3SiOTf or Me_3SiI induce the reaction of
allylsilanes with acetals and related compounds (ref. 5, 21, 22).

The allylation under strong acidic condition can be avoided by use of the catalytic method.   Aldehydes and ketones do not react with allylsilanes under such mild conditions (ref. 23). The catalytic cycle is shown in **6**.   Coordination of Me_3Si^+ to the heteroatom presumably produces the silylonium salt, which reacts with allylsilanes to give the allylated compound together with Me_3SiY and Me_3SiX.   The regenerated Me_3SiX again can be used as the activator of acetals.

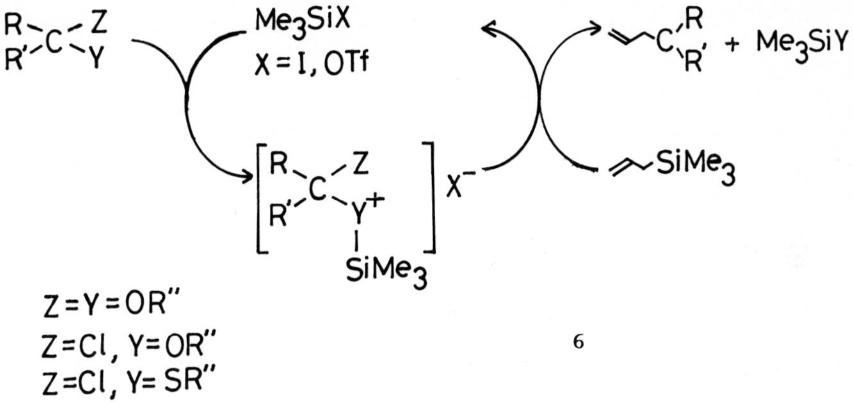

$$Z=Y=OR''$$
$$Z=Cl, Y=OR''$$
$$Z=Cl, Y= SR''$$

6

The regioselective γ-attack of allylic silanes (**1**, **2**) is synthetically very useful, since the synthesis of straight homo-allyl alcohols (**2**) with other allylic organometallic compounds is difficult.   Most of allylic organometals (M=Li, MgX, ZnX, CdX, BLn, AlLn, CrLn, TiLn, ZrLn.....) undergo rapid metallo-tropic shifts even at low temperature (**7**).   Therefore, it is difficult to prepare the unstable α-substituted isomer in pure form.   Of course, the metallotropic rearrangement can be controlled by proper choice of the ligand substituent, Ln.   For example, crotyldialkyl boron (L_2 = R_2) compounds rearrange

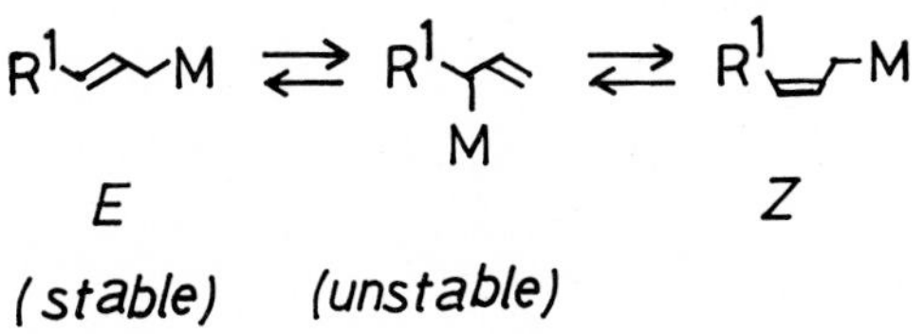

7

E   Z
(stable)   (unstable)

rapidly at room temperature, while crotylboronate ($L_2 = (OR)_2$) derivatives are sufficiently thermally stable. Allylic tri-methyl silanes are stable enough to isolate independently both regioisomers, γ- and α-substituted allylic derivatives. Consequently, either straight or branched homoallyl alcohols can be prepared selectively (**1, 2**). It is interesting that even crotyltrimethyltins undergo the 1,3-metallotropic shift and thus the preparation of α-methylallyltin becomes difficult (ref. 24a,b).

The reaction of α-alkyl or α-aryl substituted allylic silanes produced E olefins (**2**, R = Ph, Me) (ref. 3). Ethyl 3-methyl-2-trimethylsilyl-3-butenoate also gave the E olefin predominantly upon treatment with aldehydes and acetals (ref. 25a). In contrast, α-halogen substituted allylsilanes reacted with various electrophiles in the presence of Lewis acids to give the corresponding alkenyl halides with high Z preference (**2**, R = Cl, Br) (ref. 25b).

The reaction of allylsilanes with carbon electrophiles is also promoted by F⁻. The bond energy of Si-F is very high (~ 140 Kcal/mol) and further F⁻ possesses a strong affinity to Si atom. Use of Bu_4NF (TBAF) as a F⁻ source produced an allyl-anion species (ref. 26), which reacts with electrophiles to give the allylated compounds. The reaction proceeds with catalytic amounts of Bu_4NF (**8**).

Other fluoride ion sources, such as CsF and KF/18-crown-6, are
also used frequently.   In contrast to the Lewis acid mediated
reaction under strongly acidic conditions, the fluoride ion
promoted reaction can be carried out under weakly basic
conditions.   Unfortunately, the reaction is lacking in the
regioselectivity since the allylic carbanion is involved as an
intermediate.

Allylsilanes easily undergo the regioselective protonolysis
at the γ-position upon treatment with protic acids (9) (ref. 27).
On the other hand, treatment with CsF in wet dimethylsulfoxide
or dimethylformamide induces the regioselective protonolysis at
the α-position via an allylanion intermediate (ref. 28).

Therefore, the regioselective synthesis of olefins is possible
starting from the same allylic silane by merely choosing the
protonolysis condition.

Although it seems that the name, "Sakurai reaction", is given
to the Lewis acid mediated reaction of allylsilanes, we include
in this review the stereochemistry of the F⁻ promoted reaction
and of the protonolysis reaction.   Stereochemically interesting
reactions via allylsilanes, other than the above three categories, are also included.   Commonly used preparative methods of
allylic silanes are following; 1) reaction of allylmetals with
ClSiR$_3$, 2) reaction of silylanions (MSiR$_3$) with allylic substrates, 3) the hydrosilylation of 1,3-dienes, 4) reductive sily-
lation of unsaturated compounds, 5) Diels-Alder cycloaddition of
1-silylated-1,3-dienes, and 6) Wittig reaction of β-silylated
Wittig reagents.   For further reading, see the review articles
(ref. 5) and a recent paper (ref. 29).

2   INTERMOLECULAR REACTIONS

The stereoselectivity of allylic organometals is classified
into three categories; (i) P + P → 2C type reaction, that is,
erythro/threo or syn/anti problem; (ii) CP + A → 2C type reaction, namely, Cram/anti-Cram problem; (iii) CP + P → 3C type
reaction which creates three chiral centers in a single reaction

(ref. 30b).   In this classification, P means a prochiral center.
C means a chiral center, and A means an achiral center.

2.1   ALDEHYDES AND KETONES

2.1.1   P + P → 2C (Simple Diastereoselectivity)

The thermal reaction of crotyl organometallic compounds
(M = Li, Mg, B, Al, Sn, Ti, Zr, Cr, etc.) with aldehydes proceeds
through a six-membered cyclic transition state (ref. 24).   It
is thus widely accepted that the trans crotylmetal gives the
threo (anti) homoallyl alcohol, while the cis derivative produces
the erythro (syn) isomer (10).   However, in 1980, Yamamoto

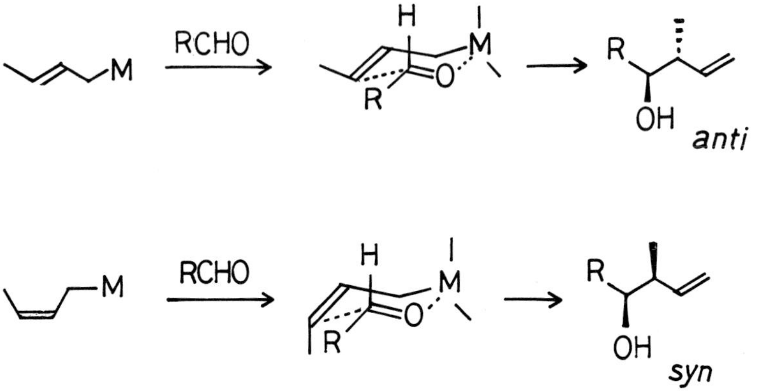

10

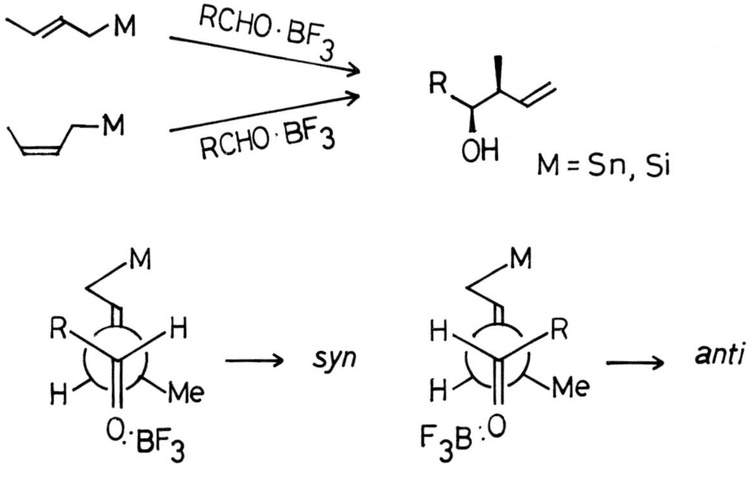

11

discovered that the Lewis acid mediated reaction of crotyltins with aldehydes produces the syn homoallyl alcohol regardless of the geometry of the crotyl unit (ref. 31). Boron trifluoride coordinates to the carbonyl oxygen, preventing the coordination of the metal to the oxygen atom. Accordingly, the acyclic transition state may be involved in the Lewis acid mediated reactions (11). It is easily determined that among several possible transition state geometries, the conformation leading to the syn isomer must be favored for steric reasons (ref. 30). The stereoconvergent syn selectivity in the presence of Lewis acids was also observed for a wide range of crotylmetals with relatively low Lewis acidity (ref. 32). Consequently, the Lewis acid coordination dramatically changes the reaction mechanism and the stereochemical outcome.

The Lewis acid mediated reaction of γ-substituted allylsilanes with aldehydes was examined in 1976 (ref. 3, 33), but they did not comment on the stereochemistry of the products. When we found the stereoconvergent syn selectivity of crotyltin-BF_3 reactions, we also investigated briefly the titanium tetrachloride mediated reaction of crotyltrimethylsilane with propanal (ref. 31). Here also, syn preference was observed; the homoallylalcohol was obtained in a ratio of 60 : 40. The syn selectivity was established unambiguously by using (E)- and (Z)-crotyltrimethylsilane and (E)- and (Z)-cinnamyltrimethylsilane (ref. 34). Both (E)- and (Z)-allylic silanes gave the syn-isomer either predominantly or exclusively (12). However, the

R^1 = Me, Ph R^2 = Me, Et, iPr, t-Bu

(E)-silanes produced higher selectivity than the corresponding (Z)-derivatives. This difference can not be explained by the anti-periplanar transition state model (11). It may be necessary to consider a modified transition state such as a synclinal model which is mentioned later.

Quite interestingly, (E)-cinnamyltributyltin produced the anti-homoallyl alcohol in the presence of Lewis acids (ref. 35), which was in a marked contrast with the syn-selectivity of the corresponding silicon derivative. Other allylic tins gave the syn-selectivity and only the cinnamyl derivative behaved differently. The difference is presumably owing to the strong covalent character of C-Si bond, compared with the C-Sn bond. The C-Sn bond of cinnamyltin may be easily polarizable, and thus a cyclic transition state might be involved in this particular compound.

The functional group substituted allylic silane exhibited different diastereoselectivity (ref. 36). The anti-homoallyl

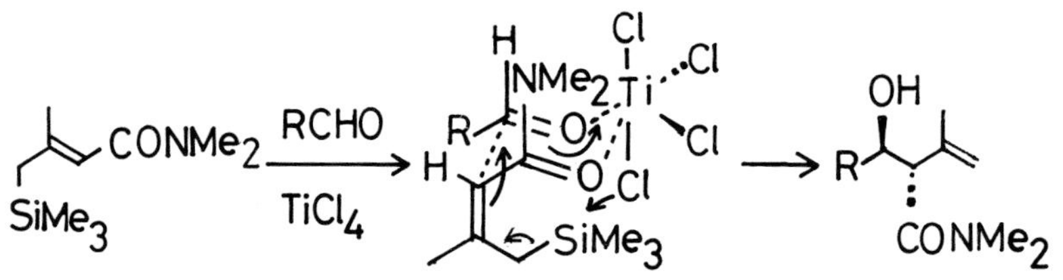

13

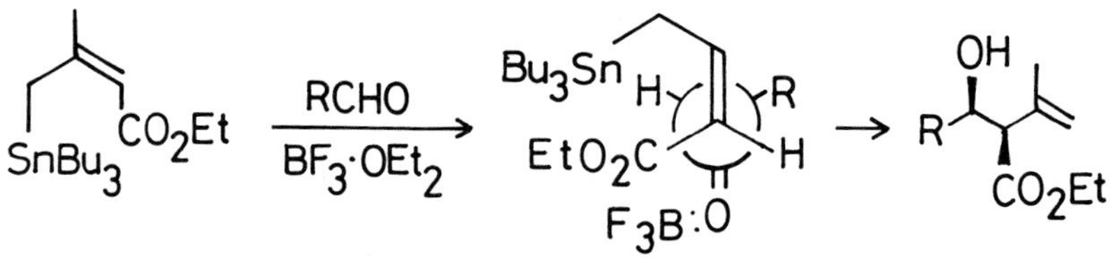

14

alcohol was produced either exclusively or predominantly. The strong affinity of titanium for oxygen and the intramolecular transfer of chloride to silicon are proposed to explain the anti-selectivity (13). The γ-silylated α,β-unsaturated amide

was prepared from the corresponding lithiated unsaturated amide.
Similarly, the γ-stannylated α,β-unsaturated ester was prepared
from the lithiated ester.   However, the tin derivative exhibited the syn-selectivity upon treatment with aldehydes in the
presence of BF_3 OEt_2 (**14**) (ref. 37).   The reaction may proceed
through the acyclic transition state, as proposed previously
(**11**).

2.1.2 CP + A → 2C (Diastereofacial Selectivity)

The reactions of ordinary chiral aldehydes having no ability
to be chelated were studied with several Lewis acids such as
$TiCl_4$, BF_3, and $SnCl_4$ (**15**) (ref. 38).   The diastereofacial
selectivity was in the range of 1.3 : 1 ~ 2.2 : 1.   As expected

$$Ph\underset{|}{\overset{}{\curlyvee}}CHO \;+\; \diagup\!\!\diagdown\!\!\diagup SiMe_3 \;\longrightarrow\; Ph\underset{\text{Cram (syn)}}{\overset{OH}{\curlyvee}}\!\!\diagdown\!\!\diagup \;+\; Ph\underset{\text{anti-Cram (anti)}}{\overset{OH}{\curlyvee}}\!\!\diagdown\!\!\diagup \qquad 15$$

from Cram and/or Felkin models, the syn-isomer was produced predominantly.   Use of $AlCl_3$ slightly enhanced the syn selectivity; 2.8 : 1 (ref. 39).

Very high diastereofacial selectivity was realized in the
Lewis acid mediated reactions with chiral alkoxy aldehydes
having a center of chirality at the α-carbon (Table 1).   Stannic
chloride (ref. 40, 41, 42) and titanium tetrachloride (ref. 42)
are excellent Lewis acids to produce the syn isomer (chelation
product).   Both are capable of forming five-membered chelate
16, and thus the reagent attacks from the less hindered side to
produce the chelation product.   In contrast, BF_3 is incapable
of chelating both oxygen atoms, causing reversal of diastereoselectivity (entries 2 and 5).   The addition of 2 equivalents
of gaseous BF_3 produced doubly complexed species **17** which reacted
with allylsilane under non-chelation control (ref. 43).   The
reaction proceeds through Cram and/or Felkin type transition
state, resulting in predominant formation of the anti-isomer
(non-chelation product).

Besides the simple α-alkoxyaldehydes, multi-functionalized
aldehydes derived from carbohydrates also exhibited very high

TABLE 1. 1,2- Asymmetric Induction

entry	X	R^1	Lewis acid	syn : anti	Yield (%)
1	OCH_2Ph	H	$SnCl_4$	35 : 1	94
2	OCH_2Ph	H	$BF_3 \ OEt_2$	1 : 1.5	50
3	OCH_2Ph	H	2eq BF_3 gas	1 : 4	85
4	OCH_2Ph	CH_3	$SnCl_4$	45 : 1	81
5	OCH_2Ph	CH_3	$BF_3 \ OEt_2$	1 : 2.6	40
6	CH_2OCH_2Ph	H	$SnCl_4$	12 : 1	92
7	CH_2OCH_2Ph	H	2eq BF_3 gas	1 : 2.3	85
8	CH_2OCH_2Ph	CH_3	$SnCl_4$	10 : 1	83

For entries 1, 2, 4, 5, 6, and 8, see ref. 40, 41, 42. For
entries 3 and 7, see ref. 43.

diastereofacial stereoselectivity. The reaction of allyltrimethylsilane with **18** or **19** in the presence of $BF_3 \ OEt_2$ produced
the non-chelation adduct with very high stereoselectivity (90
: 1 or 20 : 1, respectively), while the $TiCl_4$ mediated

reaction gave the chelation product with high selectivity (20
: 1 or 10 : 1, respectively) (ref. 44).   The $MgBr_2$ mediated
allylation of **20** produced the chelation adduct in a ratio of
> 98 : 2, clearly indicating that the α-chelation played a more
important role than the β-chelation (ref. 45).   Of course,
other bi-dentate Lewis acids such as $TiCl_4$, $SnCl_4$ and $ZnCl_2$,
were also effective in producing the chelation adduct.   2-O-
Benzyl-3-O-(t-butyldimethylsilyl)glyceraldehyde **21** gave the
syn (chelation) isomer with allylsilane-$TiCl_4$, while it produced
the anti (non-chelation) adduct with allylsilane-BF_3 OEt_2
(ref. 46) (For the transition state geometry, see **30**).

Chelation control is also feasible with the chiral ketone **22**
and acyl cyanide **23**.   The attack of allylsilane took place
exclusively from the equatorial side; the ratio of axial/equa-
torial alcohol was > 99/< 1 (ref. 47).   Allyl-MgCl or -Ti
$(NEt_2)_3$, which possesses low Lewis acidity, gave only 3 : 1
mixtures of the axial and equatorial alcohols.   With **23**, the
chelation product was obtained in the ratio of 96 : 4 upon treatment with allylsilane-$TiCl_4$ (ref. 48).

> 99   :   < 1

Chelation and non-chelation control in 1,3-asymmetric induction of **24** was studied with various Lewis acids.   With $SnCl_4$ and
$TiCl_4$, the chelation product **25** was obtained predominantly;
25a : **26a** = 9 ~19 : 1 (ref. 40, 41, 42, 49).   With BF_3 OEt_2,
24a gave again **25a** predominantly (ref. 42), in contrast to the
1,2-asymmetric induction of α-alkoxyaldehydes.   Nucleophilic
attack from the less hindered side in **28** simulates chelation
control (**27**).   The β-chiral acyl cyanide **29** also produced the
chelation adduct exclusively (ref. 48).

Consequently, excellent level of chelation control in 1,2-
and 1,3-asymmetric induction of α- and β-alkoxy-carbonyl com-

24a; R^1 = Me   chelation   nonchelation

b; R^1 = Bu   25a, b   26a, b

27   28   29

30

pounds is realized by proper choice of Lewis acids (ref. 50).
With BF_3, the 1,2-asymmetric induction proceeds in non-
chelation manner. However, non-chelation controlled 1,3-
asymmetric induction has not been realized yet.

Asymmetric synthesis of homoallyl ethers was achieved with
chiral dialkoxy-dichlorotitanium (ref. 51). The alkoxy group
presumably adds to aldehydes to produce the titanated hemiacetal
31, which undergoes substitution by the allyl group. The 1,4-

R' = CH(Me)Ph   **31**   90 % ds

32   33   Me Ph   34

asymmetric allylation with allylsilane-$TiCl_4$ was investigated in
1976 by using **32** (ref. 52).   Although the allylation of **32** did
not produce high e.e. (55%), **33** produced 89% d.s. by using
allylsilane-$SnCl_4$ (ref. 53).   Further, the allylation of **34**
with $TiCl_4$ produced > 98% e.e. (ref. 54).

Although P + P → 2C type and CP + A → 2C type stereoselectivities via allylic silanes have been extensively investigated
as mentioned above, CP + P → 3C type stereoselection is unknown.
Fortunately, this type of stereoselectivity has been studied in
detail with crotyltributyltin (ref. 30).   Crotyl- and related
γ-substituted allylic silanes must exhibit similar diastereoselectivity as the corresponding tin derivatives.

2.1.3.   Regioselectivity

As mentioned earlier, γ-substituted allylic silanes react
with electrophiles at the γ-position to produce branched homoallyl alcohols (**1**, **2**).   On the other hand, γ-trimethylsilylated
3-dialkylaminocrotonates **35** reacted with a number of electrophiles, such as aldehydes, ketones, acid chlorides and ortho-
esters, in the presence of $TiCl_4$ to give products with substitution at the α-position (ref. 55).   Compound **36** which possesses

both allylsilane and enamine functionality might react with
electrophiles at the β-carbon or at one of the terminal α- or
γ-carbon atoms.   It exhibited enamine-type reactions with
aldehydes and acetals in the presence of Lewis acids leading to
37 (ref. 56).   In contrast, an allylsilane type reaction occurs

upon treatment with ketones in the presence of F⁻. As mentioned in **8**, both α- and γ-regioisomers were obtained. Compound
38 did not react with electrophiles in the presence of Lewis
acids, but F⁻ activation promoted reaction with carbonyl compounds to give a mixture of both α- and γ-regioisomers.

2.2 α,β-UNSATURATED CARBONYL COMPOUNDS

The conjugate allylation of α,β-enones proceeded quite
smoothly with allylsilanes-$TiCl_4$ to give the corresponding δ,ε-
enones in good to high yields (ref. 19). Again, the carbon-
carbon bond formation took place at the γ-position of allylsilanes (**39**). Treatment of the enolate intermediate with

R = H or Me

carbon electrophiles, instead of proton, gave the double alkylated products (ref. 57).

Diastereofacial selectivity, CP + A → 2C type, was investigated in the conjugate addition. The addition to **40** gave the
anti-isomer preferentially; the ratio of anti/syn was 4/1 (ref.
38). In analogy with the addition to the aldehyde **15**, the
stereoselectivity can be accounted for by the Cram or Felkin
model. 1,2-Asymmetric conjugate addition of γ-alkoxy-α,β-enone
41 produced the anti isomer predominantly in the ratio of 7 : 1.
Trajectory of attack of allylsilane **42** was proposed to explain
the anti selectivity. However, it seems to us that the attack
shown in **42** is hindered by the methyl group and this model is
opposite to the so-called extended Felkin model in which H is in
the position of Me. An another model, where OCH_2Ph is in the
position of H and H is in the position of Me, may be applicable
to γ-alkoxy-α,β-unsaturated carbonyl compounds (ref. 58). In
contrast to **41**, cis isomer **43** gave the syn adduct preferentially
(10 : 1) in which the chelate intermediate **44** would play an

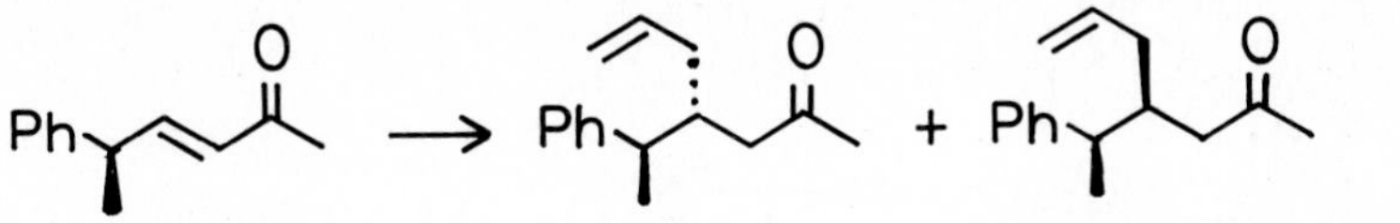

40 → anti(Cram) + syn(anti-Cram)

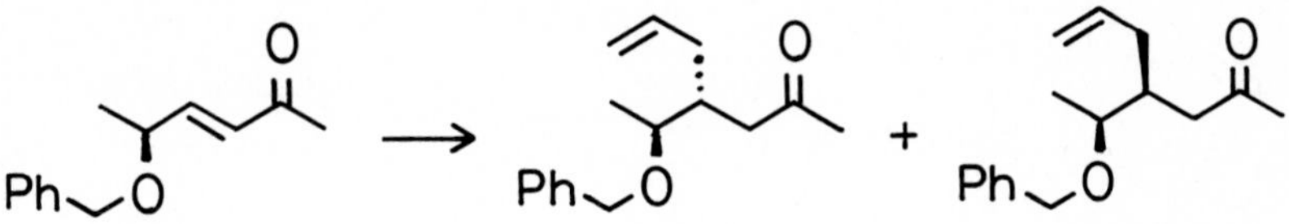

41 → anti + syn

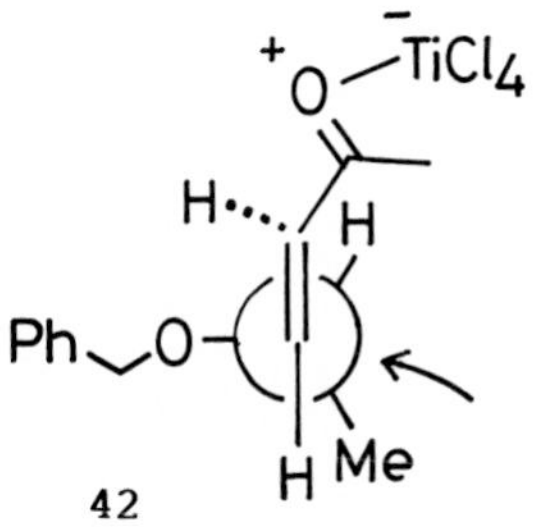

42 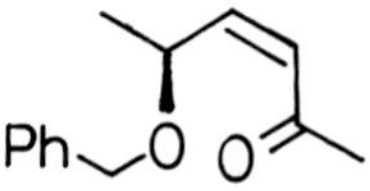   43 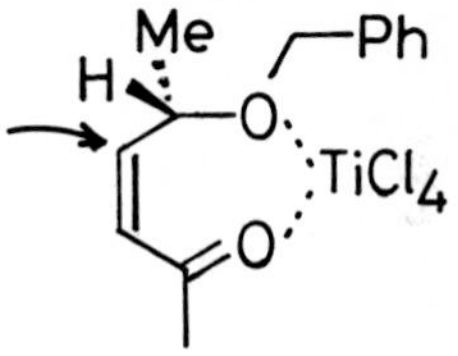   44

important role in the diastereoselection.

Conjugate addition to the bicyclic enone **45** gave **46** as a
single product (ref. 19).   4-Methylcyclohexanone **47** produced
the cis-isomer **50** predominantly; trans : cis = 32 : 68 (ref. 59).
On the other hand, the 5-methyl derivative **48** gave the trans
isomer **52** predominantly; trans : cis = 98 : 2.   The reaction
may proceed through the more stable reactant conformer and the
incoming nucleophile may attack from the antiparallel direction
to the axial hydrogen on the C-4.   Therefore, predominant for-

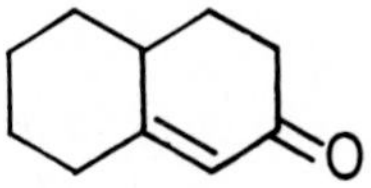

45 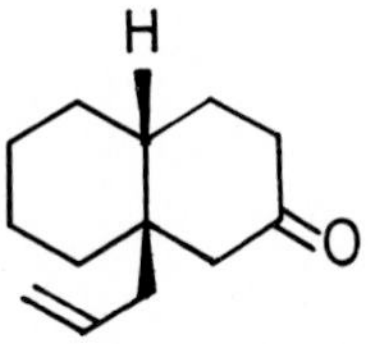   46 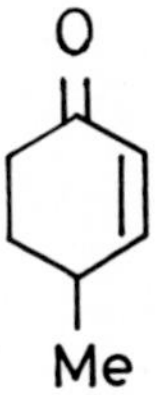   47 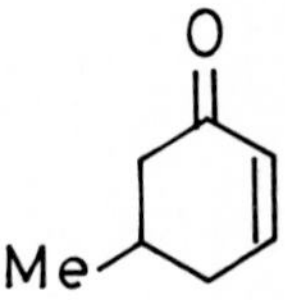   48

49   50   51   52

mation of **50** can be accounted for by **49** and that of **52** can be
explained by **51**.   The conjugate addition to cycloheptenone
series exhibited essentially the same diastereoselectivity as
cyclohexenone derivatives.   Total synthesis of (±)-fawcettimine
was accomplished with the aid of Sakurai reaction (ref. 60).
   Simple diastereoselectivity, P + P → 2C type, was studied in
the conjugate addition of **53** to α,β-unsaturated acyl cyanide **54**
(ref. 61).   The syn isomer **56** was predominantly produced via
55.   An another transition state **57**, which give the anti adduct,
is destabilized in comparison with **55** owing to the steric repulsion between Me and CH_2SiMe_3 groups.   It should be noted that
the conjugate addition of crotylmetals, such as crotyl-MgCl,
-9-BBN, -Ti(OiPr)$_3$ and -ZrCp$_2$Cl, to diethyl ethylidenemalonate

53   54   syn

                                                            56

57   anti   59
                      58

produces the anti adduct **58** with high stereoselectivity (ref. 62). The addition of crotyltin in the presence of TiCl$_4$ also produced **58** predominantly. In contrast to **55**, the acyclic transition state **59** is proposed to explain the anti selectivity.

α,β-Unsaturated acylsilanes underwent a facile conjugate addition of allylsilanes in the presence of TiCl$_4$ (ref. 63). The reactivity of acylsilanes exceeded that of analogous α,β-unsaturated ketones. The stereoselectivity was not investigated in this addition. The reaction of allenylsilanes with α,β-unsaturated acylsilanes in the presence of TiCl$_4$ can be directed to produce either five or six-membered carbocyclic compounds (ref. 64). The acylsilanes with R = H gave [3 + 2] annulation product **60**. The annulation products derived from

2-alkyl substituted acylsilanes underwent a novel rearrangement to **61** upon treatment with TiCl$_4$. This rearrangement could be prevented by using t-butyldimethylsilyl acylsilanes.

Allylation of 2-alkanoyl-1,4-quinones with allyltrimethylsilane in the presence of BF$_3$ OEt$_2$ gave the corresponding allylhydroquinone in 40% yield along with several other products (ref. 65). It seems that the conjugate addition to quinones proceeds more smoothly with allyltins than with allylsilanes.

Michael additions of allyl- and methallyl-trimethylsilane to α-nitro olefins in the presence of $AlCl_3$ proceeded smoothly at low temperature to give the corresponding nitronic acids, which were transformed into the γ,δ-enones with the aid of aqueous Ti(III) (ref. 20). The diastereoselectivity was not investigated in this addition reaction. Transformation of the γ,δ-enones into the 1,4-diketones via the palladium catalyzed oxydation (O_2, $PdCl_2$, CuCl), followed by the intramolecular aldol condensation provided cyclopentenone derivatives.

Chemoselectivity in the conjugate addition of allylsilane to various Michael acceptors was investigated (ref. 66). Although the Lewis acid mediated conjugate addition was highly efficient for allylation of enones (ref. 19), nitro olefins (ref. 20) and acysilanes (ref. 63), the process failed for enoates and α,β-unsaturated nitriles. However, doubly activated Michael acceptors such as ethylidenemalonate derivatives readily underwent allylation. The allylic carbanion species generated with the aid of F^- underwent highly chemoselective addition to enoates and α,β-unsaturated nitriles. Unfortunately, the F^- induced addition to enones was accompanied frequently by the 1,2-addition (62) (ref. 26). The F^- mediated addition to enoates and nitriles did not produce such 1,2-adducts.

	96 %	0 %
$TiCl_4$	96 %	0 %
Bu_4NF	25 %	50 %

2.3 ACETALS, 1,3-DIOXOLAN-4-ONES, THIOACETALS, AND RELATED COMPOUNDS.

The reaction of allylsilanes with acetals in the presence of $TiCl_4$ afforded homoallyl ethers in good yields (ref. 10). The chemoselectivity of ketoacetals was investigated (ref. 67). The reaction of 1,1-dimethoxybutan-3-one proceeded selectively on the acetal moiety regardless of the kind of Lewis Acid, while

the chemoselectivity of the corresponding reaction with 1,1-
dimethoxypropan-2-one was changed dramatically by the type of
Lewis acid employed (63).   Even when two equivalents of allyl-

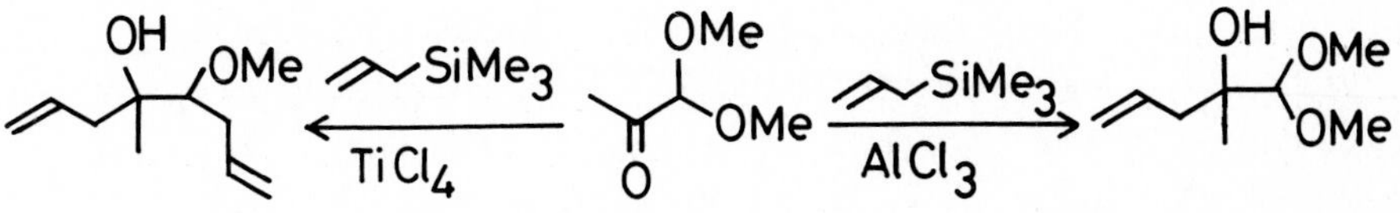

63

silane-AlCl$_3$ were used, the mono-allylated homoallyl alcohol was
obtained selectively.   As mentioned earlier, catalytic amounts
of Me$_3$SiOTf, Me$_3$SiI and Ph$_3$CClO$_4$ (ref. 21, 22) as well as
stoichiometric amounts of Lewis acids are normally used as a
homogeneous promoter.   The allylation of acetals, aldehydes and
ketones could be catalyzed by solid acids in a heterogeneous
system (ref. 68).   Clay montmorillonite such as Al-Mont, H-Mont
and Mont-K10 were effective for the allylation.

   Regioselective cleavage of unsymmetrical acetals was investigated in the reaction of 2-methoxyethyl hemiacetals 64 with
allylsilanes-TiCl$_4$ (ref. 69).   The selective elimination of the
2-methoxyethoxy (ME) group was achieved presumably owing to the
selective coordination of TiCl$_4$ to the ME function.   The reaction will serve as a new synthetic method for carbon homologative
etherification of alcohols.

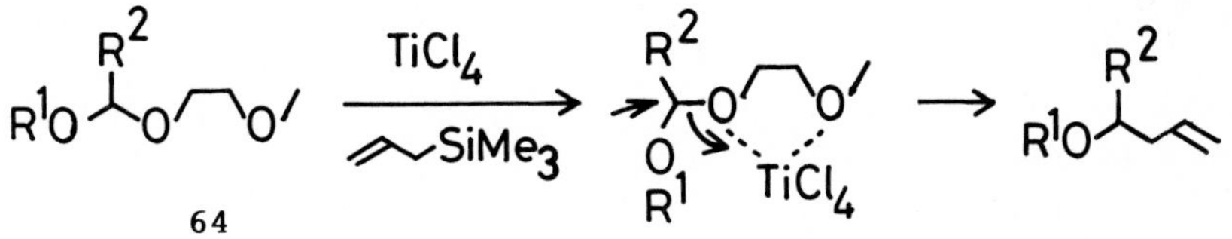

64

   As mentioned previously, γ- and α-substituted allylic
silanes also reacted smoothly with acetals (ref. 69, 25).
Simple diastereoselectivity, P + P → 2C type, was studied with
crotylsilane (ref. 70).   Aliphatic acetals exhibited syn-selectivity regardless of the geometry of crotylsilanes, as expected
from the acyclic transition state 11.   In contrast, the dia-

stereoselectivity of aromatic acetals was controlled not only by
the geometry of crotylsilanes but also by the substituent on the
aromatic nucleus.   Benzaldehyde dimethylacetal reacted with (Z)-
crotylsilane with high anti-preference, though the syn-selectivity was observed with (E)-isomer.   Increasing anti-selectivity
was observed in the case of (Z)-isomer with increasing electron
withdrawing character of the substituent in the order of p-CN >
p-Me >p-MeO.   On the other hand, syn-selectivity increased with
electron withdrawing substituent in the case of (E)-isomer.

65   *anti*   67

X = CN, Me, MeO

66   *syn*   68

In contrast to aliphatic acetates, aromatic acetates will generate benzylic cation species and thus the p-CN group may produce
the partial double bond nature between C and O (65 and 68).
However, such double bond nature may disappear or decrease in
extent by the electron donating p-MeO group, resulting in the
formation of transition states 66 and 67.   With p-CN, 65 is
more stable than 66 and 68 is more stable than 67.   With p-MeO,
65 is destabilized in comparison with 66 and 68 is much more
destabilized in comparison with 67.

    Syn-selectivity was observed in the Lewis acid mediated reac-
tions of allylsilane with β-alkoxy acetals (69) (ref. 71).   The
selectivity was not so high (2.8 : 1), but a striking difference
in the sense of the selectivity in comparison with that of

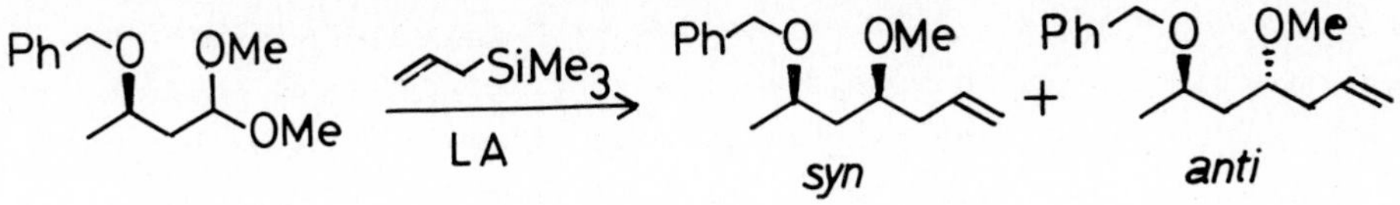

Ph–O OMe → (allyl)SiMe₃ / LA → Ph–O OMe (syn) + Ph–O OMe (anti)

69

	syn : anti
TiCl₄	1.55 : 1
TiCl₄–Ti(OiPr)₄	2.80 : 1

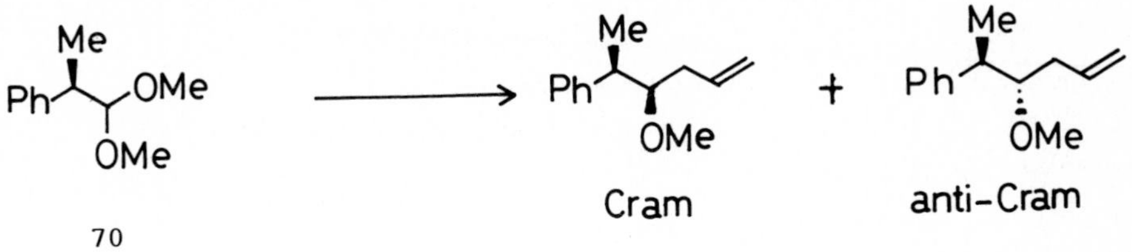

Me, Ph, OMe, OMe → Me, Ph, OMe (Cram) + Me, Ph, OMe (anti-Cram)

70

β-alkoxyaldehydes **24** was observed. Chelation product (anti isomer) was produced either predominantly or exclusively via **24**. Therefore, the selectivity of **69** is dictated by non-chelation control. Presumably, coordination of $TiCl_4$ to oxygen of the oxonium ion $>C=\overset{+}{O}-Me$ is blocked by the Me group. The Cram isomer (syn) was obtained predominantly (3.5 : 1) in the reaction of **70** with $SnCl_4$.

Very high asymmetric induction was realized via the chiral acetal template **71** (ref. 72). The Lewis acid catalyzed reaction

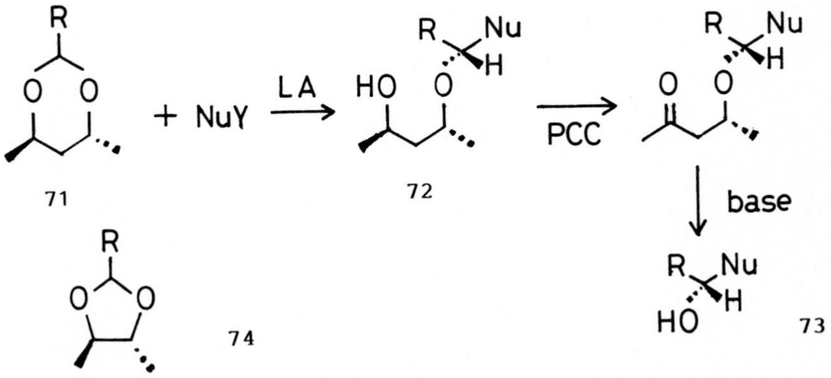

71 + NuY → (LA) → **72** → (PCC) → ... → (base) → **73**

74

of **71** derived from R,R-2,4-pentanediol with various nucleophiles proceeded highly diastereoselectively to give adduct **72** in high yields.   Removal of the chiral auxiliary from **72** via the ketone produced **73** in high yield and normally in e.e. of 90% or over. The selectivity depended upon (a) the nature of the substituent R of **71**, (b) the structure of the nucleophile NuY, (c) the type of Lewis acids, and (d) the reaction conditions.   Among the nucleophiles examined, allylsilanes proved to be the most susceptible to variations in the selectivity of the addition reaction.   Normally, $TiCl_4$ is used in the allylation reaction.   In some cases $TiCl_4$-$Ti(OiPr)_4$ provided higher selectivity than $TiCl_4$ itself (ref. 72b).   The five membered acetal **74** was also effective for the asymmetric allylation via allylsilane-$SnCl_4$ (ref. 73).

An S_N2-like transition state is proposed to explain both high asymmetric induction and direction of chiral induction (ref. 72c).   The transition state A is stabilized by a lengthening of the 2,3 bond (process a) of the ground state B with consequential relief of the relatively large 2,4-diaxial H/Me interaction.   Such interaction is not removed in the process b involving 1,2 bond lengthening in the transition state C. Therefore, the reaction proceeds through path a.

A   B   C

The acetal **75** was treated with methallyltrimethylsilane in the presence of $6TiCl_4$ $5Ti(OiPr)_4$ (ref. 74).   The product **76** was obtained in 98.5 : 1.5 (23S : 23R) diastereomeric ratio. This compound was further converted to the key intermediate for producing calcitriol lactone, a major metabolite of vitamin D_3. It should be noted that the chiral center at C-21 may exert an influence upon the chiral induction at C-23, since both the acetal template and chiral induction due to C-21 are a sort of 1,3-

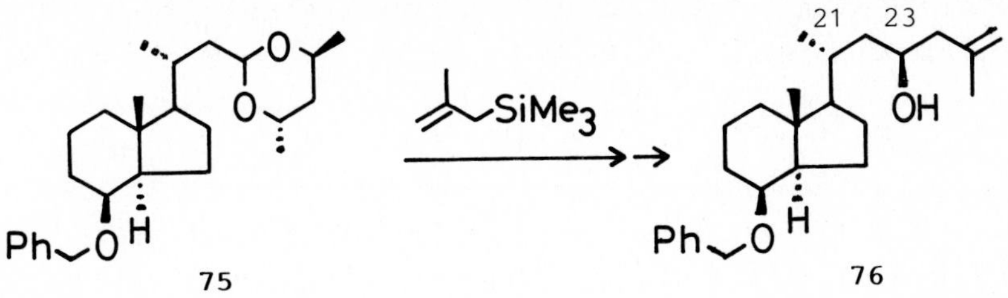

75 → 76

asymmetric induction.

The synergistic effect in chiral induction was investigated in steroidal acetals (ref. 75).   Treatment of the chiral steroidal acetal **77** (S-R,R isomer), prepared from **79** and (2R,4R)-(-)-pentanediol, with allylsilane in the presence of TiCl$_4$ followed by the usual workup gave **80** exclusively.   On the other

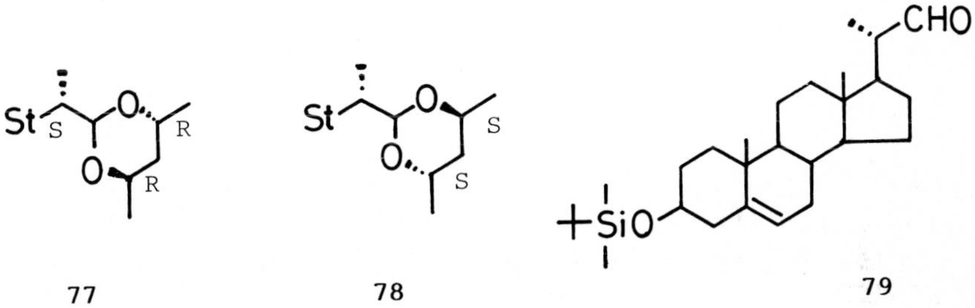

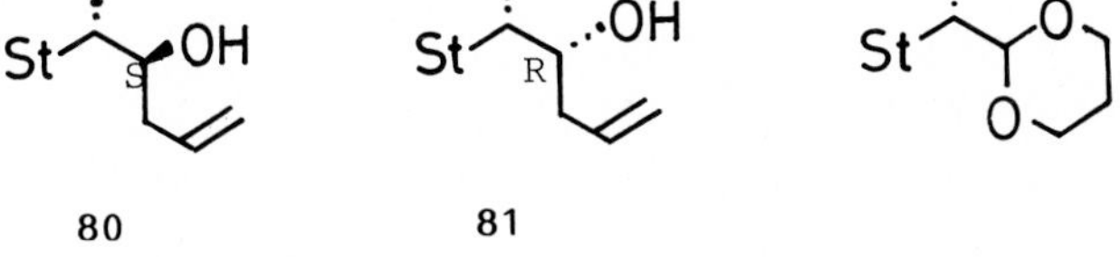

77   78   79

80   81   82

hand, the reaction of the S-S,S isomer **78**, prepared from **79** and (2S,4S)-(+)-pentanediol, with allylsilane produced a mixture of **80** and **81** in a ratio of 90 : 10, respectively.   This result was unexpected, since the SS acetal normally induces R chirality at the carbon bearing both oxygen atoms. Quite interestingly, the reaction of **78** with allyltributyltin in the presence of BF_3 OEt_2 gave **81** predominantly. With allyltin, the direction of asymmetric induction is dictated primarily by the acetal template and a violation of Cram's rule takes place in **78**. In contrast, the direction is dictated essentially by Cram rule in the reaction of **78** with allylsilane. In fact, **82** produced **80** preferentially upon treatment with allylsilane-$TiCl_4$ (see also **70**). Mechanistically, the above results clearly indicate an importance of the timing of bond cleavage and bond formation of acetal template. Allylsilane, having low nucleophilicity, presumably reacts after the bond cleavage process and thus the chiral induction is dictated primarily by Cram's rule. In contrast, allyltributyltin possesses higher nucleophilicity than allylsilane and therefore reacts simultaneously as the bond cleavage takes place.

The 1,3-dioxolan-4-ones **83** prepared by condensation of (S)-(+)- or (R)-(-)-mandelic acid with aldehydes and ketones reacted with allylsilane in the presence of BF_3 OEt_2 or $ZnBr_2$ to produce the allylated compound **84** (ref. 76). The chiral auxi-

liary can be removed with $Pb(OAc)_4$ followed by acid hydrolysis to furnish without racemization the homoallyl alcohols. The

diastereoisomer ratio of **84** was nearly independent of the starting geometry of **83**; cis-**83a** produced the ratio of 72 : 28, and
trans-**83b** gave that of 75 : 25.   Presumably, the bond breaking
of the dioxolan ring takes place before the attack of allylsilane, resulting in relatively low diastereoselectivity.   The
stereochemical results are consistent with nucleophilic attack
from the less hindered side on the less hindered oxonium ion **85a**.
Instead of mandelic acid, 3-hydroxybutanoic acid was used to prepare 1,3-dioxolan-4-one derivatives (ref. 77).   Both (R)- and
(S)-3-hydroxybutanoic acids are easily available from inexpensive
sources, and the dioxolanes may be prepared in good yields from
aliphatic aldehydes.   Further, removal of the chiral auxiliary
after nucleophilic ring opening of the acetal requires simple
β-elimination from a 3-alkoxy acid.   The diastereoselectivity
is sufficiently high in the allylation reaction; 87 ~ 97% ds.

Cyclic hemiacetals (γ-lactols) reacted with allylsilane in
the presence of $BF_3 OEt_2$ to produce the allylated tetrahydrofuran derivatives in good yields with high diastereoselectivity
(**86**) (ref. 78).   Since γ-lactols are readily available and the

allyl group is easily converted to other functional groups, the
method allows access to a variety of tetrahydrofuran derivatives.

The S-methyl monothioacetals (**87a**) reacted with allylic silanes in the presence of $SnCl_4$ to give predominantly the homoallyl
ethers with generally regioselective cleavage of the C-S bond
(ref. 79).   This selective cleavage was interpreted to be due to
the strong affinity of the tin atom for the sulfur atom in a
soft acid - soft base interaction.   In contrast, use of $TiCl_4$
induced selective C-O bond cleavage.   The S-phenyl thioacetals
87b gave the homoallyl sulfides in the presence of $SnCl_4$, indicating that C-O bond fission took place owing to the stabilization of the thionium cation by the phenyl group.   Treatment of
cyclic monothioacetals with $SnCl_4$ gave cyclic ethers in good

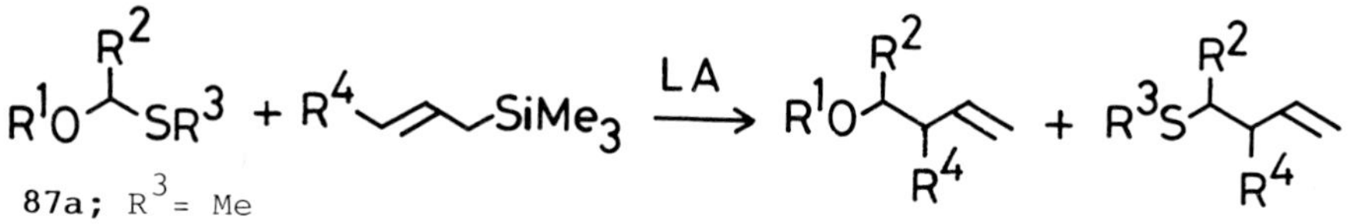

87a; R^3 = Me

b; R^3 = Ph

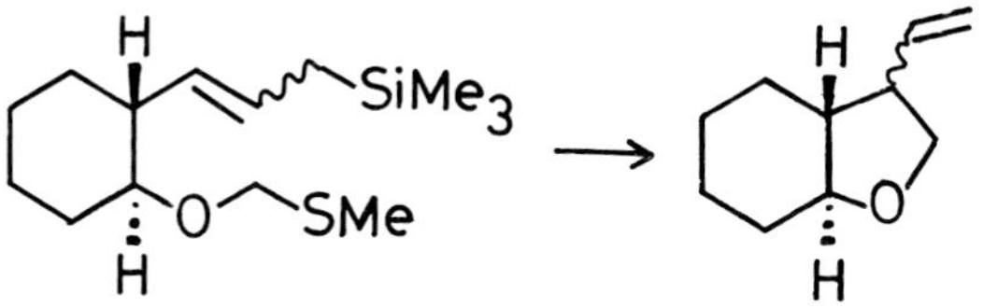

88

yields (88).

Treatment of allylsilane with triethyl orthoformate-$TiCl_4$ gave only bishomoallyl ether and no homoallyl acetal (ref. 10). Presumably, the reaction of allylsilane with the resulting acetal is faster than the monoallylation of the orthoformate. Thioacetals are generally more stable to Lewis acids than the corresponding O-analog.   Treatment of 89 with allylsilane

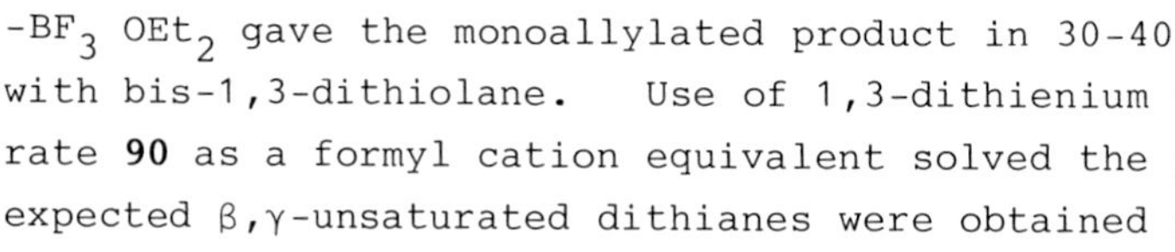

90

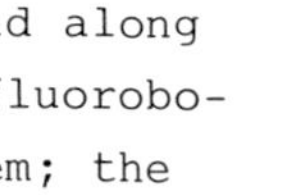

89

$-BF_3$ OEt_2 gave the monoallylated product in 30-40% yield along with bis-1,3-dithiolane.   Use of 1,3-dithienium tetrafluoroborate 90 as a formyl cation equivalent solved the problem; the expected β,γ-unsaturated dithianes were obtained in good yields (ref. 80).   Lewis acid assisted condensations between a 5-methoxyisoxazolidine and silicon-based nucleophiles were also investigated (ref. 81).

2.4 CARBOHYDRATES

C-Nucleosides have become interesting synthetic targets since many of naturally occurring C-nucleosides exhibit a broad spect-

rum of biological activity.   The anomeric centers in tetrasub-
stituted pentose sugars are acetals, and thus it was expected
that suitably substituted sugars might undergo the allylation
via allylsilane-Lewis acids.   Treatment of **91** with allylsilane
(1.5 eq) in the presence of 1.4 eq Me_3SiOTf gave **92** in 83%
yield (ref. 82).   The ratio of **92a** : **92b** was 10 : 1, indicating

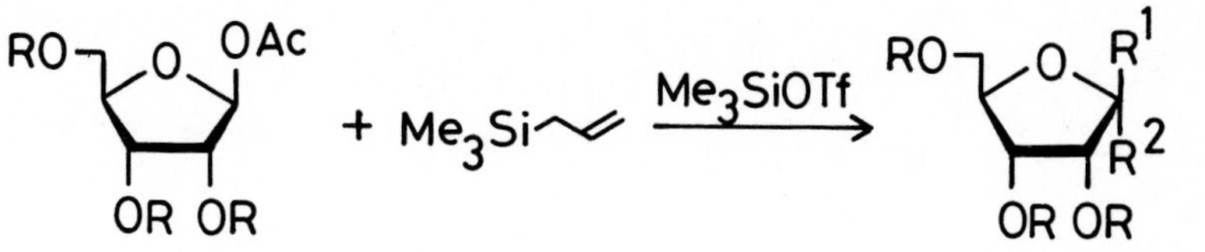

91   R = COPh   **92a**; R^1=H, R^2=allyl

b; R^1=allyl, R^2=H

that the reagent attacked predominantly from the α face super-
ficially via an S_N2 process.   The allylation of **91** in the presence of 2.5 eq $ZnBr_2$ gave a 4 : 1 mixture of **92a** and **92b** in
85% combined yield (ref. 83), while the allylation in the presence of 1.1 eq $BF_3 OEt_2$ in acetonitrile produced a 7 : 1 mixture in 93% combined yield (ref. 84).

The diastereoselectivity is dependent upon both the nature
of the Lewis acids and the nature of the leaving group used to
activate the carbohydrate.   The protected lyxose derivative
93a, having p-nitrobenzoate as the leaving group, gave exclusively the desired α-product **94** under the $BF_3 OEt_2$-CH_3CN condition
(ref. 84).   If the acetate derivative **93b** was used, a 4 : 1

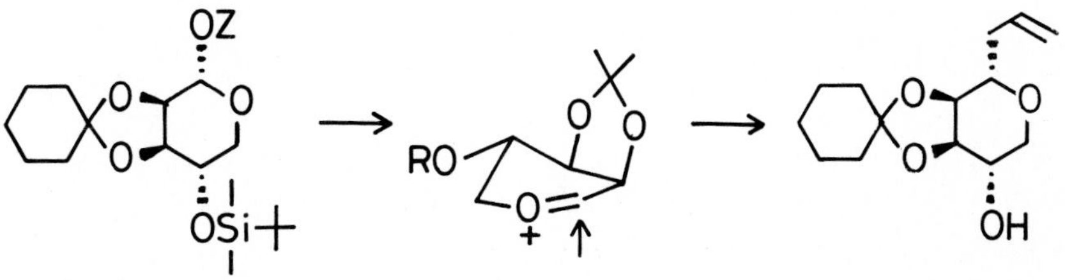

93a; Z = PNB   **94**

b; Z = Ac

mixture of the α and β products was obtained in the presence of
$ZnBr_2$.   This same acetate in acetonitrile with $BF_3 OEt_2$ afford-

ed a 19 : 1 mixture of the α and β products.   The better leaving ability of the PNB group must produce an open oxonium ion,
which subsequently undergoes the stereoelectronically and sterically favored axial attack.   When **93a** was treated with ZnBr$_2$,
a 5 : 1 mixture of the α- and β-isomers was obtained.   The zinc
bromide method may involve substitution via both S$_N$1 and S$_N$2
processes (ref. 84).

When the tetrabenzoate derivative of L-lyxose **95** was reacted
with allylsilane in BF$_3$ OEt$_2$ - acetonitrile, a 1 : 5 mixture of
the α and β products was obtained (ref. 84).   This was quite

| **95** | α product | β product |

unexpected in comparison with **93**.   The β-preference was also
observed in the ribose derivative **91**.   Consequently, the stereoselectivity (inversion or retention) is also dependent upon the
structure of carbohydrates.

The reaction of 2,3,4,6-tetrabenzylglucopyranose (**96a**)
with allylsilane and BF$_3$ OEt$_2$ in acetonitrile gave a 10 : 1
mixture of the α and β allylglucopyranoside in 55% combined yield
(ref. 85).   The chemical yield was improved by activating the
C-1 position; **96b** gave a 10 : 1 mixture of α and β products in
80% combined yield.   Treatment of **96b** with allylmagnesium

| β product | **96a;** Z = H | α product |
| | **b;** Z = PNB | |

bromide gave the corresponding hemiketal, which was then reduced

with triethylsilane and boron trifluoride etherate to afford the
β product in 85% overall yield.   Here again, the Lewis acid
promoted allylation proceeds through the pyran oxonium ion,
which preferentially accepts the nucleophile from the α (axial)
side due to the anomeric effect from the ring oxygen.   The
compound **97** also produced a 10 : 1 mixture of the α and β

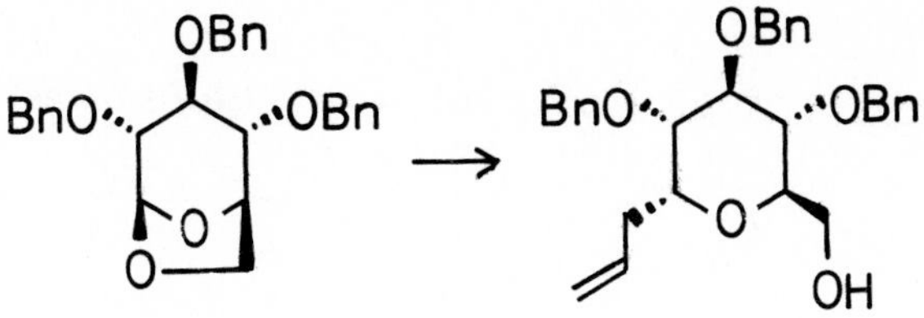

97

isomers.

Methyl α-D-glucopyranoside, methyl α-D-mannopyranoside and
α-D-glycopyranosyl chloride (**98**) readily underwent the allylation with allylsilanes in the presence of catalytic amounts of
trimethylsilyl triflate or iodotrimethylsilane (ref. 86).   As
expected, the α anomer was produced either exclusively or predominantly.   Various allylic functional groups, such as allyl,
methallyl, 2-bromoallyl and 3-butenyl, could be introduced into
the pyranoside structure.   The simple diastereoselectivity in
the reaction of crotylsilane was not clarified.

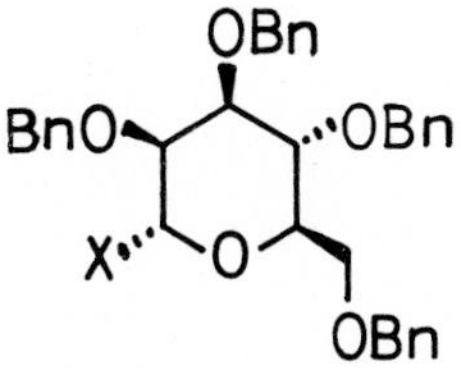

98a; X = OMe   **99**   **100**

b; X = Cl

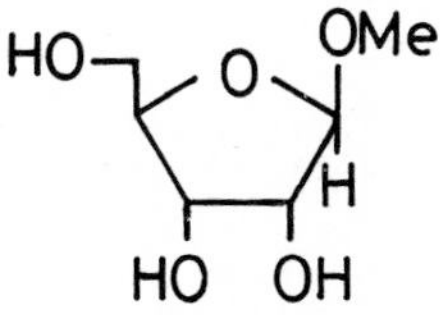

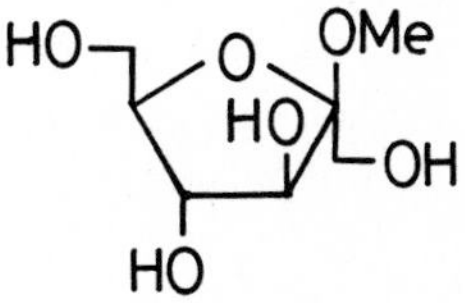

Compounds **99** and **100** also underwent the allylation with allyl-
silane-TMSOTf after treatment with bis(trimethylsilyl)trifluoro-
acetamide to protect the OH groups (ref. 87).

5,6-Dihydro-γ-pyrones **101** reacted with allylsilane in the
presence of TiCl$_4$ to give the C-allyl glycosides **102** in good to

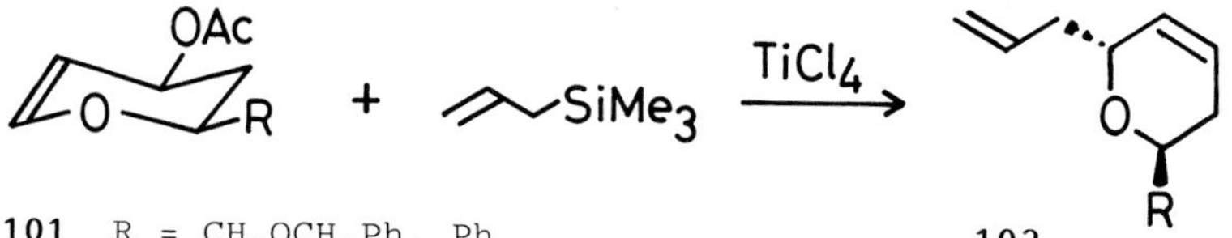

101   R = CH$_2$OCH$_2$Ph, Ph
          CH$_2$SPh, Pr

102

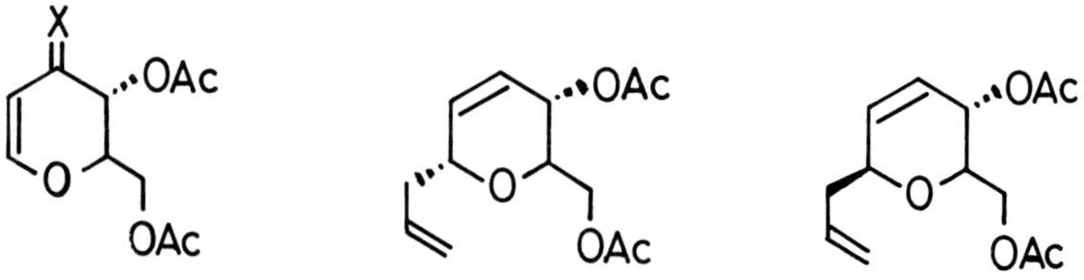

103a; X = β-OAc, α-H

   b; X = α-OAc, β-H

high yields (ref. 88).   The only regioisomer observed was the one arising from entry of the nucleophile at C$_1$, with transposition of the double bond to C$_2$-C$_3$.   This process was applied to glycal derivatives obtained from natural hexoses.   The reaction of D-glucal triacetate **103a** gave an 85% yield of a 16 : 1 mixture of **104a/104b**.   When the same reaction was carried out with D-allal triacetate **103b**, the same products were produced in 95% yield but in a 6 : 1 ratio of **104a/104b**.   No indication for direct displacement (S$_N$2 type) was evident as before.   Therefore, in all cases the reactions proceed via axial attack by the allyl nucleophile, and the most axial allylation results in an anti S$_N$2' displacement (**101**, **103a**).   It should be noted that the allylations of **101** and **103** are mechanistically different from those of other carbohydrate derivatives (**91**, **93**, **95**, **96**, **97**); the former involves S$_N$2' type substitution and the latter proceeds through an oxonium ion intermediate or through S$_N$2 substitution in some cases.

2.5   ETHERS, HALIDES, EPOXIDES, OXETANES, AND ACID HALIDES

Allylic ethers and halides regioselectively reacted with ethyl 3-methyl-2-trimethylsilyl-3-butenoate without allylic rearrangement (105) (ref. 12).   α-Phenylethyl ethers and α-phenyl-

105

ethyl bromide also underwent the allylation under similar conditions.   Acid halides, acetals and chloromethylphenylsulfide as
well as aldehydes and ketones were also reacted with the same
allylic silane to produce the corresponding α,β-unsaturated
esters with mostly E-preference (ref. 25a)(See also ref. 25b on
the olefin geometry).

Allylic couplings between allyl- or prenyltrimethylsilane and
allylic substrates such as halides, ethers or acetates were
investigated in detail (ref. 89).   In contrast with **105**, the
coupling took place both at the α-position and at the γ-position of allylic substrates.   Needless to say, allylic silanes
reacted at the γ-position.   Primary allylic substrates and
benzyl halides, gave the coupling product in very low yields,
though the coupling of secondary allylic substrates was efficient.   Although the α (S$_N$2 reaction) or γ substitution (S$_N$2'
reaction) with respect to the allylic compounds can not be
sufficiently controlled, the regioselectivity may be reasonably
explained by the steric effect around the reaction sites where
positive charge develops to a small extent by the complexation
of a leaving group X with TiCl$_4$.   The γ-position of **106**,

106   **107**

though a less electron deficient center, is more reactive than
the α-position since the secondary α site may be more hindered
than the γ-position.   On the other hand, the α-position of **107**
is more reactive because of both electronic and steric reasons.

The coupling between **108** and the secondary benzylic substrates
109 took place in the presence of $TiCl_4$ to give **110** in good
yields (ref. 90).   The diastereomer ratios of the products were
1 : 1 in **110a**, 3 : 1 in **110b**, and 7 : 1 in **110c**.   The syn
isomers ($3R^*$, $4R^*$) were produced predominantly in **110b** and **110c**.
The anti isomer ($3R^*$, $4S^*$) is a precursor of hexestrol derivatives, affinity labeling and tumor imaging agents.   The benzylic
hydroxy group of the chromium complex could be substituted with
allyl group, because the benzylic cation was extremely stabilized
by $Cr(CO)_3$ complexation (ref. 91).   The reaction proceeded
stereoselectively to creat the quarternary chiral center.
Adamantyl chloride reacted with allylsilane under the influence
of $AlCl_3$ to yield 1-allyladamantane derivatives (ref. 9, 92).

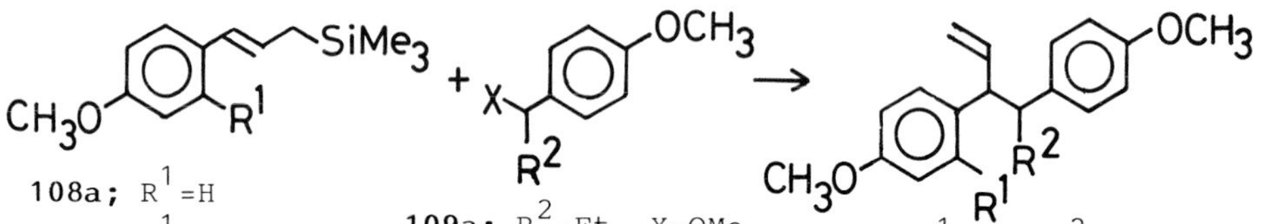

108a; R^1 =H

b; R^1 =NO$_2$

109a; R^2 =Et, X=OMe

b; R^2 =Et, X=Cl

c; R^2 =Me, X=OMe

110a; R^1 = H, R^2 =Et

b; R^1 = NO$_2$, R^2 =Me

c; R^1 = NO$_2$, R^2 =Et

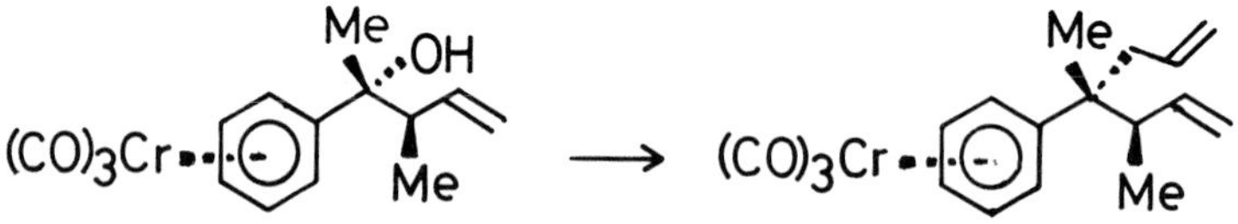

γ-Alkenyl-γ-butyrolactones reacted regio- and stereoselectively with allylsilanes in the presence of trimethyloxonium tetrafluoroborate to afford methyl (E)-4,8-alkadienoates in high
yields (ref. 93).   Synthetic utility of the substitution reac-

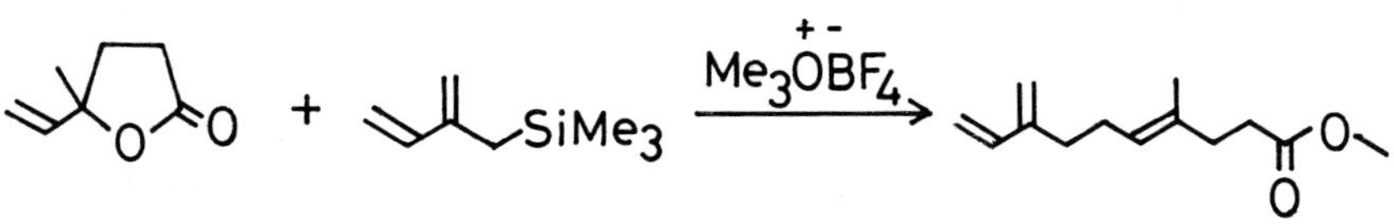

tion was demonstrated by the synthesis of β-sinensal.

Cyclic β-bromo ethers could be allylated with allyltrimethylsilanes in the presence of silver tetrafluoroborate via the
corresponding carbocations (ref. 94).   The initially formed
cation may undergo 1,2-hydrogen shift to give the more stable
carboxonium ions.   The former must afford the direct substitution product, while the latter should produce the branched
isomer.   2-(Bromomethyl)tetrahydrofuranes (n=3) gave usually

111a; R^1=H, R^2=CH$_2$Br   112   113
  b; R^1=CH$_2$Br, R^1=H

both isomers, but tetrahydropyranyl derivatives (n=4) produced
exclusively the branched isomer.   From both **111a** and **111b**, the
allylated ether **112** was obtained as a single isomer.   The
stereoselectivity did not depend upon the geometry of the starting β-bromo ethers.   The high degree of stereoselection was
accounted for by the exclusive axial attack on the half-chair
conformation of the six-membered ring **113**.

Iodotrimethylsilane and trimethylsilyl triflate activated
selectively the C-Cl bond rather than the C-O bond of α-chloro
ethers and catalyzed the allylation with allylsilanes to give
the corresponding homoallyl ethers effectively in good yields
(ref. 95).   The catalytic reaction can be explained by the

cycle shown in **6**.   The initial formation of an silylhalonium
ion (Y = Cl, Z = OR), stabilized by an adjacent alkoxyalkyl
group, may take place selectively at low temperature.   Chloro-
methyl methyl ether reacted with allylic silanes under $SnCl_4$
catalysis (ref. 14, 96), and this reaction was used as a key
step in the preparation of a precursor to prostaglandin A and F.

β-Chloroalkyl phenyl sulphides prepared via the trans addition
of PhSCl to alkenes, underwent the allylation with allylsilane
in the presence of catalytic amounts of $ZnBr_2$ to give the α-
allylated products in good yields (ref. 97).   Small amounts of
allylphenylsulphide were produced as a by-product (<10%).   The
reaction proceeded stereoselectively; cyclic olefins gave the
trans adducts.   Moreover, the stereospecificity of addition was

Y = CH$_2$ or O

confirmed for cis- and trans-2-butene.   The syn adduct was
produced from cis-2-butene and the anti adduct was afforded from
trans-2-butene.   This stereospecificity is presumably a reflec-
tion of the nucleophilic attack in an intermediate episulphonium
ion.   Although the reaction proceeded quite smoothly with mono-
and di-substituted olefins, tri- and tetra-substituted olefins

produced allylphenyl sulphide as a major product.

α-Chlorosulfides containing a carbonyl group either at α, β, or γ position reacted with allylsilanes in the presence of Lewis acids to produce exclusively the corresponding α-allylsulfides in high yields (ref. 15). The chemoselective allylation is also demonstrated in **63** and **64**. In contrast to **63**, the α-chlorosulfides underwent substitution for the chlorine atom

n = 0, 1, 2
R^1 = Me, Hex, OEt

regardless of the type of Lewis acids and regardless of the carbon number n. With γ-substituted allylic silanes, the coupling product was obtained in very low yield. Without the need for high dilution conditions, medium-membered lactones (8 ~11 membered rings) were obtained in good yields by $EtAlCl_2$ -promoted intramolecular condensation of α-chlorosulfides bearing an ester group and allylsilane moiety (ref. 98).

The reaction of allylic silanes with epoxides in the presence of $TiCl_4$ produced 4-penten-1-ol derivatives (ref. 9, 99). Intramolecular version of this reaction was applied to the synthesis of karahanaenol (ref. 100). The cyclization product was obtained in 71% yield upon treatment with $BF_3 OEt_2$, though use of

karahanaenol

TiCl$_4$ gave a ring opening product, chlorohydrin, with no formation of the cyclic product.   The titanium tetrachloride promoted reaction of allylic silanes with oxetanes gave 5-hexen-1-ol
derivatives in good yields (ref. 101).   The reaction was limited to cyclic ethers that possessed high ring strain, since similar reaction with tetrahydrofuran failed.

1,4-Bis-(trimethylsilyl)-2-cycloctene reacted with acetyl
chloride and AlCl$_3$ to give 3-acetyl-4-trimethylsilyl cyclooctene
(ref. 102).   The diastereoselectivity was not reported in this

paper.   The reaction with acid chlorides normally proceeded with
γ-regioselectivity (ref. 103).   However, the reaction of 5-tri-
methylsilyl-1,3-pentadiene with pivaloyl chloride afforded a
mixture of products derived from electrophilic attack at both the
γ and ε carbons of the pentadienyl system (ref. 104).   In the
case of (2,4-hexadienyl)trimethylsilanes, the γ-adduct was produced preferentially.   The reaction of 1-chloro-3-trimethylsilyl
propene with acid chlorides in the presence of AlCl$_3$ gave α-chlo-
ro-β,γ-unsaturated ketones (ref. 105).   α-Trimethylsilylallyl
phenyl sulfides underwent regiospecific γ acylation with acid
chlorides - AlCl$_3$ to give γ-phenylthio-β,γ-unsaturated ketones
(ref. 106).   Homoallylic trimethylsilanes reacted with acid chlo-
rides-TiCl$_4$ to produce mixtures of β-chloro ketones, cyclopropyl
carbinyl ketones, and homoallylic ketones (ref. 107).

2.6 NITROGEN COMPOUNDS AND OTHERS

The N, O-acetal, a 1 : 1 mixture of diastereomers, reacted
with allylsilane in the presence of TiCl$_4$ to give the amino alco-

hol derivative in 85% yield; the isomer ratio of (R,R)/(R,S) was
4 : 1 (ref. 108, 109).   Quite similarly, a mixture of two dia-

stereoisomers (1.2 : 1) derived from the anodic oxidation of the
alanyl-valine dipeptide afforded the allylation product in a
ratio of 1.1 : 1 upon treatment with allylsilane.   Presumably,
the reaction proceeds through the iminium intermediate generated
by the coordination of $TiCl_4$ to OMe group (6).   Alkoxylactams
and alkoxycarbamates underwent the allylation reaction with
allylsilane in the presence of $SnCl_4$ or $BF_3 OEt_2$.   Here also, the
iminium ion intermediate must be involved.   The resulting allylated compound was transformed into the carbapenem ring system

(ref. 110).   This transformation is important, since only a few
successful carbon-carbon bond formation reactions at C-4 in N-
unsubstituted azetidinones are known (ref. 111).
    The reaction of N-carbethoxy-4-hydroxy-1,2,3,4, -tetra-
hydropyridine with allylsilane in the presence of $SnCl_4$, $TiCl_4$,
$BF_3 OEt_2$, or Me_3SiOTf gave 2-allyl-substituted-Δ^3-piperidine in

good yields (ref. 11b).   The Lewis acid must produce a conjugated iminium ion system, which is attacked by the carbon electro-
phile at the 2 position.

Simple iminium salts, generated in situ under Mannich-like
conditions, upon exposure to allylsilanes in the presence of
excess formaldehydes gave rise to piperidines via a novel amino-
methano desilylation-cyclization process (ref. 112).   Despite
the relatively acidic conditions in order to generate iminium
ion in water, protodesilylation of the allylsilane was not a
competitive process.

The iminum ions **114** gave 1,2,5,6-tetrahydropyridines in
excellent yield (ref. 113).   Two mechanisms could be considered
for the cyclization reaction.   The simplest was direct cyclization via a β-silyl cation intermediate (path a).   Alternatively, **114** could undergo cationic aza-Cope rearrangement to the
allylsilane intermediate which then could react with the iminium ion (path b).   The following experiment clarified that path
b took place more rapidly than path a.   Treatment of **115** with

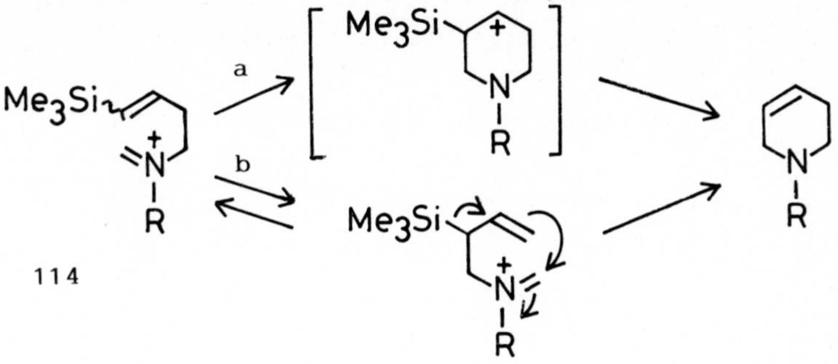

114

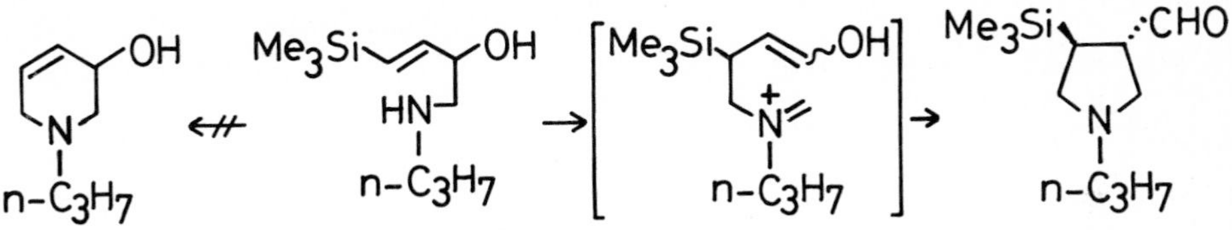

116   115   117

paraformaldehyde and camphorsulfonic acid in refluxing ethanol
did not give tetrahydropyridine **116** but rather pyrrolidine **117**.
The photochemical reaction of iminium salts with allylsilanes
gave homoallylamines presumably through an electron transfer
from allylsilanes to the excited singlet state of iminium salts
(ref. 114).

Boron trichloride served as a useful activator of nitriles
and under this condition nitriles were converted to β,γ-unsatu-
rated ketones in high yields (ref. 115). The reaction of substituted allylsilanes and intramolecular reaction of allylic

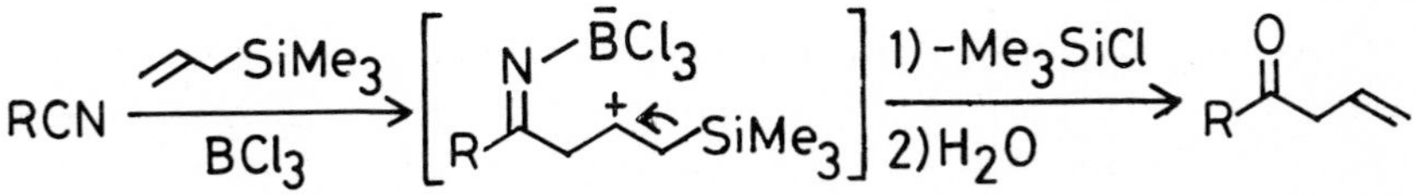

silanes with nitriles proceeded similarly in good yields. The
reaction did not proceed with other Lewis acids. Allylsilanes
added to nitrones and nitrile oxides regioselectively to give
the corresponding isoxazolidenes and isoxazolines in good

yields (ref. 116).   Hydrogenative cleavage of the N-O bond of
[3 + 2] cycloadducts with allylsilanes afforded homoallylamines.

2-Substituted allylsilanes on treatment with iodosobenzene
and boron trifluoride etherate in dioxane gave conjugated enals
in good yields (ref. 117).   Without BF_3, the oxidation reaction
did not occur and thus the activation of iodosobenzene by BF_3
seemed to be important.   The reaction may proceed through the

118

hypervalent organoiodine intermediate which might be attacked by
the nucleophilic oxygen of another molecule of iodosobenzene.
The allylsilanes **118**, containing suitably substituted hydroxy
groups, gave 5- or 6-membered β-methylene cyclic ethers in good
yields upon treatment with PhIO-BF_3 OEt_2 in an ether solvent
(ref. 118).   Oxidative desilylation of the cyclic allylsilane
at the γ-position was examined with a number of peracids under
various conditions, but the effort was unsuccessful (ref. 119).
However, such oxidative desilylation was achieved by a two step
sequence, OsO_4 catalyzed dioxidation of the double bond to produce the diol followed by β-elimination of the Me_3Si - OH unit.

Diphenylseleninic anhydride reacted with the allylsilane at
the γ-position to give the γ-selenylated olefin which underwent
a further migration reaction (ref. 120). A combination of allylsilane and thallium trifluoroacetate gave rise to an allyl cationic species which reacted with aromatic compounds to give allylated aromatics (ref. 121).   Therefore, the usual reactivity of
allylsilane as a nucleophile changed to electrophilic character.

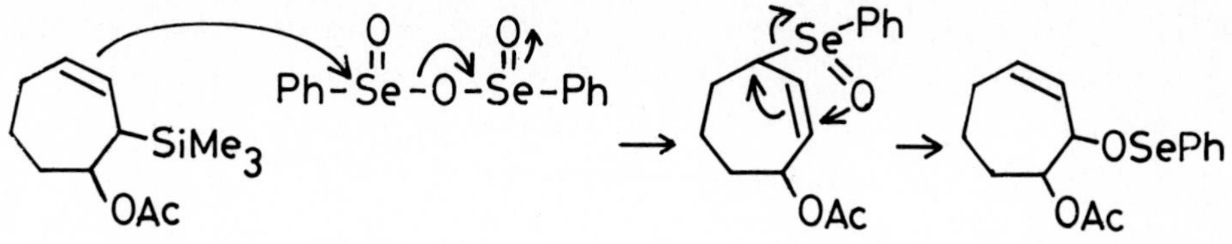

3. INTRAMOLECULAR REACTIONS

3.1 ALDEHYDES AND KETONES

The compound **119a** gave the cyclization product upon treatment with Lewis acids (ref. 122). The ratio of axial/equatorial alcohols were 59/41 with $SnCl_4$ and 85/15 with $BF_3 \cdot OEt_2$. The axial alcohol must be produced via the transition state geometry **120** which is more stable than the geometry leading to the equa-

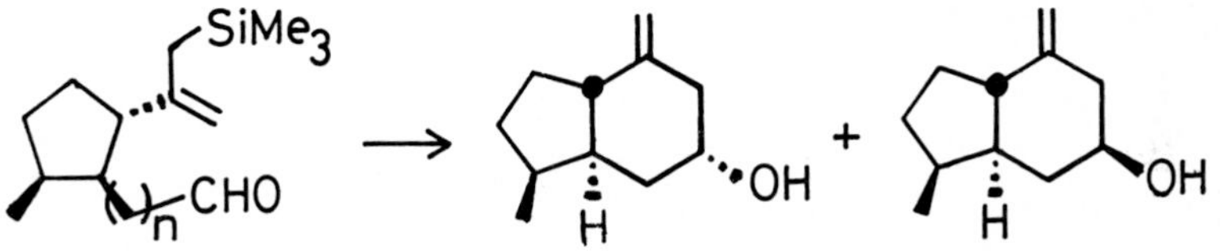

119a; n = 1, **b**; n = 2   axial-OH   equatorial-OH
    c; n = 0

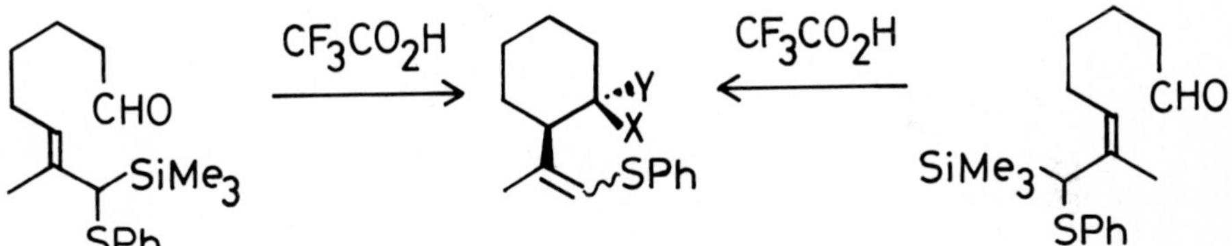

121a

122a; X=OH, Y=H
    b; X=H, Y=OH

121b

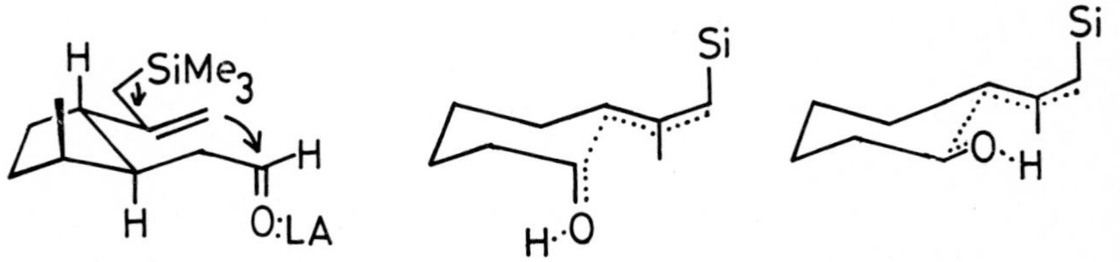

120   **123a**   **123b**

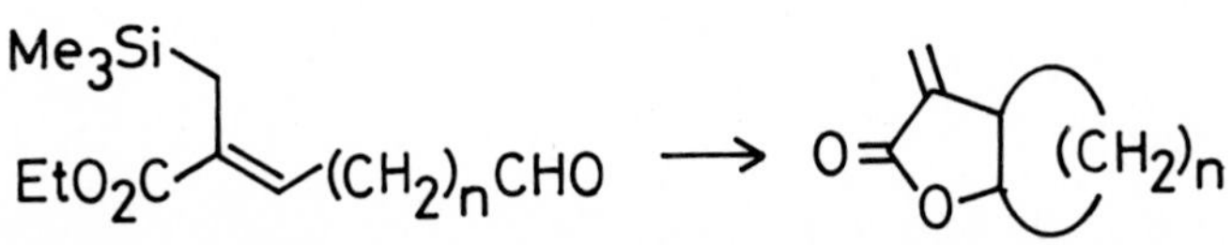

torial alcohol.   While **119c** did not afford the cyclization product under the similar conditions, **119b** produced the corresponding bicyclic alcohols.   Treatment of **121a** with CF_3CO_2H or
CF_3CH_2OH gave the cis alcohol **122a**, while the reaction of **121b**
with the same acid produced the trans alcohol **122b** (ref. 123).
Cyclization of **121a** with $SnCl_4$, $TiCl_4$, BF_3 OEt_2 or Bu_4NF gave a
mixture of **122a** and **122b**.   Exclusive formation of **122a** from
121a may be explained by the steric and stereoelectronic effect
of the transition state geometry; **123a** may be more stable than
123b.   Bicyclic α-methylene-γ-lactones were prepared by intramolecular reaction of the aldehyde having α,β-ethoxycarbonyl-
allylsilane function with the aid of $TiCl_4$ (n = 3, 4) (ref. 124).

The ketones also underwent the cyclization in high yields
upon treatment with $EtAlCl_2$ (ref. 125).   Fluoride ion and titanium tetrachloride induced cyclization of these ketones suffered
from protodesilylation (ref. 126).   Ethylaluminum dichloride
avoided this complexation, which presumably arised from the presence of HX.   Treatment of **124** with $EtAlCl_2$ gave spirocycles
(n = 3, 4 and 8) with greater than 95% yield (ref. 127).   The
reaction proceeded through the normal cyclization **125**, followed

| 124 | 125 |

by a pinacol type rearrangement in which the sulfone group
served as a leaving group in the presence of a Lewis acid.

The $TiCl_4$ catalyzed chelation controlled cyclization of **126**
gave a single diastereomeric product in good yields (ref. 128).

126a; X=OEt
   b; X=NR_2^2

cis   trans

Although use of TiCl$_4$ gave exclusively the cis product, a 1 : 4.8 mixture of the cis/trans products was obtained with a non-chelating BF$_3$ OEt$_2$ (X = OEt, R = R' = Me).   In contrast, the ratio was 2 : 1 with Bu$_4$NF.

Although syn-stereoselectivity in the Lewis acid mediated intermolecular condensation of γ-substituted allylic organometallics with aldehydes has been accounted for by the anti-periplanar transition state geometry **11**, an alternative model, synclinal geometry, is proposed in the intramolecular condensation reaction (ref. 129).   Cyclization of **127b** gave the syn isomer either exclusively or predominantly regardless of the Lewis acids used.   The compound **127a** again afforded the syn isomer predominantly with BF$_3$ OEt$_2$.   Since the syn isomer arises from

127a; MLn=SiMe$_3$
  b; MLn=SnBu$_3$

syn   anti

synclinal   antiperiplanar I   antiperiplanar II

the synclinal transition geometry and the anti isomer arises from the antiperiplanar I, a preference for the synclinal over the antiperiplanar geometry is assumed.   However, the originally proposed model **11** for intermolecular condensation corresponds to the antiperiplanar II in the reaction of **127**.   Such a geometry can not be assumed in **127** owing to the restriction of the carbon chain of -CH$_2$CHO.   It seems to us that intramolecular steric congestion forces the reactions to take synclinal geometry.

3.2 α,β-UNSATURATED CARBONYL COMPOUNDS

When **128** was treated with BF$_3$ OEt$_2$ in ether, intramolecular cyclization occurred readily to produce 4-methyl-4-vinylcyclohexanone in 73% yield (ref. 130). The reaction of **129** (E/Z = 3/2) with TiCl$_4$ gave a 3 : 2 mixture of **130a** and **130b** in 90%

combined yield (ref. 131). The stereochemistry between C-5 and C-6 did not depend upon the double bond geometry of the allylic silane **129**. Therefore, the cyclization of **129** must proceed through **131a** (antiperiplanar type) rather than **131b** (synclinal type).

Cyclization of **132** with EtAlCl$_2$ in toluene at -78°C produced a 7 : 1 mixture of **133a** and **133b** in 72% combined yield (ref.

132).   Use of other Lewis acids such as BF_3 OEt_2 and $SnCl_4$ lead only to protodesilylation, while decomposition occurred with $TiCl_4$.   The synclinal transition state is more favorable than the antiperiplanar geometry, since the former gives **133a** and the latter produces **133b**.   Allylsilanes of type **134** and propargyl-silanes of type **135** underwent smooth cyclization upon treatment with $EtAlCl_2$ to give stereoselectively functionalized hydrindano-nes and spiro[4.5]decanes, respectively (ref. 133).   The compound **136** cyclized only in 1,6 fashion leading to the bicyclo [4, 5, 0]undecenone as a single diastereomer (ref. 134, 135).

134   135   136

In some cases, the 1,6 addition was accompanied with the 1,4 addition leading to bicyclo[4, 3, 0]nonanone derivatives.

   The diastereoselectivity of γ-substituted allylic silanes was investigated with the aid of Lewis acids and F^- (ref. 136). Cyclization of **137** with $EtAlCl_2$ or $TiCl_4$ produced **138a** predomi-nantly.   The synclinal geometry may permit the developing car-

[A]   138a   138b   [B]

bonium ion to be stabilized by the nearby π-electrons of the titanium (or aluminium) enolate [A], while such a π-overlap stabilization is not expected in the cationic intermediate [B] derived from the antiperiplanar transition state. On the other hand, the F⁻ induced cyclization produced **138b** preferentially. The mechanistic details will be discussed in **5.1**.

As mentioned in **136**, the regioselectivity was also investigated in **139** (ref. 137). With EtAlCl$_2$, the 6-7 bicyclic ring **140a** was obtained as a sole product. Here again, the 1,6-addition took place exclusively. The F⁻ induced cyclization gave a mixture of **140b** (1,4-addition) and **140c** (1,2-addition) along

139   140a   140b   140c

with **140a**, and the regioselectivity highly depended upon the structure of substrates (ref. 138, 139).

3.3 ACETALS AND RELATED COMPOUNDS

The allylsilane **141** cyclized cleanly to the exocyclic product **142** under the influence of SnCl$_4$ (ref. 10). This concept was extended to the cyclization of epoxyallylsilanes (ref. 99). The compound **143** (Z : E = 4 : 1) produced cis-cyclopentane **144** (cis : trans = 4 : 1) stereoselectively upon treatment with TiCl$_4$. The MEM-substituted allylsilane **145** was treated with 1.1 eq TiCl$_4$ in CH$_2$Cl$_2$ at low temperature to give the tetrahydropyran derivative **146** (ref. 69a). Tetrahydrofuran derivatives were also prepared by a similar procedure. Cyclization of monothioacetals produced cyclic ethers upon treatment with SnCl$_4$ (88). Treatment of the acetal **147** with SnCl$_4$ in pentane

141   142   143   144

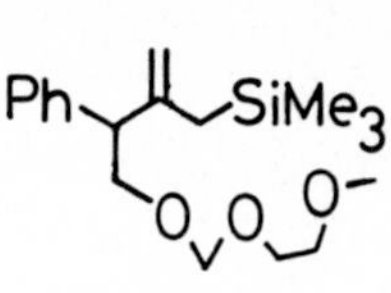

145 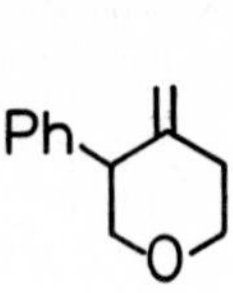 **146** 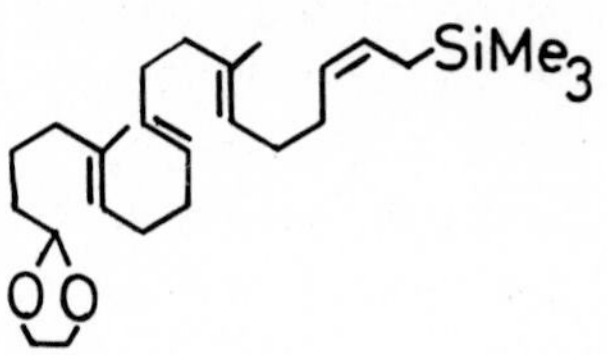 **147** 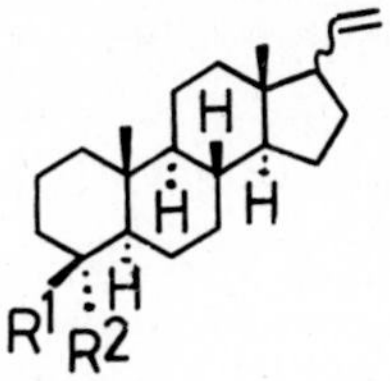 **147a;** $R^1 = OCH_2CH_2OH$, $R^2 = H$
b; $R^1 = H$, $R^2 = OCH_2CH_2OH$

at 0°C afforded a 2 : 1 mixture of **147a** and **147b** (ref. 140).
Both isomers were obtained as mixtures of their 17α and 17β epi-
mers.   Without separation of isomers, the products could be
converted to progesterone.

Cyclization of **148** in the presence of $SnCl_4$ gave an epimeric
mixture of bicyclic products **149** in 51% yield and a ratio of
9α/9β of 1 : 3 for R = iPr (ref. 141).   In the chair-chair transition state there is a gauche interaction between the ester
CO_2R and both the C-10 methyl and the C-1 methylene (**150a**).
In the chair-boat transition state **150b** the ester is eclipsed

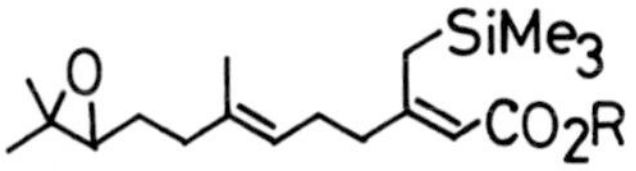

148 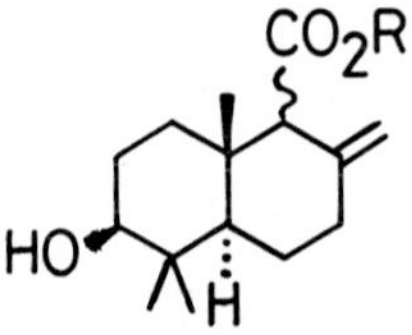 **149**  **150a**

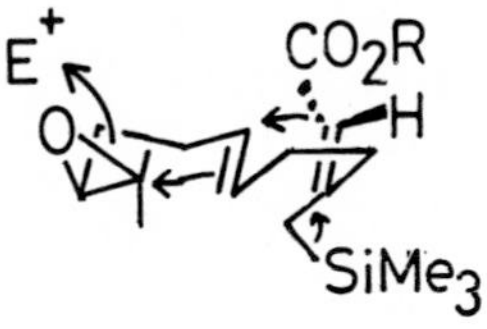

150b 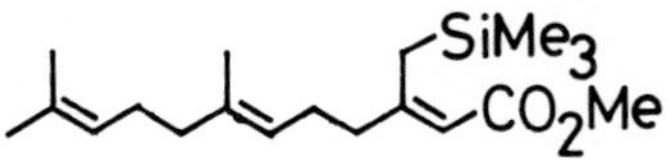 **151** 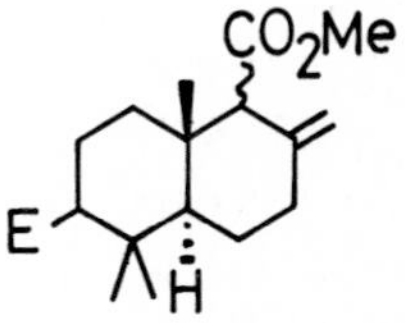 **152**

with the C-1 methylene.   Therefore the chair-chair geometry
150a leading to the 9β epimer is preferred (ref. 99, 141).
The polyolefinic allylsilane **151** was cyclized with Lewis acids
or mercuric trifluoroacetate to give **152** (E = H or HgCl) which
were converted into albicanyl acetate and isodrimenin (ref. 142,
143).   With $SnCl_4$, **152** (E=H) was obtained in 95% yield and the

ratio of the 9α : 9β isomers was 1 : 4.   With Hg^{2+}, the ratio
of the 9α : 9β isomers was 1 : 3.

The intramolecular [3 + 2] cycloaddition of 2-trimethylsilyl
substituted allyl cations to alkenes lead to bicyclo[3.3.0]
octane derivatives (ref. 144).   When $TiCl_4$/N-methylaniline
(1 : 1) was allowed to react with **153** in CH_2Cl_2 at -15°C, the
bicyclooctane was obtained in 70% yield.   The reaction of cobalt

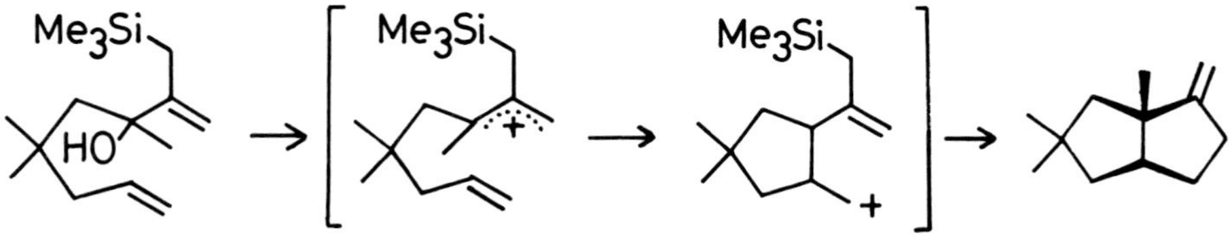

153

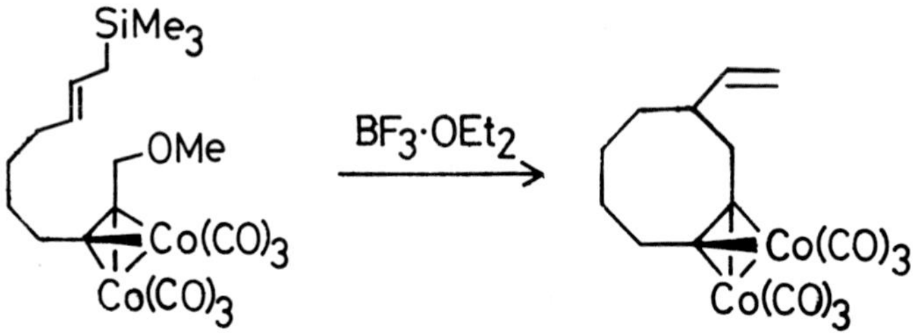

154

complexed propargylic alcohols with HBF_4 provides a cobalt
stabilized carbocation that can be treated with a variety of
carbon nucleophiles to give alkylated products (Nicholas reaction) (ref. 145).   Lewis acid mediated version of the Nicholas
reaction was shown in **154** (ref. 146).   By using this procedure,
six-, seven-, and eight-membered cyclic compounds were prepared.

3.4 NITROGEN COMPOUNDS

Alkylation of the lithium enolates derived from ω-alkoxy
lactams with trimethylsilyl-substituted unsaturated iodides
afforded **155** in high yields.   The N-acyliminium ion cycliza-

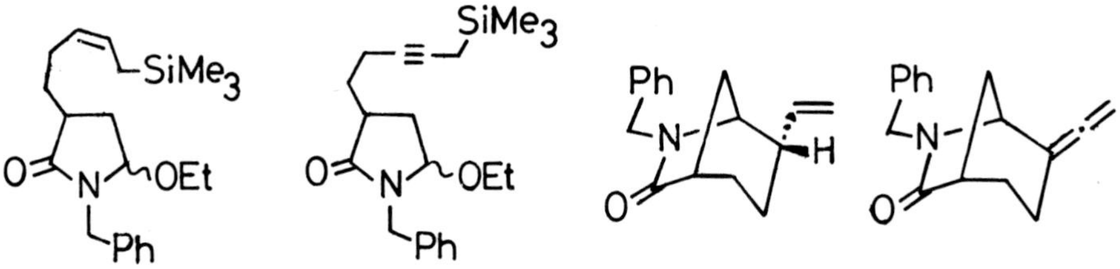

155a   **155b**   **156a**   **156b**

tion reaction was carried out in the presence of CF_3CO_2H in CH_2Cl_2. Thus **155a** gave **156a** in 73% yield, and **155b** produced **156b** in 88% yield (ref. 147). Quite similarly, intramolecular reactions of acyclic N-acyliminium ions with propargyl silanes, induced by protic or Lewis acid, gave α-allenic amides of carbamates, i.e. derivatives of 3-vinylidenepyrrolidine, 3-vinylidenepiperidine or 1-amino-2-vinylidenecyclopentane (ref. 148). The allylsilane analog **157a** produced a 2 : 1 isomer mixture of

157a; n=3. b; n=2   158a   158b   159

158a (cis) and **158b** (trans) in 60% combined yield (ref. 149). In contrast, **157b** gave the trans pyrrolidine as a single product. The reaction must proceed through the most stable transition state geometry **159**. Similar intramolecular allylsilane cyclization via acyliminium ions afforded indolizidine and quinolizidine compounds in high yields (ref. 150).

As mentioned previously, the intramolecular version of the iminium salt cyclization (ref. 112) gave five, six, seven, and eight-membered rings containing nitrogen (ref. 151). By using

n = 2, 3, 4

this procedure, (±)yohimbone was synthesized in 10% overall yield (ref. 152). The intramolecular reaction of allylsilane with nitrile under the influence of BCl_3 was investigated (ref. 115) in the course of the study on the intermolecular reaction.

4.   OPTICALLY ACTIVE ALLYLSILANES

4.1   STEREOCHEMISTRY OF ELECTROPHILIC SUBSTITUTION

Electrophilic substitution of allylic silanes with electrophiles proceeds through either syn S_E' (ref. 153) or anti S_E' manner (ref. 154). The stereoselectivity was highly dependent upon the structures of allylic silanes. For example, in the cyclopente-

nyl system electrophilic substitution took place with syn manner (ref. 153) but in the cyclohexyl system it occurred with anti fashion (ref. 154). These stereochemistries seemed to be controlled not by the inherent nature of allylsilanes but by the stereochemical bias in the cyclic system.

The S_E' reactions of **160** with a range of electrophiles were predominantly anti in the allylsilane portion of the molecule, but this was offset by axial or equatorial preference in the ring system (ref. 155). The epoxidation of **160a** with MCPBA followed by the ring opening with Bu_4NF gave the axial alcohol as a sole product. The same reaction of **160b** produced a 96 : 3 mixture on the equatorial and axial alcohol. All the reaction were anti overall, and, although there was a small preference for axial attack, the overall stereochemistry was mostly dictated by the allylsilane group. Protonation and protodesilylation were also very largely anti. Trifluoroacetolysis of various cyclo-hex-2-enylsilanes proceeded with γ-regiospecific and highly anti stereoselective manner (ref. 154a).

The stereochemistry of S_E' reactions in acyclic systems was studied with optically active allylsilanes (ref. 156). Tert-butylation, acetylation and hydroxymethylation of **161** proceeded via anti-S_E' fashion (ref. 157). The anti stereochemistry is explained as follows. The allylsilane **161** must exist in confor-

mation A with the C-Si bond overlapping with the π lobes of the C-C double bond, due to the $\sigma-\pi$ conjugation (3). The electro-phile attacks from the side opposite to $SiMe_3$ group to form cationic intermediate B in which the cation is stabilized by the $\sigma-\pi$ conjugation. Displacement of the silyl group produces (E)

olefin C.   The anti-S_E' attack is also predicted from the theoretical studies (ref. 159, 154a).   Optically active allylsilanes, **161a** and **161b**, were treated with aldehydes in the presence
of $TiCl_4$.   As expected, optically active syn-homoallyl alcohols
were produced in high yields (ref. 160).   The syn-preference is
in good agreement with the acyclic transition model **11**. Interestingly, the E olefin **161a** produced very high diastereoselectivity
and excellent enantiomer excess, but the Z olefin **161b** gave
lower diastereoselectivity and very low ee.   Reaction of aldehydes with an optically active allylsilane **161** (R = H), (R)-3-
phenyl-3-(trimethylsilyl)propene, in the presence of $TiCl_4$ gave
homoallyl alcohols with high enantioselectivity (up to 91% ee)
(ref. 158).   Reaction of optically active cyclic allylsilanes,
(S)-3-(trimethylsilyl)cyclopentene and -cyclohexene, with ethylene oxide in the presence of $TiCl_4$ proceeded with anti stereochemistry to give (R)-3-(2-hydroxyethyl)cyclopentene and -cyclo-
hexene, respectively (ref. 161).   The reaction of lithium
palladates with **161** gave complexes through anti attack of palladium (II) (ref. 162).

64 - 91 % ee

4.2 ASYMMETRIC SYNTHESIS

Chirality transfer from asymmetric silicon to carbon is an
interesting problem.   As mentioned above, the optically active
allylsilanes **161** gave homoallyl alcohols with high ee upon
treatment with aldehydes (ref. 158, 160).   The chirality of
161 exists in the α-carbon, that is, "C-centered chirality".
(-)-α-Naphthylphenylmethylallylsilane **162** underwent the BF_3
mediated condensation with benzaldehyde dimethyl acetal to produce low enantioselectivity (3.9 ~5.5 ee) (ref. 163).   In contrast to the "C-centered" chirality, the "Si-centered" chirality
created only low enantioselectivity.   The "C-centered"
optically active allylsilane **163** reacted with carbonyl compounds
in the presence of Lewis acids to give the corresponding homoallyl alcohols with ee varying from 21 to 56% (ref. 164).

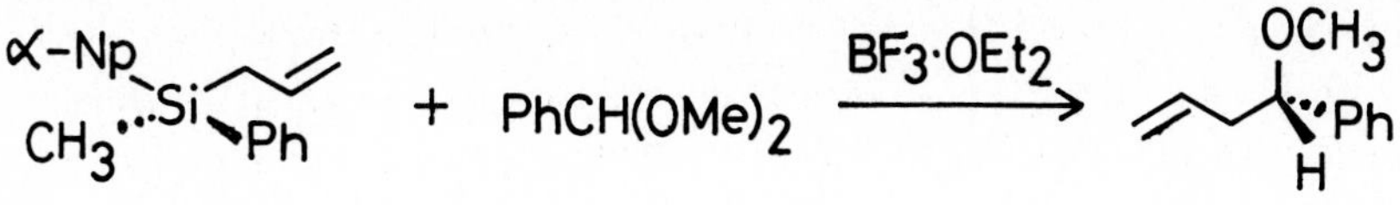

162

(-)-(S)

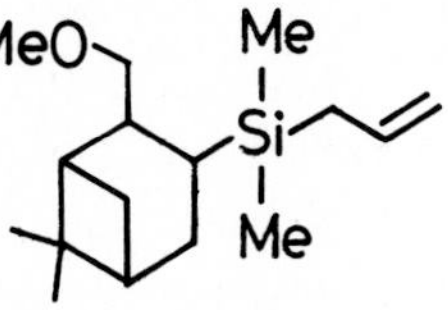

163    164a    164b

Optically active **164a** and **164b** also produced relatively low ee (ref. 161).

A highly enantiospecific synthesis of (1S,2S)-(+)-2-methyl-3-cyclopenten-1-ol was achieved via asymmetric allylsilane mediated carbocyclization (ref. 165). The compound (R,E)-**165**

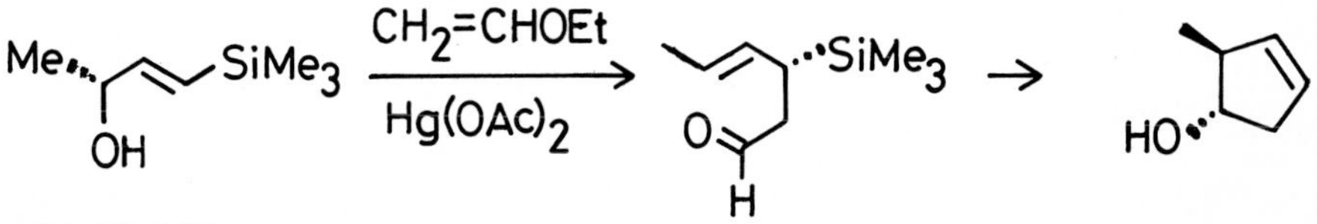

(R,E) **165**

underwent the enol ether Claisen rearrangement upon treatment with $CH_2=CHOEt/Hg(OAc)_2$ to give the allylic silane with moderate (E)-selectivity (E/Z = 83 : 17). Treatment with $TiCl_4$ at -78°C gave the trans cyclopentenol with > 98% ee.

5. OTHER REACTIONS

5.1 F⁻ MEDIATED REACTIONS

Fluoride ion exhibits a marked nucleophilic affinity for silicon due to the high Si-F bond energy. As mentioned in **8**, an allyl carbanion is involved as an intermediate (ref. 26). The allylation of nitriles, epoxides and esters did not take place via the F⁻ induced reaction. Isoprenylation of aldehydes and ketones with 2-[(trimethylsilyl)-methyl]-1,3-butadiene was achieved by a catalytic amount of TBAF (ref. 168). Ipsenol and

ipsdienol were synthesized via this procedure.   Cinnamyltri-
methylsilane reacted with electrophiles at the α-position in the
presence of KF/18-crown-6 or silica-TBAF under mild conditions
(ref. 166).   Allyltrimethylsilane underwent conjugate allylation
(1,4-addition) to cyclohexenone upon treatment with CsF, while
the 1,4-addition took place along with the 1,2-addition upon
treatment with TBAF-silica (ref. 167).

As shown in the above examples, the regio- and chemo-selecti-
vity of F⁻ mediated reaction is different from the Lewis acid
mediated reaction.   Prenylsilane reacted at the α position upon
treatment with TBAF (ref. 26).   Crotylsilane gave both α - and
γ-adducts.   The regioselectivity of the carbonyl addition and

the chemoselectivity in α,β-enones are in good agreement with an
allyl cation intermediacy (See also **62**, ref. 66).

Although the Lewis acid mediated conjugate addition to α,β-
enones proceeded very smoothly (**39**), this allylation procedure
failed with monoactivated α,β-unsaturated esters, nitriles, or
amides regardless of the choice of Lewis acids.   Doubly activated Michael acceptors underwent the conjugate allylation with
$TiCl_4$.   Under F⁻ catalysis conditions, only 1,4-addition took
place for α,β-unsaturated esters and nitriles.   Therefore, the
F⁻ mediated method is superior to the Lewis acid promoted method
for conjugate allylation of these substrates (ref. 169).   Even

by using the F⁻ method, the 1,2-addition took place for the amide (R = t-Bu, R' = H, EW = CONEt$_2$).

With α,β-unsaturated aldehydes, the F⁻ procedure gave only the 1,2-addition product in high yield. The BF$_3$ OEt$_2$ promoted method afforded again the 1,2-addition product in about 50% yield. With α,β-unsaturated ketones, the Lewis acid mediated method gave only the 1,4-addition product in high yield. In contrast, the F⁻ promoted procedure produced both the 1,2- and 1,4-adducts, whose ratio depended upon the steric circumstance of carbonyl group.

A free allyl anion was suggested as an intermediate in the F⁻ promoted reaction (8) (ref. 26). As an alternative mechanism, a nonbasic hypervalent silicon intermediate may be involved (ref. 169). Such pentacoordinate silicon intermediate is isolated; [Ph$_4$As]⁺[SiF$_5$]⁻, RSiF$_4$⁻ and R$_2$SiF$_3$⁻.

The intramolecular cyclization via the fluoride ion method was compared with that through the Lewis acid mediated procedure (ref. 136, 170, 171). The reaction of **166** with TBAF produced **167a** normally as a sole product in good yields. When two methyl groups were present in R^1, R^2 or R^3 substituents, small amounts of **167b** were obtained as a by-product. On the other hand, the reaction with EtAlCl$_2$ or TiCl$_4$ gave normally **167b** and/or **167c**. The reaction of **168** (R^2 = H) with F⁻ produced **169a**

166   167a   167b   167c

168   169a   169b   169c

in good yield. However, when R^3 = Me, the yield of **169a** dramatically decreased and **169b** and/or **169c** were afforded as a major
product. The Lewis acid mediated reaction again produced **169b**
and/or **169c** (ref. 172).

Two mechanisms, S_E2' and S_E2, were conceivable for the above
cyclization. Cyclization may occur at the γ-carbon of the
allylsilane (path a, S_E2' process), or take place at the α-carbon
(path b, S_E2 process). Investigation with certain unsymmetrical

allylsilanes revealed that both mechanisms were in operation in
the cyclization reaction.

As mentioned in **3.2**, the F^- mediated cyclization gave **138b**
predominantly (ref. 136). The stereochemistry was explained in
terms of the geometries of the olefinic components prior to
cyclization. The synclinal geometry must suffer from the steric
repulsion between R^1 and R^2 group, thus the reaction presumably
proceed through the antiperiplanar transition state geometry.

The regioselectivity of the F^- induced cyclization of **139** was

$$137 \xrightarrow{F^-} \left[\text{synclinal} \rightleftarrows \text{antiperiplanar} \right] \rightarrow 138b$$

synclinal   antiperiplanar

dependent upon the substituents R and R[1] (ref. 137). With
R = Me, **140a** was not obtained at all and a major product was
140b. With R = H, **140a** was afforded as a major product. The
substrate dependency was accounted for by the conformational
and/or steric hindrance effects. Non-bonded steric interactions between the C-4 substituent and the C-3 vinyl group may force
to take a cisoid conformation which produces **140b** via 1,4-addi-
tion. In contrast, the absence of the C-4 substituent allows

transoid   cisoid   1,4-addition   **140b**

transoid   1,6-addition   **140a**

the more stable transoid conformer to predominate, leading to
140a through 1,6-addition.

Although EtAlCl$_2$ mediated cyclization of **136** gave only **170a**
via 1,6-γ-addition in high yields (ref. 135, 138), the F$^-$ mediated reaction produced the 1,6-α-adduct **170b** and the 1,2-γ-
adduct **170c**.

Muscone (ref. 126) and coriolin (ref. 173) were prepared by
the F$^-$ mediated cyclization of **171** and **172**, respectively. The
indole derivative **173** underwent 1,4-fragmentation upon treatment

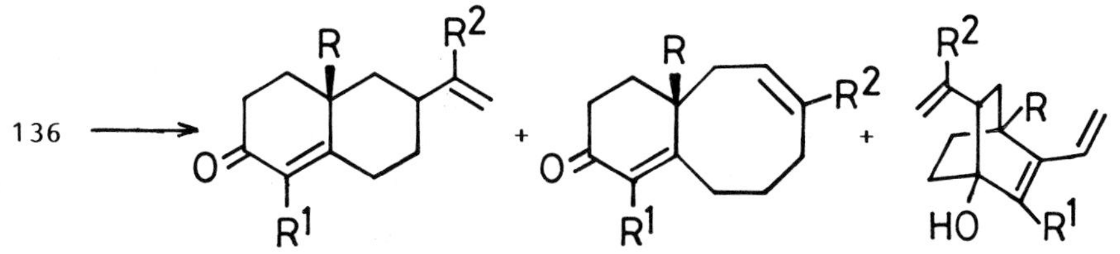

136 ——→

170a   170b   170c

with fluoride ion to give the desired indole-2,3-quinodimethane.
The intramolecular [4 + 2] cycloaddition of this quinodimethane
provided a highly convergent way for making indole alkaloids
(ref. 174). The F$^-$ mediated reaction of 13 was also investiga-

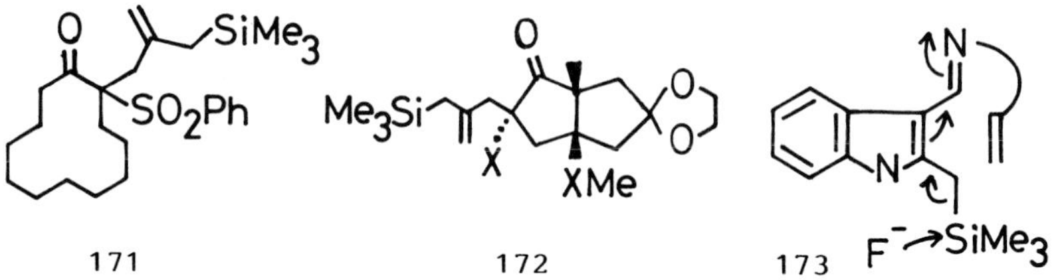

171   172   173

ted (ref. 36).

Aromatic heterocyclic N-oxides were readily converted into
α-alkylated heterocycles by allylsilane and fluoride ion (ref.
175). The reaction presumably proceeded through the initial
1,2-adduct followed by elimination of trimethylsilanol to 2-
allylpyridine which was isomerized during the reaction to 2-
propenylpyridine.

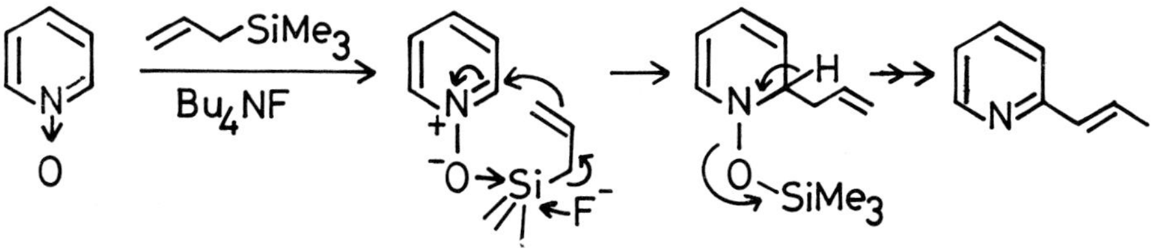

The compound 174 reacted with aldehydes and ketones to give
175 in high yields which was converted into 176 upon treatment
with BrCN followed by NaH (ref. 176).

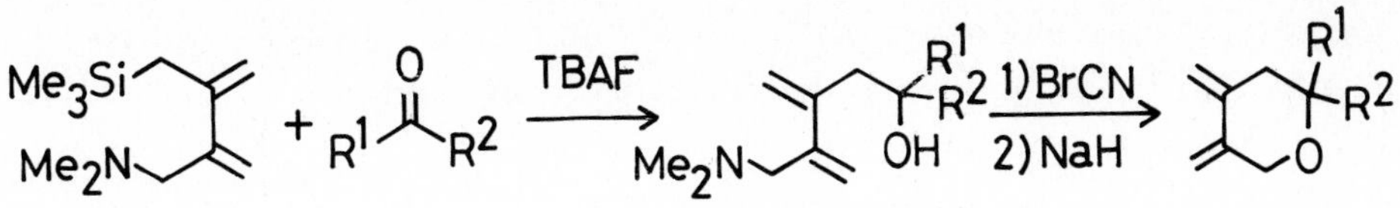

174   175   176

5.2 PROTODESILYLATION REACTIONS.

3,3- Dialkylallylsilanes reacted with protons by two delicately balanced mechanisms. The allylsilane **177** reacted directly by protonation on the γ-position of the allyl group followed by loss of the silyl group. On the other hand, **178** reacted, at least in part, by protonation of the β-position, followed by hydride shift and loss of the silyl group. The overall result is the same, but the difference in pathways was revealed by using D^+ instead of H^+ (ref. 177).

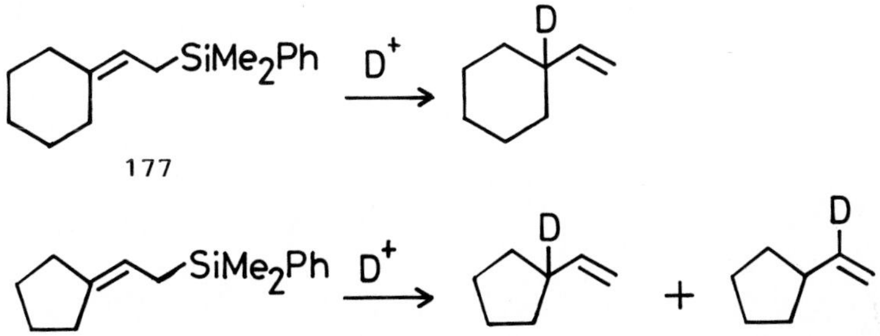

177

178

S_E' reaction of **161a** (R = Ph) with trifluoroacetic acid-d_1, proceeded with anti stereochemistry to give (R)-(E)-1,3-diphenyl-propene-3-d (ref. 178). The stereochemistry of protodesilylation reactions of cis- and trans-3,4-bis(trimethylsilyl)cyclohexene with CF_3CO_2D were also investigated (ref. 154a). Here again, the protonation took place in anti-S_E' manner.

Treatment of **179** with BF_3/AcOH in CH_2Cl_2 gave **180** in 85% yield (ref. 130). Protonation of **179** proceeded intramolecularly leading to 1,3-asymmetric induction as shown in **180**, (2S)(4R)-2,4-dimethyl-5-hexenoic acid (> 8 : 1 ratio). The reaction

426

presumably takes place via a six-membered chair transition state.
The intramolecular protonation was also investigated in similar
alcohol derivatives **181** (ref. 179).   Protonolysis of **181a**
gave **182a** selectively, while protonolysis of **181b** produced **182b**.
The reverse selectivity was observed.   This interesting stereo-

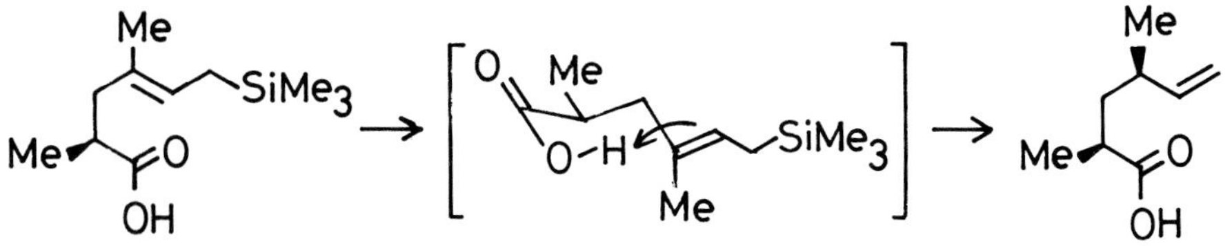

Me

179   180

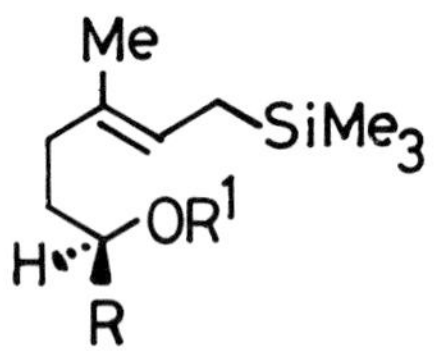

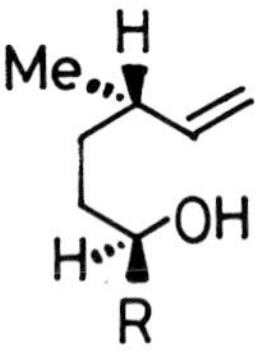

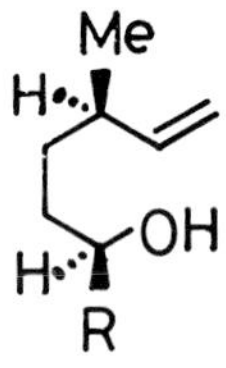

181a; R¹=H   182a   182b
   b; R¹=COMe

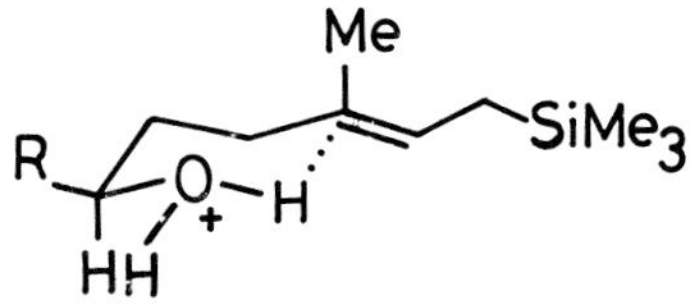

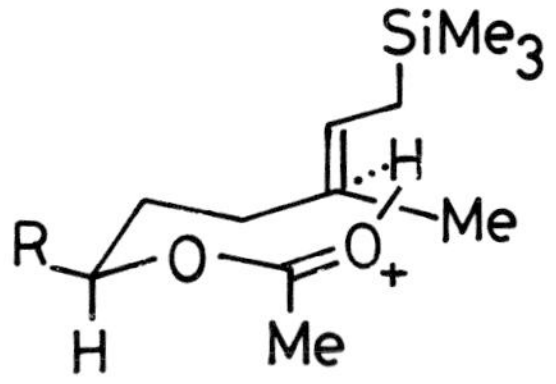

[A]   [B]

chemistry can be accounted for by intramolecular protonation
via a six-membered transition state A for **181a** or an eight-
membered transition state B for **181b**.

6.   CONCLUDING REMARKS

The Lewis acid mediated reactions of allylsilanes with a
number of electrophiles have occupied an important position in
modern organic synthesis, despite its short history. The fluoride
ion catalyzed reactions of allylsilanes are becoming an equally
important methodology.   The two methods are often complementary
from the synthetic point of view.   Perhaps the most important
aspect of allylic silane chemistry is that allylsilanes are
stable, storable, and are able to be handled under open atmos-

phere without any special cautions. Therefore, an allylic silane moiety can be incorporated in complex organic molecules and the intramolecular cyclization can be carried out by addition of Lewis acids or fluoride ion. This type of reaction is not possible via classical allylic organometallic compounds. Undoubtedly, the tremendous potential of the Sakurai reaction will result in greater application in organic synthesis. Finally, a number of interesting reactions of allylsilanes other than the Sakurai reaction are not cited; for example, the photochemical reaction with iminium salts (ref. 180) or with quinones (ref. 181), ene reaction (ref. 182), and cycloaddition reaction (ref. 183).

We would like to thank Professors Hideki Sakurai and Akira Hosomi for helpful discussions and Miss Harumi Hiratsuka for her help in the preparation of the manuscript.

REFERENCES and NOTES

1) R. Calas, J. Dunoguès, G. Deleris and F. Pisciotti, J. Organomet. Chem., 69 (1974) C 15. G. Deleris, J. Dunoguès and R. Calas, ibid., 93 (1975) 43.

2) E. W. Abel and R. J. Rowley, J. Organomet. Chem., 84 (1975) 199.

3) A. Hosomi and H. Sakurai, Tetrahedron Lett., (1976) 1295. H. Sakurai, Pure & Appl. Chem. 54 (1982) 1; ibid., 57 (1985) 1759. Later, Calas et al. reported the $AlCl_3$-mediated allylation of ordinary aldehydes, G. Deleris, J. Dunogues and R. Calas, Tetrahedron Lett., (1976) 2449.

4) T. Mukaiyama, K. Banno and K. Narasaka, J. Am. Chem. Soc. 96 (1974) 7503.

5) a) E. W. Colvin, Silicon in Organic Synthesis, Butterworths, London, 1981. b) W. P. Weber, Silicon Reagents for Organic Synthesis, Springer-Verlag, Berlin, 1983. c) P. D. Magnus, T. Sarker and S. Djuric, "Comprehensive Organometallic Chemistry", (1982), Vol. 7, 515. d) T. H. Chan and I. Fleming, Synthesis, (1979), 761. e) I. Fleming, Chem. Soc. Rev., 10 (1981) 83.

6) a) A. Schweig, U. Weidner and G. Manuel, J. Organomet. Chem., 67 (1974) C 4. b) R. S. Brown, D. F. Eaton, A. Hosomi, T. G. Traylor and J. M. Wright, ibid., 66 (1974) 249.

7) L. H. Sommer, L. J. Tyler and F. C. Whitmore, J. Am. Chem. Soc., 70 (1948) 2872.

8) a) J. -P. Pillot, J. Dunoguès and R. Calas, Tetrahedron Lett., 17 (1976) 1871. b) A. Hosomi, H. Hashimoto and H. Sakurai, J. Org. Chem. 43 (1979) 2551.

9) I. Fleming and I. Paterson, Synthesis, (1979) 446.

10) a) A. Hosomi, M. Endo and H. Sakurai, Chem. Lett., (1976) 941. b) I. Fleming, A. Pearce and R. L. Snowden, J. Chem. Soc. Chem. Commun., (1976) 182.

11) a) D. J. Hart and Y. -M. Tsai, Tetrahedron Lett. 22 (1981) 1567. b) A. P. Kozikowski and P.Park, J. Org. Chem. 49 (1984) 1674.

12) Y. Morizawa, S. Kanemoto, K. Oshima and H. Nozaki, Tetrahedron Lett., 23 (1982) 2953.

13) T. Fujisawa, M. Kawashima and S. Ando, Tetrahedron Lett., 25 (1984) 3213.

14) B. Au-Yeung and I. Fleming, J. Chem. Soc. Chem. Commun., (1977) 79.

15) M. Wada, T. Shigehisa and K. Akiba, Tetrahedron Lett., 24 (1983) 1711.

16) H. Nishiyama, S. Narimatsu, K. Sakuta and K. Itoh, J. Chem. Soc. Chem. Commun., (1982) 459.

17) J. E. O'Boyle and K. M. Nicholas, Tetrahedron Lett, 21 (1980) 1595.

18) A. J. Birch, L. F. Kelly and D. J. Thompson. J. Chem. Soc. Perkin I, (1981) 1006.

19) A. Hosomi and H. Sakurai, J. Am. Chem. Soc. 99 (1977) 1673; Org. Syn., 62 (1984) 86. A. Hosomi, H. Kobayashi and H. Sakurai, Tetrahedron Lett., 21 (1980) 955.

20) M. Ochiai, M. Arimoto and E. Fujita, Tetrahedron Lett., 22 (1981) 1115.

21) T. Tsunoda, M. Suzuki and R. Noyori, Tetrahedron Lett., 21 (1980) 71.

22) H. Sakurai, K. Sasaki and A. Hosomi, Tetrahedron Lett., 22 (1981) 745. Catalytic amounts of Ph_3CClO_4 and Ph_2BOTf are also effective; T. Mukaiyama, H. Nagaoka, M. Murakami and M. Ohshima, Chem. Lett., (1985) 977.

23) A. Hosomi, Y. Sakata and H. Sakurai, Chem. Lett., (1983) 405.
    H. Sakurai, Y. Sakata and A. Hosomi, ibid., (1983) 409.

24) a) R. W. Hoffmann, Angew. Chem. Int. Ed. Engl., 21 (1982)
    555. b) Y. Yamamoto and K. Maruyama, Heterocycles, 18 (1982)
    357.

25) a) P. A.-Robertson and J. A. Katzenellenbogen, Tetrahedron
    Lett., 23 (1982) 723. b) A. Hosomi, M. Ando and H. Sakurai,
    Chem. Lett., (1984) 1385.

26) A. Hosomi, A. Shirahata and H. Sakurai, Tetrahedron Lett.,
    19 (1978) 3043. See also, 26a) T. K. Sarker and N. H.
    Andersen, Tetrahedron Lett., 19 (1978) 3513.

27) a) E. Frainnet and R. Calas, C. R. Acad. Sci. Paris, 240
    (1955) 203. b) M. J. Carter and I. Fleming. J. Chem. Soc.
    Chem. Commun., (1976) 679.

28) A. Hosomi, H. Iguchi, J. Sasaki and H. Sakurai, Tetrahedron
    Lett., 23 (1982) 551.

29) A. Hosomi, H. Iguchi and H. Sakurai, Chem. Lett., (1982) 223.

30) a) Y. Yamamoto, H. Yatagai, Y. Ishihara, N. Maeda and K.
    Maruyama, Tetrahedron Sym. in print, 40 (1984) 2239. b) Y.
    Yamamoto, Accounts Chem. Res., 20 (1987) 243. c) Y.
    Yamamoto, Aldrich Chim. Acta, 20 (1987) 45.

31) Y. Yamamoto, H. Yatagai, Y. Naruta and K. Maruyama, J. Am.
    Chem. Soc., 102 (1980) 7107.

32) Y. Yamamoto and K. Maruyama, J. Organomet. Chem., 284 (1985)
    C 45.

33) G. Deleris, J. Dunoguès and R. Calas, Tetrahedron Lett.,
    (1976) 2449.

34) T. Hayashi, K. Kabeta, I. Hamachi and M. Kumada, Tetrahedron
    Lett., 24 (1983) 2865.

35) M. Koreeda and Y. Tanaka, Chem. Lett., (1982) 1297, 1299.

36) J. R. Green, M. Majewski, B. I. Alo and V. Snieckus,
    Tetrahedron Lett., 27 (1986) 535.

37) Y. Yamamoto, S. Hatsuya and J. Yamada, J. Chem. Soc. Chem.
    Commun., (1987) 561.

38) C. H. Heathcock, S. Kiyooka and T. A. Blumenkopf, J. Org.
    Chem. 49 (1984) 4214.

39) Y. Yamamoto, T. Komatsu and K. Maruyama, J. Organomet. Chem.
    285 (1985) 31. See also, C. H. Heathcock and L. A. Flippin,
    J. Am. Chem. Soc., 105 (1983) 1667.

40) S. Kiyooka and C. H. Heathcock, Tetrahedron Lett., 24 (1983)
    4765.

41) C. H. Heathcock, S. Kiyooka and T. A. Blumenkopf, J. Org. Chem., 49 (1984) 4214.

42) M. T. Reetz, K. Kesseler and A. Jung, Tetrahedron Lett., 25 (1984) 729.

43) M. T. Reetz and K. Kesseler, J. Chem. Soc. Chem. Commun., (1984) 1079.

44) S. Danishefsky and M. DeNinno, Tetrahedron Lett., 26 (1985) 823.

45) D. R. Williams and F. D. Klingler, Tetrahedron Lett., 28 (1987) 869.

46) M. T. Reetz and K. Kesseler, J. Org. Chem., 50 (1985) 5434. In case of BF_3 mediated allylsilane addition to 20 and 21, non-chelation control was realized presumably via a Conforth-type diplar transition state 30. For diastereo-selectivity of α-alkoxyaldehydes with allenylsilanes, see R. L. Danheiser, C. A. Kwasigroch and Y. -M. Tsai, J. Am. Chem. Soc., 107 (1985) 7233.

47) M. T. Reetz, K. Kesseler, S. Schmidtberger, B. Wenderoth and R. Steinbach, Angew. Chem. Int. Ed. Engl., 22 (1983) 989.

48) M. T. Reetz, K. Kesseler and A. Jung, Angew. Chem. Int. Ed. Engl., 24 (1985) 989.

49) M. T. Reetz and A. Jung, J. Am. Chem. Soc., 105 (1983) 4833.

50) M. T. Reetz, Angew, Chem. Int. Ed. Engl., 23 (1984) 556.

51) R. Imwinkelried and D. Seebach, Angew. Chem. Int. Ed. Engl., 24 (1985) 765. See also, H. Sakurai, K. Sasaki, J. Hayashi and A. Hosomi, J. Org. Chem., 49 (1984) 2808. T. Mukaiyama, M. Ohshima and N. Miyoshi, Chem. Lett., (1987) 1121.

52) I. Ojima, Y. Miyazawa and M. Kumagai, J. Chem. Soc. Chem. Commun., (1976) 927.

53) K. Soai and M. Ishizaki, J. Chem. Soc. Chem. Commun., (1984) 1016.

54) K. Soai, M. Ishizaki and S. Yokoyama, Chem. Lett., (1987) 341.

55) T. H. Chan and G. J. Kang, Tetrahedron Lett., 23 (1982) 3011.

56) R. J. P. Corriu, V. Huynh and J. J. E. Moreau, J. Organomet. Chem., 259 (1983) 283.

57) A. Hosomi, H. Hashimoto, H. Kobayashi and H. Sakurai, Chem. Lett., (1979) 245.

58) Y. Yamamoto, S. Nishii and T. Ibuka, J. Chem. Soc. Chem. Commun., (1987) 464.

59) C. H. Heathcock, E. F. Kleinman and E. S. Binkley, J. Am. Chem. Soc., 104 (1982) 1054. H. O. House, C. C. Yau and D. VanDerveer, J. Org. Chem., 44 (1979) 3031.

60) C. H. Heathcock, K. M. Smith and T. A. Blumenkopf, J. Am. Chem. Soc., 108 (1986) 5022. For the synthesis of (+) nootkatone, see T. Yanami, M. Miyashita and A. Yoshikoshi, J. Org. Chem., 45 (1980) 607.

61) D. El-Abed, A. Jellal and M. Santelli, Tetrahedron Lett., 25 (1984) 1463.

62) Y. Yamamoto, S. Nishii and K. Maruyama, J. Chem. Soc. Chem. Commun., (1985) 386.

63) R. L. Danheiser and D. M. Fink, Tetrahedron Lett., 26 (1985) 2509.

64) R. L. Danheiser and D. M. Fink, Tetrahedron Lett., 26 (1985) 2513.

65) a) Y. Naruta, H. Uno and K. Maruyama, Tetrahedron Lett., 22 (1981) 5221. b) H. Uno, J. Org. Chem., 51 (1986) 350.

66) G. Majetich, A. M. Casares, D. Chapman and M. Behnke, Tetrahedron Lett., 24 (1983) 1909.

67) I. Ojima and M. Kumagai, Chem. Lett., (1978) 575.

68) M. Kawai, M. Onaka and Y. Izumi, Chem. Lett., (1986) 381.

69) a) H. Nishiyama and K. Itoh, J. Org. Chem., 47 (1982) 2496. b) M. C. Pirrung and P. M. Kenney, J. Org. Chem., 52 (1987) 2335.

70) A. Hosomi, M. Ando and H. Sakurai, Chem. Lett., (1986) 365.

71) S. Kiyooka, H. Sasaoka, R. Fujiyama and C. H. Heathcock, Tetrahedron Lett., 25 (1984) 5331.

72) a) P. A. Bartlett, W. S. Johnson and J. D. Elliott, J. Am. Chem. Soc., 105 (1983) 2088. b) W. S. Johnson, P. H. Crackett and J. D. Elliott, Tetrahedron Lett., 25 (1984) 3951. c) V. M. F. Choi, J. D. Elliott and W. S. Johnson, Tetrahedron Lett., 25 (1984) 591.

73) J. M. McNamara and Y. Kishi, J. Am. Chem. Soc., 104 (1982) 7371.

74) W. S. Johnson and M. F. Chan, J. Org. Chem., 50 (1985) 2598.

75) Y. Yamamoto, S. Nishii and J. Yamada, J. Am. Chem. Soc., 108 (1986) 7116.

76) S. H. Mashraqui and R. M. Kellogg, J. Org. Chem., 49 (1984) 2513.

432

77) D. Seebach, R. Imwinkelried and G. Strucky, Angew. Chem.
    Int. Ed. Engl., 25 (1986) 178.
78) C. Bruckner, H. Lorey and H. Reissig, Angew. Chem. Int. Ed.
    Engl., 25 (1986) 556.
79) H. Nishiyama, S. Narimatsu, K. Sakuta and K. Itoh, J. Chem.
    Soc. Chem. Commun., (1982) 459.
80) C. Westerlund, Tetrahedron Lett., 23 (1982) 4835.
81) A. P. Kozikowski and P. D. Stein, J. Am. Chem. Soc., 107
    (1985) 2569.
82) T. L. Cupps, D. S. Wise and L. B. Townsend, J. Org. Chem.,
    47 (1982) 5115.
83) A. P. Kozikowski and K. L. Sorgi, Tetrahedron Lett., 23
    (1982) 2281.
84) A. P. Kozikowski, K. L. Sorgi, B. C. Wang and Z. Xu,
    Tetrahedron Lett., 24 (1983) 1563.
85) M. D. Lewis, J. K. Cha and Y. Kishi, J. Am. Chem. Soc.,
    104 (1982) 4976.
86) A. Hosomi, Y. Sakata and H. Sakurai, Tetrahedron Lett., 25
    (1984) 2383.
87) J. A. Bennek and G. R. Gray, J. Org. Chem., 52 (1987) 892.
88) S. Danishefsky and J. F. Kerwin, Jr., J. Org. Chem., 47
    (1982) 3803.
89) A. Hosomi, T. Imai, M. Endo and H. Sakurai, J. Organomet.
    Chem., 285 (1985) 95.
90) R. Mohan and J. A. Katzenellenbogen, J. Org. Chem., 49 (1984)
    1238.
91) M. Uemura, T. Minami, K. Isobe, T. Kobayashi and Y. Hayashi
    Tetrahedron Lett., 27 (1986) 967.
92) T. Sasaki, A. Usuki and M. Ohno, Tetrahedron Lett., 19 (1978)
    4925; J. Org. Chem., 45 (1980) 3559.
93) T. Fujisawa, M. Kawashima and S. Ando, Tetrahedron. Lett.,
    25, (1984) 3213.
94) H. Nishiyama, T. Naritomi, K. Sakuta and K. Itoh, J. Org.
    Chem., 48 (1983) 1557.
95) H. Sakurai, Y. Sakata and A. Hosomi, Chem. Lett., (1983) 409.
96) I. Fleming and B. -W. Au-Young, Tetrahedron Supplement, 37
    (1981) 13.
97) R. P. Alexander and I. Paterson, Tetrahedron Lett., 24
    (1983) 5911.
98) M. Wada, T. Shigehisa and K. Akiba, Tetrahedron Lett., 26

(1985) 5191.

99) R. J. Armstrong and L. Weiler, Can. J. Chem., 61 (1983) 214. T. S. Tan, A. N. Mather, G. Procter and A. H. Davidson, J. Chem. Soc. Chem. Commun., (1984) 585.

100) D. Wang and T. -H. Chan, J. Chem. Soc. Chem. Commun., (1984) 1273.

101) S. A. Carr and W. P. Weber, J. Org. Chem., 50 (1985) 2782.

102) R. Calas, J. Dunogues, J. -P. Pillot, C. Biran, F. Pisciotti and B. Arreguy, J. Organomet. Chem., 85 (1975) 149.

103) J. -P. Pillot, G. Déléris, J. Dunoguès and R. Calas, J. Org. Chem., 44 (1979) 3397. I. Ojima, M. Kumagai, Y. Miyazawa, Tetrahedron Lett., (1977) 1385. M. Laguerre, J. Dunogues and R. Calas, ibid., 19 (1978) 57. A. Hosomi, M. Saito and H. Sakurai, ibid., 20 (1979) 429.

104) A. Hosomi, M. Saito and H. Sakurai, Tetrahedron Lett., 21 (1980) 3783.

105) M. Ochiai and E. Fujita, Tetrahedron Lett., 21 (1980) 4369.

106) K. Hiroi and L. -M. Chen, J. Chem. Soc. Chem. Commun., (1981) 377.

107) H. Sakurai, T. Imai and A. Hosomi, Tetrahedron Lett., 18 (1977) 4045.

108) P. Renaud and D. Seebach, Angew. Chem. Int. Ed. Engl., 25 (1986) 843.

109) P. J. Sinclair, D. Zhai, J. Reibenspeies and R. M. Williams, J. Am. Chem. Soc., 108 (1986) 1103.

110) G. A. Kraus and K. Neuenschwander, J. Chem. Soc. Chem. Commun., (1982) 134.

111) C. W. Greengrass and D. W. T. Hoople, Tetrahedron Lett., 22 (1981) 1161.

112) S. D. Larsen, P. A. Grieco and W. F. Fobare, J. Am. Chem. Soc., 108 (1986) 3512.

113) L. E. Overman, T. C. Malone and G. P. Meier, J. Am. Chem. Soc., 105 (1983) 6993.

114) K. Ohga, U. C. Yoon and P. S. Mariano, J. Org. Chem., 49 (1984) 213.

115) H. Hamana and T. Sugasawa, Chem. Lett., (1985) 921.

116) A. Hosomi, H. Shoji and H. Sakurai, Chem. Lett., (1985) 1049. 116a) For the reaction of nitrones with allylsilane which gives [3 + 2] dipolar cycloadducts, see P. DeShong,

434

J. M. Leginus and S. W. Lander, Jr., J. Org. Chem., 51
(1986) 574.

117) M. Ochiai, E. Fujita, M. Arimoto and H. Yamaguchi, Tetrahedron Lett., 24 (1983) 777.

118) M. Ochiai, E. Fujita, M. Arimoto and H. Yamaguchi, J. Chem.
Soc. Chem. Commun., (1982) 1108.

119) M. Koreeda and M. A. Ciufolini, J. Am. Chem. Soc., 104
(1982) 2308.

120) P. Magnus, "Organic Synthesis Today and Tomorrow", B. M.
Trost, C. R. Hutchinson, Ed., Pergamon Press, New York,
1981, p97.

121) M. Ochiai, M. Arimoto and E. Fujita, Tetrahedron Lett.,
22 (1981) 4491.

122) T. K. Sarker and N. H. Andersen, Tetrahedron Lett., (1978)
3513.

123) A. Itoh, K. Oshima and H. Nozaki, Tetrahedron Lett., 20
(1979) 1783.

124) K. Nishitani and K. Yamakawa, Tetrahedron Lett., 28 (1987)
655.

124a) For the synthsis of 4-halotetrahydropyrans from allylsilanes and aldehydes, see L. Coppi, A. Ricci and M. Taddei,
Tetrahedron Lett., 28 (1987) 973.

125) B. M. Trost and B. P. Coppola, J. Am. Chem. Soc., 104
(1982) 6879.

126) B. M. Trost and J. E. Vincent, J. Am. Chem. Soc. 102 (1980)
5680.

127) B. M. Trost and B. R. Adams, J. Am. Chem. Soc. 105 (1983)
4849.

128) G. A. Molander and S. W. Andrews, Tetrahedron Lett., 27
(1986) 3115.

129) a) S. E. Denmark and E. J. Weber, Helv. Chim. Acta, 66
(1983) 1655. b) S. E. Denmark and E. J. Weber, J. Am.
Chem. Soc., 106 (1984) 7970. c) S. E. Denmark, B. R. Henke
and E. Weber, J. Am. Chem. Soc., 109 (1987) 2512.

130) S. R. Wilson and M. F. Price, J. Am. Chem. Soc., 104 (1982)
1124.

131) T. Tokoroyama, M. Tsukamoto and H. Iio, Tetrahedron Lett.,
25 (1984) 5067.

132) D. Schinzer, Angew. Chem. Int. Ed. Engl., 23 (1984) 308.

133) D. Schinzer, S. Solyom and M. Becker, Tetrahedron Lett.,

26 (1985) 1831.

134) D. Schinzer, J. Steffen and S. Solyom, J. Chem. Soc. Chem. Commun., (1986) 829.

135) G. Majetich, M. Behnke and K. Hull, J. Org. Chem., 50 (1985) 3615.

136) G. Majetich, J. Defauw, K. Hull and T. Shawe, Tetrahedron Lett., 26 (1985) 4711.

137) G. Majetich, K. Hull, J. Defauw and R. Desmond, Tetrahedron Lett., 26 (1985) 2747.

138) G. Majetich, K. Hull and R. Desmond, Tetrahedron Lett., 26 (1985) 2751.

139) G. Majetich, K. Hull, J. Defauw and T. Shawe, Tetrahedron Lett., 26 (1985) 2755.

140) W. S. Johnson, Y. Chen and M. S. Kellogg, J. Am. Chem. Soc. 105 (1983) 6653.

141) R. J. Armstrong and L. Weiler, Can. J. Chem., 64 (1986) 584.

142) R. J. Armstrong, F. L. Harris and L. Weiler, Can. J. Chem., 60 (1982) 673.

143) R. J. Armstrong, F. L. Harris and L. Weiler, Can. J. Chem., 64 (1986) 1002.

144) J. Ipaktschi and G. Lauterbach, Angew. Chem. Int. Ed. Engl. 25 (1986) 354.

145) K. M. Nicholas and J. Siegel, J. Am. Chem. Soc., 107 (1985) 4999.

146) S. L. Schreiber, T. Sammakia and W. E. Crowe, J. Am. Chem. Soc., 108 (1986) 3128.

147) H. Hiemstra, W. J. Klaver and W. N. Speckamp, J. Org. Chem. 49 (1984) 1149.

148) H. Hiemstra, H. P. Fortgens, S. Stegenga and W. N. Speckamp, Tetrahedron Lett., 26 (1985) 3151.

149) H. Hiemstra, H. P. Fortgens and W. N. Speckamp. Tetrahedron Lett., 26 (1985) 3155.

150) J. -C. Gramain and R. Remuson, Tetrahedron Lett., 26 (1985) 327.

151) P. A. Grieco and W. F. Fobare, Tetrahedron Lett., 27 (1986) 5067.

152) P. A. Grieco and W. F. Fobare, J. Chem. Soc. Chem. Commun., (1987) 185.

153) a) ref. 96. b) B. W. Au-Yeung and I. Fleming, J. Chem. Soc. Chem. Commun., (1977) 79, 81. c) H. Wetter, P. Scherer and

W. B. Schweizer, Helv. Chim. Acta, 62 (1979) 1985.

154) a) G. Wickham and W. Kitching, J. Org. Chem., 48 (1983) 612; Organomet., 2 (1983) 541. b) M. J. Carter and I. Fleming, J. Chem. Soc. Chem. Commun., (1976) 679.

155) I. Fleming and N. K. Terrett, Tetrahedron Lett., 24 (1983) 4153.

156) T. Hayashi, Chemica Scripta, 25 (1985) 61.

157) T. Hayashi, M. Konishi, H. Ito and M. Kumada, J. Am. Chem. Soc., 104 (1982) 4962.

158) T. Hayashi, M. Konishi, M. Kumada, J. Org. Chem., 48 (1983) 281.

159) a) N. T. Anh, J. Chem. Soc. Chem. Commun., (1968) 1089. b) K. Fukui and H. Fujimoto, Bull. Chem. Soc. Jpn., 39 (1966) 2116.

160) T. Hayashi, M. Konishi and M. Kumada, J. Am. Chem. Soc., 104 (1982) 4963.

161) T. Hayashi, K. Kabeta, T. Yamamoto, K. Tamao and M. Kumada, Tetrahedron Lett., 24 (1983) 5661.

162) T. Hayashi, M. Konishi and M. Kumada, J. Chem. Soc. Chem. Commun., (1983) 736.

163) S. J. Hathaway and L. A. Paquette, J. Org. Chem., 48 (1983) 3351.

164) L. Coppi, A. Mordini and M. Taddei, Tetrahedron Lett., 28 (1987) 969.

165) K. Mikami, T. Maeda, N. Kishi and T. Nakai, Tetrahedron Lett., 25 (1984) 5151.

166) A. Ricci, A. Degl'Innocenti, M. Fiorenza, M. Taddei and M. A. Spartera, Tetrahedron Lett., 23 (1982) 577.

167) A. Ricci, M. Fiorenza, M. A. Grifagni, G. Bartolini and G. Seconi, Tetrahedron Lett., 23 (1982) 5079.

168) A. Hosomi, Y. Araki and H. Sakurai, J. Org. Chem., 48 (1983) 3122.

169) G. Majetich, A. Casares, D. Chapman and M. Behnke, J. Org. Chem., 51 (1986) 1745.

170) For the F⁻ mediated intramolecular cyclization via the reaction of an aldehyde with allylic silane, see C. Kuroda, S. Shimizu and J. Y. Satoh, J. Chem. Soc. Chem. Commun., (1987) 286. See also ref. 26a.

171) G. Majetich, R. Desmond and A. M. Casares, Tetrahedron Lett., 24 (1983) 1913.

172) G. Majetich, R. W. Desmond, Jr. and J. J. Soria, J. Org. Chem., 51 (1986) 1753.

173) B. M. Trost and D. P. Curran, J. Am. Chem. Soc., 103 (1981) 7380. For the F⁻ induced intramolecular cyclization of 2-substituted 1,3-bis(trimethylsilyl)propene, see M. Ochiai, K. Sumi, E. Fujita and M. Shiro, Tetrahedron Lett., 23 (1982) 5419.

174) P. Magnus, T. Gallagher, P. Brown and P. Pappalardo, Acc. Chem. Res., 17 (1984) 35. See also, Y. Ito, M. Nakatsuka and T. Saegusa, J. Am. Chem. Soc., 102 (1980) 863.

175) H. Vorbruggen and K. Krolikiewicz, Tetrahedron Lett., 24 (1983) 889.

176) A. Hosomi, K. Hoashi, S. Kohra, Y. Tominaga, K. Otaka and H. Sakurai, J. Chem. Soc. Chem. Commun., (1987) 570.

177) I. Fleming, D. Marchi and S. K. Patel, J. Chem. Soc. Perkin I, (1981) 2518. For protodesilylation of 1-trimethylsilyl-2,3-dienes see, C. Nativi, A. Ricci and M. Taddei, Tetrahedron Lett., 28 (1987) 2751.

178) T. Hayashi, H. Ito and M. Kumada, Tetrahedron Lett., 23 (1982) 4605.

179) S. R. Wilson and M. F. Price, Tetrahedron Lett., 24 (1983) 569.

180) P. S. Mariano, Acc. Chem. Res., 16 (1985) 130.

181) M. Ochiai, M. Arimoto and E. Fujita, J. Chem. Soc. Chem. Commun., (1981) 460.

182) J. Dubac and A. Laporterie, Chem. Rev., 87 (1987) 319.

183) H. Sakurai, A. Hosomi, M. Saito, K. Sasaki, H. Iguchi, J. Sasaki and Y. Araki, Tetrahedron, 39 (1983) 883.

8 ADDED IN PROOF

Recent advances in the reaction of allylsilanes are included in this Appendix, which covers the literature from January 1987 to September 1987.   The condensation of alkoxyallylsilanes with carbonyl compounds in the presence of Lewis acids gave all cis-4-chloro-2,6-disubstituted-tetrahydropyranes (ref. 184). The reaction was found to proceed through a mixed silyl acetal intermediate.   The reactivities of the allyl-silicon bonds in two kinds of pentacoordinated allylsilanes (**183** and **184**) were compared (ref. 185).   These two allylsilicates exhibited quite different reactivity in the allylation of carbonyl compounds under nucleophilic conditions (Bu$_4$NF, KF or NaOMe) and

electrophilic conditions (TiCl$_4$, AlCl$_3$ or BF$_3$).   The allylation via **183** took place under nucleophilic conditions in DMSO, while the allylation via **184** occurred under electrophilic conditions.   These results indicate that the reactivities of these silicates depend on the overall charge and geometry. Quite interestingly, triethylammonium bis(pyrocatecholato)allylsilicates (**185**), readily prepared from allyltrimethoxysilane, pyrocatechol, and triethylamine, reacted with aldehydes regiospecifically and chemoselectively <u>without catalyst</u> under mild conditions (ref. 186).   This provides the first example of direct transfer of an organic ligand from a pentacoordinate silicate to a carbon electrophile without catalyst.   The allylation took place with allylic transposition as usual, and ketones did not undergo the allylation reaction.   The allylation of aldehydes occurred even in protic solvents such as ethanol. The pentacoordinated allylsilicate, generated in situ from allyltrialkoxysilane, pyrocatechol and triethylamine, also exhibited the similar reactivity as **185** (ref. 187).

Treatment of allyltrifluorosilane with CsF produced the corresponding pentacoordinated allylsilicate, which underwent chemoselective, regiospecific and highly stereoselective allylation of aldehydes (ref. 188). The allylation occurred exclusively at the γ-carbon of the allylsilanes, though the fluoride ion mediated reaction of allyltrialkylsilanes took place both at the γ-position and at the α-position (see **5.1**). Here also, the allylsilicate did not react with ketones.

$$R^1R^2C=CHCH_2SiF_3 \ + \ R^3CHO \ \xrightarrow{F^-/THF} \ H_2C=CHCR^1R^2\underset{\underset{OH}{|}}{C}HR^3$$

(E)-Crotylsilicate gave the anti homoallyl alcohol and (Z)-crotyl derivative produced the syn-isomer, indicating that the reaction proceeded via six-membered chair transition state.

In the presence of a catalytic amount of trityl perchlorate, secondary and tertiary allyl ethers reacted with allylsilanes to give the corresponding 1,5-dienes in good yields (ref 189). Normally, a stoichiometric amount of a Lewis acid is required in the allylsilane-Lewis acid system. Accordingly, use of the catalyst may enhance synthetic usefulness of allyltrimethylsilanes. The conjugate addition of allylsilanes to α,β-unsaturated ketones was effectively promoted by a catalytic amount of trityl perchlorate to give the corresponding Michael adducts in good yields (ref. 190). 2-Alkenyl-1,3-dithiolan or dithian-2-ylium cations were produced by hydride abstraction of cyclic ketene dithioacetals with trityl tetrafluoroborate. These cations reacted with allylsilane completely regioselective-

ly to give the corresponding allylated ketene dithioacetals
(ref. 191). In the presence of a catalytic amount of diphenyl-
boryl triflate, aldehydes reacted with (S)-1-phenyl-1-trimethyl-
silyloxyethane and allyltrimethylsilane to produce the chiral
homoallyl ethers in good yields with high stereoselectivities

(ref. 192). Allyl ethers were oxidized by DDQ to produce the
corresponding cationic species, which reacted with allylsilane
in the presence of a catalytic amount of lithium perchlorate
to give the allylated ether in a one-pot procedure (ref. 193).
Allylation of 6-methoxy-1, 2, 3, 6-tetrahydro- and 4-
methoxy-1, 2, 3, 4-tetrahydropyridines with allylsilane as a
carbon nucleophile by using electrogenerated acid as an acid
catalyst proceeded regioselectively at the C-6 position via
vinylogous acyliminium ion (ref. 194). Cyclization of
acyclic N-acyliminium ions, generated from the adducts of
formaldehyde, n-butyl glyoxylate or methyl glyoxylate with
amines containing an allyl or propargyl silane as terminator,
produced 3-ethenyl and 3-ethenylidene substituted pyrrolidines
and piperidines (ref. 195). Electrochemically generated

triphenylphosphine radical cation reacted with allylic silanes
to give the corresponding allyltriphenylphosphonium tetra-
floroborates (ref. 196).   Theoretical studies on the regio-
and stereochemistry of electrophilic addition to chiral double
bonds in allylic silanes were reported (ref. 197).

References
184) Z. Y. Wei, J. S. Li, D. Wang and T. H. Chan, Tetrahedron
     Lett., 28 (1987) 3441.
185) G. Cervean, C. Chuit, R. J. P. Corriu and C. Reye, J.
     Organomet. Chem., 328 (1987) C17.
186) a) A. Hosomi, S. Kohra and Y. Tominaga, J. Chem. Soc.
     Chem. Commun., (1987) 1517. b) M. Kira, K. Sato,
     M. Kobayashi and H. Sakurai, Abstracts of the
     31st Symposium on Organometallic Chemistry, Kyoto (1987).
187) A. Hosomi, S. Kohra and Y. Tominaga, Chem. Pharm. Bull.,
     35 (1987) 2155.
188) M. Kira, M. Kobayashi and H. Sakurai, Tetrahedron Lett.,
     28 (1987) 4081.
189) M. Murakami, T. Kato and T. Mukaiyama, Chem. Lett., (1987)
     1167.
190) M. Hayashi and T. Mukaiyama, Chem. Lett., (1987) 289.
191) Y. Hashimoto and T. Mukaiyama, Chem. Lett., (1986) 755.
192) T. Mukaiyama, M. Ohshima and N. Miyoshi, Chem. Lett.,
     (1987) 1121.
193) Y. Hayashi and T. Mukaiyama, Chem. Lett., (1987) 1811.
194) S. Torii, T. Inokuchi, S. Takagishi, F. Akahoshi and K.
     Uneyama, Chem. Lett., (1987) 639.
195) H. H. Mooiweer, H. Hiemstra, H. P. Fortgens and W. N.
     Speckamp, Tetrahedron Lett., 28 (1987) 3285.
196) T. Takanami, K. Suda, H. Ohmori and M. Masui, Chem. Lett.,
     (1987) 1335.
197) S. D. Kahn, C. F. Pau, A. R. Chamberlin and W. J. Hehre,
     J. Am. Chem. Soc., 109 (1987) 650.

SUBJECT INDEX

(grouped by chapter)

CHAPTER 1 --(pp. 1-138)

CO Hydrogenation

448

Immobilized Mixed Metal Cluster Catalysts on Functionalized Supports... 73-74

Hetero-Metallic Clusters in Heterogeneous Catalysis 76-123

Heterogeneous Catalysis 76

Chapter 2 -- (pp. 139-223)

Stereoelectronic Effects

Conformational Analysis for [CpFe(CO)(PPh₃)C(OMe)R]⁺

CpFe(CO)(PPh₃)C(OMe)R] R = Me, Et

Conformational Analysis for [CpFe(CO)(PPh₃)COR]

Stereoselectivity of Ligands Attached to the Chiral Auxiliary CpFe(CO)(PPh₃)

CHAPTER 3 --(pp 225-300)

Compound of Type MM'(CO)$_{10}$ With M, M' = Mn, Re

Photo Disproportionations in Metal Carbonyls

Photo Substitutions in Metal Carbonyls

Structural and Spectroscopic Properties

Mn(CO)$_5$

$Mn_2(CO)_9$

Primary Photoprocess in

$HMn(CO)_5$

$Mn_2(CO)_{10}$

CHAPTER 4 -- (pp. 301-362)

Abbreviations: TMCC = Transition Metal Carbonyl Cluster
EAN = Equivalent Atomic Number
PESP = Polyhedral Skeletal Electron Pair Approach

464

CHAPTER 5 (pp. 363-441)